Monographs in Mathematical Economics

Volume 6

This series, consisting of research monographs and advanced textbooks for graduate students, is designed to bring together mathematicians interested in obtaining challenging new stimuli from economic theories and economists seeking effective mathematical tools for their research.

The scope of the series includes but is not limited to:

- Economic theories in various fields based on rigorous mathematical reasoning
- Mathematical methods (e.g., analysis, algebra, geometry, probability) motivated by economic theories
- Mathematical results of potential relevance to economic theory
- Historical study of mathematical economics

Comparable, existing monographs series are mainly organized from the viewpoint of "users" of mathematics, and are thus of limited interest to mathematicians. This series in contrast aims at genuine interaction between economists and mathematicians.

Most economic phenomena are described by (1) optimizing behaviors of agents (consumers, firms, governments, etc.) and (2) equilibria generated through the interactions of these agents. Consequently, the most basic mathematical subjects for economics fall into two categories:

- Optimization theory, static and dynamic. Basic linear / nonlinear programming, calculus of variations, optimal control, dynamic programming etc. provide key mathematical tools. Modern developments in convex analysis, nonlinear functional analysis, set-valued analysis, and non-smooth analysis form indispensable foundations.
- Equilibrium theory. Existence in finite- / infinite-dimensional settings, characterization, computational algorithms, dynamic adjustment processes, correspondence with economic structures.

Some fixed-point theorems and variational inequalities were created to solve the existence problem of equilibria. The stability theory and the viability theory of differential equations play important roles in the dynamic aspects of the equilibrium theory. Further, differential topology is a basic tool for the geometric analysis of equilibrium manifolds.

Economic phenomena significant in the real world often have a dynamic character. For instance, business cycles and movements of asset prices are governed by dynamic laws expressed in terms of differential or difference equations. Nonlinear analysis, harmonic analysis, and stochastic calculus play significant roles in solving dynamic economic problems. Econometric methods, including time series analysis and forecasting, also require sophisticated mathematical tools. Finally, significant developments in game theory and interaction with mathematical logic should be mentioned.

All of these topics and phenomena come within the scope of this series, aimed at cooperation between economists and mathematicians and the development of their respective disciplines.

Toru Maruyama
Editor

Marginal Revolution in Economics

A Reappraisal

Springer

Editor
Toru Maruyama
Keio University
Tokyo, Japan

ISSN 2364-8279 ISSN 2364-8287 (electronic)
Monographs in Mathematical Economics
ISBN 978-981-99-4341-8 ISBN 978-981-99-4342-5 (eBook)
https://doi.org/10.1007/978-981-99-4342-5

This Springer imprint is published by the registered company Springer Nature Singapore Pte Ltd.
The registered company address is: 152 Beach Road, #21-01/04 Gateway East, Singapore 189721, Singapore

Preface

In 1971, several academic conferences were held on the occasion of the 100th anniversary of the Marginal Revolution in economics. These meetings, as any economist agrees, exerted the stimulating influence upon the historical research into the "revolution". While W. Jaffé's masterpiece, *Correspondence of Léon Walras and Related Papers* (3 Vol's, 1965) and C. Menger's *Gesammete Werke* (2 Auflage. 4 Vol's, 1970) edited by F. A. von Hayek had been already published, various important books and articles were released during the 1970s, which were motivated by the 100th anniversary. We may count, among others, *Papers and Correspondence of William Stanley Jevons* (7 Vol's, 1972–1981) R. D. C. Black ed., J. R. Hicks and W. Weber eds., *Carl Menger and the Austrian School of Economics* (1973), and the special issue of *History of Political Economy* (Vol. 4, No.2, 1972).

The history of economics is filled with dramas of succession to, dependence upon and opposition to doctrines. It may be regarded as a gradual and continuous process in the long run. However, we also admit that a few major and fundamental turning points are clearly discernible in the process. The major turning points should be distinguished from "relatively minor controversies and amendments" (T. W. Hutchison).

I am not sure if the Marginal Revolution is a real "revolution" or not. Nevertheless, the simultaneous appearance of C. Menger, W. S. Jevons and L. Walras is, at least, well-qualified for the title "Fundamental Turning-Point".

To the best of my belief, the foremost achievement of the Marginal Revolution is the completion of the general equilibrium theory as a basic framework of economic analysis. In two decades preceding 1971, the mathematical structure of this theory was rigorously scrutinized by means of carefully chosen methods. The highly formal problems such as the existence of general equilibrium (by the fixed-point theorems), the fundamental theorem of welfare economics (by the separation theorem of convex sets), the dynamic stability of price adjustment processes (by Ljapunov's second method), etc. were successfully solved by many mathematical economists including K. J. Arrow, G. Debreu, H. Nikaido and L. Hurwicz.

We have to keep in mind that the participants of 1971 conferences had learned, more or less, these theoretical achievements in 1960–1970 as their commonsense background.

Now more than fifty years have passed since then. During this period, economic theory has experienced further substantial changes in researchers' central interest, the way of reasoning and the styles of description, even if we restrict our sight to narrow micro-economic theory. (1) The results obtained in 1960–1970 were extended to much wider fields (say, imperfect competition, uncertainty, externalities, ...). (2) The concept of an economic structure (i.e. a distribution of the determinants of equilibrium over the space of agents) was given a rigorous mathematical expression. Analysis of a (set-valued) function which associates to each economic structure the corresponding equilibrium paved a way to the general comparative statics. (3) The theory of games provided quite effective frameworks of economic analysis other than general equilibrium theory. It has been exerting a dominant influence upon the world of economics. (4) Various disciplines in mathematics have extended the scope of their applications to economic theory. Moreover, some novel realms of mathematics have been cultivated in response to the requirements from economic problems. For instance, variational analysis, set-valued analysis and computational algorithm.

In view of the new achievements acquired in recent 50 years, it seems an indispensable task for us to review and reevaluate the Marginal Revolution based upon the present status of economics.

We also keep in mind the "renaissance phenomena"; that is, some concepts and doctrines once discarded may reappear in a later stage of history in a more or less transfigured form.

This volume is thus devoted to the reappraisal of the Marginal Revolution on the occasion of its 150th anniversary. I requested both the historians of economic thought and the economic theorists to contribute their works. It was a great pleasure and honor for me that they willingly accepted my offer and joined the project. I endeavored to compose a coherent book, not just a collection of articles. For that purpose, I myself add a somewhat lengthy introductory chapter, in which I give brief expositions about what aspect of the Marginal Revolution each of our authors is trying to clarify. I am also pleased if our project fortunately attracts mathematicians' interest as well as economists'. I expect my exposition to be helpful as a guide for readers from mathematics world. This book is scheduled to be included in the Springer series *Monographs in Mathematical Economics*. We chose this publication policy exactly because we wish to address many mathematicians, not only economists.

In the summer of 1971, I stayed in Bellagio. One of the conferences to celebrate the 100th anniversary of the Marginal Revolution was held in a tasteful villa on a hilltop, from where we could look down on Lake Como. I worked as the secretary to Professor T. Matsuura during the session, and enjoyed ample opportunities to learn quite a lot from the participants. Professors T. W. Hutchison, W. Jaffé and G. J. Stigler—all these principal members of the conference already passed away. Wonderful memories of our rich conversations as well as the lovely landscape of Como vividly come up to my mind.

I devote my warmest gratitude to all of our authors for their faithful cooperation. I am also grateful to the Oak Society Inc. for providing us with very efficient secretarial service.

Tokyo, Japan
March 2023

Toru Maruyama

Contents

Prelude

Toru Maruyama

Abstract This introductory chapter aims at an exposition of the contents and the evolution of the Marginal Revolution, which became a turning-point in the history of economics. I present a short history of the drama, restricting myself to the most basic issues: (i) the general equilibrium (Sect. 3), (ii) the utility-demand theory (Sect. 4), and (iii) the production-distribution theory (Sect. 5). Some aspects of the prehistory of the revolution are discussed in Sect. 2. And the final section is devoted to the philosophical foundations. In order to attract the interest of mathematicians, I talk about even common-sense concepts which most economists may be acquainted with very well. Some brief comments on the following chapters are also given as a preview of what kinds of problems are to be discussed by our authors.

1 Introduction

A crucial event which occurred in the history of economics 150 years ago has been treated as a scientific "revolution". The achievements of the trio of revolutionists are fairly well-known among economists. William S. Jevons (London). Carl Menger (Vienna) and Léon Walras (Laussanne) published their masterpieces almost simultaneously:

Jevons, *Theory of Political Economy*, 1871;
Menger, *Grundsätze der Volkswirtschaftslehre*, 1871;
Walras, *Eléments d'économie politique pure*, 1874–7.

Although these works were completely independently written, they shared some peculiar features in common, which were to leave definite vestiges in the course of economics after them. The simultaneous appearance of these three treatises in different countries (and in different languages) has been celebrated as the Marginal Revolution in the history of economics.

T. Maruyama (✉)
Keio University, Tokyo, Japan
e-mail: maruyama@keio.ac.jp

T. Maruyama (ed.), *Marginal Revolution in Economics*, Monographs in Mathematical Economics 6, https://doi.org/10.1007/978-981-99-4342-5_1

Reading through the works of the trio, nobody disagrees with the following aspects as their common features.

(i) All the three theories are based upon subjective elements (utility) of human wants.
(ii) Mathematical analysis (differential calculus, in particular) was made use of explicitly in the works of Jevons and Walras, and implicitly by Menger.

However, the most fundamental achievement of the Marginal Revolution is, among others, the completion of general equilibrium system as a basic skeleton of economic analysis. From this viewpoint, we may safely justify the special status occupied by Walras superior to others, because it is definitely Walras who constructed the system by himself.

According to *Concise Oxford Dictionary*, the term "revolution" is defined as follows:

> a forcible overthrow of a government or social order, in favour of a new system, a dramatic and wide-reaching change.

It seems to connote a radical sudden change. Nevertheless it was not until the middle of the next century that Walras' system was accepted and became rooted in the academic world of economics.[1] In this sense, Walras had to fight a very long revolution.

The COD definition tells us that the term "revolution" means a forcible substitution by a new ruler for the old. Who then are the new and old rulers in the case of the Marginal Revolution?

We can easily find an answer for Jevons' revolution. The old ruler is, of course, the English Classical School represented by D. Ricardo–J. S. Mill, which was exclusively dominant in that country until the 1850s. Jevons regarded himself as a challenger to the hegemony of the Classical School. On the other hand, there was no dominant hegemony that Menger or Walras bid defiance to. Ample economic theories had been accumulated by outstanding German economists (I christened the group "the German Classical School") prior to Menger. Menger and his school succeeded the genealogy of these forerunners.[2] Although Menger got involved in the "Methodenstreit" with the German Historical School, it was a drama after the establishment of his authority in Vienna. G. Schmoller and other historians were never the old rulers for Menger. Walras always praised two names as his predecessors; Auguste Walras (his father) and A. A. Cournot. We command, behind these names, a magnificent view of two intellectual traditions. One is the old Italian subjective value theory, which was brought about to France in the eighteenth century. Another one is the tradition of "mathématique sociale" founded by Marquis de Condorcet. Walras' historical standpoint is exactly at the confluence of these two great rivers. The succession of these traditions to Walras was achieved without any historical discrepancy. The novelty of Walrasian economics was claimed to the academic world in general, not against any

[1] Walras' system rooted firmly in the community of economic theory since J. R. Hicks' *Value and Capital* (1939) and P. A. Samuelson's *Foundations of Economic Analysis* (1947) were published.

[2] See Chipman [16] and Streissler [115].

specific old ruler. Walras' revolution (if it may be called so) was rather a long and quiet revolution than a radical over–turn.

I expect this collection of articles to invoke mathematicians' as well as economists' interests in the Marginal Revolution. I start with recollecting briefly a rough outline of the drama and previewing what kinds of problems are to be discussed in the following chapters. Subsections marked by ♣ are my comments on the following chapters.

I restrict myself to the most basic issues in the Marginal Revolution; that is, the general equilibrium (Sect. 3), the utility–demand theory (Sect. 4) and the production–distribution theory (Sect. 5). Other important topics such as capital, money, or market adjustment are all omitted here. The Marginal Revolution could not happen suddenly without any preceding background. Hence I discuss the prehistory of the revolution (Sect. 2). In the final section, I cast a glance at the philosophical foundations of economics.[3]

2 Traditions Behind Walras

2.1 Mathématique Sociale

The eighteenth century in the history of science is characterized as the age of mechanics and astronomy. L. Euler, D. Bernoulli, J. L. Lagrange and P. S. Laplace, all following Newton's *Principia*, piled up enormous works which founded modern mathematical analysis. Lagrange's *Mecanique analytique* (1788) and Laplace's *Traité de mécanique celéste* (1792–1825) are the shining stars of the scientific cosmos of the age.

J. B. J. Fourier, S. D. Poisson and A. L. Cauchy were the core members of the next generation of the French school of mathematics. A. A. Cournot came after them. The mathematics world of Cournot's generation was starred with marvolous figures like G. P. L. Dirichlet, N. H. Abel, E. Galois and so on. (See Table 1.) We must admit that Cournot's name as a mathematician seems to have faded in comparison with the splendid achievements of such ingenious mathematicians as those listed above.

One reason why he could not reach the summit comparable to his contemporary colleagues is that his main concern was transmuted from analysis and probability to economic theory. It must have been a serious decision in his life as a mathematician. From a viewpoint of economics, his *Recherches sur les principes mathématiques de la théorie des richesses* (1838) has been evaluated as a path-breaking venture to mold economics in the Lagrange–Laplace tradition. Although we can learn a lot of concepts and reasonings from Cournot, the most important message he left is the methodological discipline, which tells us how to reason and express economic problems mathematically. Then we can not help asking a question—why did Cournot

---

[3] Some of this chapter is based upon my former monograph [69]. The generous permission of its publisher Keisoshobo is gratefully acknowledged.

Table 1 Mathematicians of the period

Timeline markers: 1735, 1789, 1814, 1830, 1848, 1870

Mathematician	From	To
Vandermonde	1735	1796
Lagrange	1736	1813
Condorcet	1743	1794
Monge	1746	1818
Laplace	1749	1827
Legendre	1752	1833
Fourier	1768	1830
Gauss	1777	1855
Poinsot	1777	1859
Poisson	1781	1840
Cauchy	1789	1857
Cournot	1801	1877
Abel	1802	1829
Dirichlet	1805	1859
Galois	1811	1832
Poincaré	1854	
Hilbert	1862	

step aside from the main street of mathematics to social science? This question reminds us of another tradition in modern mathematics in France.

It was founded by M. J. N. C. Condorcet.[4] After achieving celebrated works in classical analysis and astronomy, he left mathematics and devoted himself to public service in partnership with A. R. J. Turgot. This blank period, *prima facie*, carried Condorcet far away from mathematics. However, he started to engage in constructing a new branch of mathematics. Its purpose is two-fold:

(i) to analyze human societies consisting of many men with free will, and
(ii) to construct rationally an ideal society.

Mathematical tools and concepts of probability theory were effectively made use of instead of classical analysis in mathematical physics. Condorcet called this new field "mathématique sociale". Many mathematicians, like Lagrange, Laplace, A. T. Vandermond, …, soon joined this new wave, expecting to discover new frontiers of mathematics.

In the midst of the Revolution, Condorcet was arrested and commited suicide. This miserable tragedy gave a motivation for R. Rolland's drama.

After Condorcet's death, France entered the stage of Napoleon's despotism which was an adverse wind blowing against the mathématique sociale. It was after the end

[4] See Ingrao and Israel [47] pp. 46–60 for Condorcet's "mathématique sociale". Mclean and Hewitt [74] as well as Williams [136] should also be consulted.

Table 2 Arbitrage

A	B	C
1	2	
	1	3
1		4

of Napoleon era that Poisson encouraged his disciple, Cournot, to develop a rigorous mathematical analysis of social economic phenomena. Thus Cournot became a successor to Condorcet's mathématique sociale.

We now summarize Cournot's principal achievements in his economic theory.

(1) *Demand function.* Adam Smith, and many economists before him, knew the close relation between prices and quantities demanded. But it is Cournot who gave an explicit expression to this vague common sense as a function of prices. Walras effectively incorporated this concept in his general equilibrium system. On the other hand, Jevons did not know Cournot when he wrote the first edition of *Theory of Political Economy*.[5]

(2) *Non–arbitrage condition.* Assume that there are three agents ($i = 1, 2, 3$) and three goods ($j = 1, 2, 3$). The initial endowment vector ω^i of agent $i (= 1, 2, 3)$ is given by

$$\omega^1 = (\omega_1, 0, 0), \quad \omega^2 = (0, \omega_2, 0), \quad \omega^3 = (0, 0, \omega_3).$$

A pair of agents exchange their intial endowment with each other. Table 2 shows that 1 unit of the first commodity is exchanged with 2 units of the second commodity, 1 unit of the second commodity is exchanged with 3 units of the third commodity; and finally, 1 unit of the first commodity is exchanged with 4 units of the third. If the first agent exchanges 1 unit of the first commodity with the third commodity, he obtains only 4 units. However, if he exchanges 1 unit of the first commodity with 2 units of the second commodity, and then gives the second commodity in exchange for the third commodity, he can get 6 units of the third commodity. The situation shown in Table 2 is unstable since such an indirect exchange (arbitrage) gives benefit to an agent. Any perfect settlement requires that no more arbitrage is admitted.[6]

(3) *Theory of monopoly.* If a market is monopolized by a single firm, its optimal behavior is decided by the condition: marginal revenue = marginal cost. Cournot solved this problem in the simplest case.[7]

---

[5] Cournot [19] Chap. IV.

[6] Cournot [19] Chap. III.

[7] Cournot [19] Chap. V.

(4) *Theory of oligopoly*. Cournot analyzed the simple case of duopoly, and created an equilibrium concept, now called Cournot–Nash equilibrium.[8]
(5) *Limit theorem*. Suppose that the number of firms in an oligopolistic market is n and that the cost function of all firms are identical, denoted by $C(x_j)$, where x_j is the output of the j-th firm. The inverse demand function of a market is expressed by $p(x) = p(x_1 + x_2 + \cdots + x_n)$. Then we obtain, by a simple calculus, that $x_1 = x_2 = \cdots = x_n$ at the Cournot–Nash equilibrium and

$$p(x) - C'(x_j) = -\frac{1}{n}\left(\frac{dp}{dx}\Big/\frac{p}{x}\right) \cdot p = \frac{p}{ne},$$

where e is the elasticity of demand. Assuming that e is constant, the gap between the price and the marginal cost $C'(x_j)$ converges to zero as the number n of firms tends to infinity. The situation $p = C'$, where every producer is a price-taker, is the competition in Cornot's sense. This concept of competition was succeeded by Walras.[9]
(6) *Method of classical analysis*. According to Cournot, the demand of the i-th individual agent depends not only upon price p but also upon some randomness. Let Ω be a probability space and P the set of prices. Then the individual demand function ξ^i is defined on $P \times \Omega$. If the number of agents is m, the mean demand function is given by

$$\frac{1}{m}\sum_{i=1}^{m} \xi^i(p, \omega). \tag{1}$$

If random variables $\xi^i : \omega \mapsto \xi^i(p, \omega)$ (for each fixed $p \in P; i = 1, 2, \ldots, m$), are independent and identically distributed ($i.i.d.$), then the mean (1) converges $a.s.$ to the expectation $\mathbb{E}\xi^i(p, \omega)$ for each $p \in P$ (this limit is independent of i) as m tends to infinity. Thus the mean demand $\mathbb{E}\xi^i(p, \omega)$ does not depend upon ω in an economy consisting of infinitely many agents. The mean demand is released from the probability world in a large economy.

When we investigate a demand function by means of classical analysis, it is required to be continuous and smooth. However, an individual demand function is very likely to be discontinuous or nonsmooth. We can not apply classical analysis to such an irregular function. Nevertheless, if the number of agents is very large and the shapes of demand functions are suitably dispersed,[10] then the aggregate demand function becomes more regular through "smoothing by aggregation".

In Fig. 1, three individual demand functions, ξ^1, ξ^2 and ξ^3, are depicted. These are neither continuous nor smooth. The graph of $\xi^1 + \xi^2 + \xi^3$ is depicted by the bold

[8] Cournot [19] Chap. VII.

[9] Cournot [19] Chap. VIII. The Cournot–Walras concept of competition is a very technical one. We have to distinguish it from the other concept of competition due to Smith, Ricardo, and many others. The most important historical research is Stigler [111]. See also McNulty [75].

[10] Intuitively, we assume that there are ample types of individual demand functions and they are heterogeneous little by little.

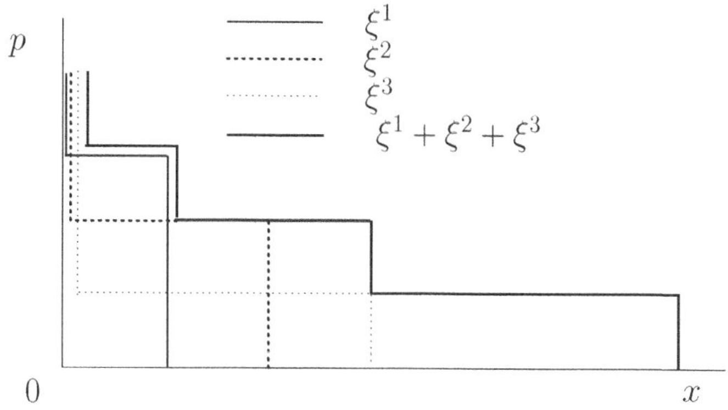

Fig. 1 Smoothing by aggregation

line. It may be seen that the aggregate function becomes smoother and smoother if the number of agents tends to infinity.

Thus, as far as the aggregate mean demand is concerned, we are permitted to regard demand as a smooth function of p independent of ω. This is why Cournot made use of aggregate (not individual) demand exclusively.

In this way, Cournot invented key tools of economic analysis so as to be well adapted to classical analysis. Walras did not consider deeply the subtle justification for the concepts and the methods he made use of.[11]

There were many authors on economic problems even before Cournot, who used numerical examples or simple algebraic illustrations to help their expositions. However, Cournot's method should be distinguished from such numerical illustrations. He specified implicitly a class of functions endowed with some required properties, and then applied a mathematical way of reasoning. He is the first mathematical economist in this sense.

Jevons' evaluation of Cournot deserves quotation.

> Even apart from its economic importance, this investigation, so far as I can venture to judge it, presents a beautiful example of mathematical reasoning, in which knowledge is apparently evolved out of ignorance. In reality the method consists in assuming certain simple conditions of the functions as conformable to experience, and then disclosing by symbolic inference the implicit results of these conditions. But I am quite convinced that the investigation is of high economic importance, and that, when the parts of political economy to which the theory relates come to be adequately treated, as they never have yet been, the treatment must be based upon the analysis of Cournot, or at least must follow his general method.[12]

---

[11] Cournot [19] Chap. IV. The role of the law of large numbers and the smoothing by aggregation behind Cournot's concept of demand function was formerly discussed in my seminar for graduate students. Based upon it, I. Muto wrote an extensive report [81] on Cournot's economics in the tradition of mathématique sociale.

[12] Jevons [54] Preface to the Second Edition, p. xxxi.

When Jevons wrote the first edition of *Theory of Political Economy*, he was not acquainted with Cournot's work. It was in 1872, when he obtained *Recherchet*.

> I procured a copy of the work as far back as 1872, but have only recently studied it with sufficient care to form any definite opinion upon its value. Even now I have by no means mastered all parts of it, my mathematical power being insufficient to enable me to follow Cournot in all parts of his analysis.[13]

Finally, we should keep in mind that Cournot, like other contemporaries, did not clearly distinguish the concept of continuity of a function from that of differentiability. Such a confusion is illustrated in the following passage.

> Si la fonction $F(p)$ est continue, elle jouira de la propriété commune à toutes les fonctions de cette nature, et sur laquelle reposent tant d'applications importantes de l'analyse mathématique: *les variations de la demande seront sensiblement proportionnelles aux variations du prix, tant que celles-ci seront de petites fractions du prix originaire*. D'ailleurs, ces variations seront de signes contraires, c'est-à-dire qu'à une augmentation de prix correspondra une diminution de la demande.[14]

No mathematician reached a full understanding of even such elementary concepts until the age of Cauchy.

2.2 ♣ *Mr. Menard on Cournot*

I summarized Cournot's achievements in economic theory and added brief remarks upon them. C. Menard's work in this volume also examined almost all of them more concretely.

While the French tradition of the "mathématique sociale" since Condorcet had primarily made use of probability theory as an analytical tool, Cournot turned to classical mathematical analysis. What is a reason for Cournot's turn around? It deserves a special notice that Menard gave an explicit answer to this important question. The point seems to be Cournot's sceptical view on the statistical method in social sciences.[15]

Once the approach of mathematical analysis is adopted, some continuity or smoothness is required for each function to be analyzed. For instance, an individual demand function does not satisfy such a requirement because it reflects the volatility and caprice of an agent.

On the other hand, a social demand function obtained by aggregating (or averaging) individual ones is expected to enjoy such properties thanks to the law of large numbers, as I discussed above. This is exactly the reason why Cournot examined aggregated demand functions exclusively.

[13] Jevons [54] p. xxx.

[14] Cournot [19] p. 39.

[15] See also Menard [76]. However, a clearer distinction between statistics and probability theory seems to be required.

Menard seems to share the same idea with us concerning this problem.

Walras, Marshall, Wicksell, ...all accepted Cournot's ideas and built their systems which incorporated them as indispensable parts. But their story occurs after the 1870s. There were some obstacles to the prompt success of Cournot in the scientific world in the nineteenth century. What are these obstacles? Menard also describes the historical background of this misery.

2.3 Italian Tradition of Value Theory

HECTOR
Brother, she is not worth what she doth cost
The keeping.
TROILUS What's aught but as'tis valued?
HECTOR
But value dwells not in particular will:
It holds his estimate and dignity
As well wherein 'tis precious of itself
As in the prizer, 'Tis mad idolatry
To make the service greater than the god;
And the will dotes that is attributive
To what infectiously itself affects,
Without some image of th'affected merit.

This conversation exchanged between the brothers in Shakespeare's *Troilus and Cressida*[16] illuminates two typical approaches to the theory of value. While Hector attaches more importance to objective factors than to subjective ones when he evaluates something, Troilus insists on the dominance of subjective factors. The former attitude reminds us of the labor– or cost–theory of value. On the other hand, the latter opinion leads to utility theory.

We should keep in mind that utility theory is never a product of modern society. Actually, the origin of the subjective view can be traced back to Aristotle.[17] A stream grew, via St. Augustine[18] into a big river and irrigated the long history of the medieval Scholastic economic doctrines of the just price up to the sixteenth century. The list of leading figures includes St. Thomas, St. Bernaldino, St. Antonino, and so on.[19]

Leaving aside the details of ancient and medieval economic thought, I restrict myself to quoting just some passages in order to show the striking examples of old subjectivism.

---

[16] W. Shakespeare. *Troilus and Cressida* (Oxford Univ. Press, Oxford) 1982, pp. 91–92.

[17] Aristotle [4]. Vol. II, *Nicomachean Ethics*, in particular Book V-5, pp. 1787–1789. See also Lowry [62, 63].

[18] St. Augustine [107] Book XI, Chap. 16, pp. 447–448.

[19] de Roover [27], Langholm [60, 61].

The first one is a passage from St. Thomas' *Summa Theologica*.

> In reply to the third argument, it is to be said that, as Augustine says (*de Civit. Dei*, XI, 16) the price of saleable things does not depend upon their rank in nature, since sometimes a horse is sold for more than a slave; but depends upon their usefulness to man. Hence a seller or buyer does not have to know the hidden qualities of a thing sold, but only those which render it fit for human use; such as the fact that a horse is strong, runs well, and so on. These qualities, however, the seller and buyer can easily recognize.[20]

During the medieval era, economic problems (including just price, usura, ...) were discussed mainly by Scholastic priests. It was the sixteenth century when a few laymen began to take part in the intellectual discussions concerning the economic problems, particularly from the viewpoint of theory of money.

The most remarkable author in Italy of this age is B. Davanzati. In his *Lezione delle moneta*, Davanzati vividly described the role of utility in the determination of prices.

> It was said in the Definition, *By the Consent of Nations made the Price and Measure of things*; because Men have agreed to fix that Value upon those Metals, for they have no such Privileges from Nature. A Natural Calf is far more noble than a Golden one, yet how much inferior in Price? An Egg that was bought for half a Grain of Gold, kept Count *Ugolino* alive in the Castle for ten days, which all the Treasure in the Universe could not do. What does more nearly concern our Lives than Corn? nevertheless ten thousand Grains thereof are sold for one of Gold.[21]

> It will not be amiss to illustrate what we have here said by some Examples. Water is excellent, said *Pindar*, and we could not well live without it: But because every one may have enough of it for nothing, *Jeremy* had reason to lament that it could not be procur'd without Price. A Mole is a vile and despicable Animal, but in the Siege of *Cassilino* the Famine was so great, that one was sold for 200 *Florins*; and yet it was not dear, for he that parted with it dy'd of Hunger, and he that bought it out-liv'd the Siege.[22]

Davanzati's subjective theory of value fortunately found excellent successors in Italy, that is G. Montanali in the seventeenth century and F. Galiani in the eighteenth century. Galiani's *Della moneta* (1750) published anonymously is the summit of the Italian tradition of the subjectivism in value theory. Abbe Galiani was a sparkling conversationalist as well as a fine writer. While he stayed in Paris as a diplomatic officer, he was loved and admired for his talent by many people in society. The Italian way of thinking was naturally transmitted to the intellectual circle in France via Galianis' vivid talk. Among men and women gathering around him, Turgot and Condillac deserve special mention. They were deeply influenced by Italian economic thought and supported the subjective theory of value. Galiani was, indeed, a messenger from Italy who brought its old economic thought to France.

Finally, Auguste and Léon Walras succeeded this theory and added a conclusion to it.

---

[20] *Summa Theologica*, Question 77. English translation is due to Monroe [79] p. 59.

[21] Davanzati [22] p. 112. English translation [23] by J. Toland p. 14.

[22] Davanzati [22] p. 114. English translation [23] by J. Toland, p. 16.

As is well-known, Walras wrote:

> Je reconnais la priorité de Gossen quant à la courbe d'utilité et celle de Jevons quant à l'équation du maximum d'utilité dans l'échange; mais je ne leur ai pas emprunté ces conceptions. En revanche, je dois à mon père, Auguste Walras, les principes fondamentaux de ma doctorine économique et à Augustin Cournot celui de l'emploi du calcul des fonctions pour l'exposition de cette doctrine. Je l'ai déclaré dès mes premiers essais et en toute occasion depuis, lors, Ici, je voudrais expliquer comment cette doctorine s'est précisée, développée et complétée dans les éditions successives du présent ouvrage.[23]

However, I can not help perceiving certain huge entities of thought behind the names of Auguste Walras and Cournot. One is the Italian traditional economic thought and another is the modern mathematics in France. Walras was standing just at the meeting point of these two great rivers.

The Italian river of subjectivism diverged into several branches after that. A branch irrigated England and resonated with theological or Benthamite utilitarianism. Other tributary poured itself into the German–speaking lands and was absorbed by the soil plowed well by the subjective philosophy of Kant.

Jevons is a representative figure in England, who combined utility theory and the Benthamite calculus of pleasure and pain.

> It will be readily conceded that pain is the opposite of pleasure; so that to decrease pain is to increase pleasure; to add pain is to decrease pleasure. Thus we may treat pleasure and pain as positive and negative quantities are treated in algebra. The algebraic sum of a series of pleasures and pains will be obtained by adding the pleasures together and the pains together, and then striking the balance by subtracting the smaller amount from the greater. Our object will always be to maximise the resulting sum in the direction of pleasure, which we may fairly call the positive direction. This object we shall accomplish by accepting everything, and undertaking every action of which the resulting pleasure exceeds the pain which is undergone; we must avoid every object or action which leaves a balance in the other direction.[24]
>
> The words of Bentham on this subject may require some explanation and qualification, but they are too grand and too full of truth to be omitted.[25]

On the Germany–Austria bank of the tributary, there ripened ample fruits of economic theories by the German Classical School (my naming) including F. B. W. von Hermann, K.H. Rau, J. H. von Thünen, H. H. Gossen, H. K. von Mangoldt and so on. It is a pity that their achievements have been concealed or, at least, underestimated behind the dominance of the Historical School.

Menger and his school came after them.

In fact, Menger's detailed footnotes clearly show his close acquaintance with, at least a part of, the German Classical School.

> ... ... der Güter (Siehe Rau, „Volkswirthschaftslehre," §.58 ff.,1863). Eine Gattung kann nützliche Eigenschaften haben, welche die concreten Güter zur Befriedigung menschlicher Bedürfnisse tauglich machen, der Grad der Nützlichkeit kann bei den verschiedenen Gattungen mit Rücksicht auf bestimmte Gebrauchszwecke ein ungleicher sein (Buchenholz

---

[23] Walras [128] Préface de la 4me édition, p. viii.

[24] Jevons [54] p. 32.

[25] Jevons [54] pp. 23–24.

und Weidenholz für Heizzwecke u. dgl. m.); weder die Nützlichkeit der Gattung, noch aber auch der verschiedene Grad derselben bei den verschiedenen Gattungen oder Species kann indess „Werth" genannt werden. Nicht die Gattungen, sondern stets nur die concreten Güter sind den wirthschaftenden Individuen verfügbar, nur diese letztern demnach Güter und nur solche: Objecte unserer Wirthschaft und unserer Werthschätzung.[26]

Menger also frequently mentioned the names of Condillac, Turgot, and even Montanari, all belonging to the old Italian tradition of subjective value theory.

Von ältern Versuchen, den allgemeinen Begriff des Werthes festzustellen, seien hier jene Montanari's, 1687, (della Moneta III, S.43, p.a. der ed. Custodi), Turgot's (Valeurs et monnaies, S.79 ff., ed. Daire), Condillac's (Le commerce et le gouvernement 1776, S.151 ff., ed. Daire), Garnier's (S.5 der Vorrede zu seiner Uebersetzung A. Smith's), Storch's (Cours d'économ. polit. 1815, I, S.56 ff.) erwähnt. Zumal ist es Condillac, dessen Begriffsbestimmung des Werthes keine geringe Aehnlichkeit mit einzelnen neuern Entwickelungen dieser Lehre in Deutschland hat.[27]

The relation between the utility theory in the German-speaking world and the philosophy of Kant is still an open question. It seems quite natural to interpret its philosophical background in the context of Kant when we take account of such a passage from Menger [77] as follows:

Der Werth ist demnach nichts den Gütern Anhaftendes, keine Eigenschaft derselben, eben so wenig aber auch ein selbstständiges, für sich bestehendes Ding. Derselbe ist ein Urtheil, welches die wirthschaftenden Menschen über die Bedeutung der in ihrer Verfügung befindlichen Güter für die Aufrechthaltung ihres Lebens und ihrer Wohlfahrt fällen, und demnach ausserhalb des Bewusstseins derselben nicht vorhanden.[28]

Nonetheless there are some difficulties in the Kanten interpretation of Menger's methodology. This problem will be discussed in the final section of this chapter.

3 General Equilibrium

3.1 Walras' Equilibrium in a Pure Exchange

The significance of the marginal revolution can be evaluated from various viewpoints. But, I believe, no one would deny the appraisal that the general equilibrium theory founded by Walras is the most influential achievement of the revolution. Indeed it is Walras' theory that determined the basic framework of economic analysis or the economists' basic manner of grasping the world since Walras.

Assume that there are l kinds of goods, discriminated by $j = 1, 2, \ldots, l$. The number of agents is m and their names are indicated by $i = 1, 2, \ldots, m$. Each agent i has

[26] Menger [77] s. 81n.

[27] Menger [77] s. 80n.

[28] Menger [77] s. 86. Chipman [16] is an invaluable study on the German School. See also Streissler [115].

a utility function u^i as well as an initial endowment vector $\omega^i = (\omega_1^i, \omega_2^i, \dots, \omega_l^i)$.[29] A consumption vector of the i-th agent is expressed by $x^i = (x_1^i, x_2^i, \dots, x_l^i)$. $p = (p_1, p_2, \dots, p_l)$ denotes a price (vector) of commodities. Each agent is assumed to act so as to maximize his utility subject to the budget constraint.

$$\text{Maximize} \quad u^i(x^i)$$
$$\text{subject to} \quad \sum_{j=1}^{l} p_j x_j^i \leqq \sum_{j=1}^{l} p_j \omega_j^i. \tag{1}$$

The market is assumed to be *competitive* in the sense of Cournot-Walras—every agent perceives p as given. If the agent's utility is non-satiated (for instance, if marginal utilities are positive), the inquality $\leqq$ in the constraint of (1) may be changed to the equality $=$.

The solution of the problem (1) must satisfy the condition[30]:

$$\frac{D_j u^i(x^i)}{D_k u^i(x^i)} = \frac{p_j}{p_k}, \quad j, k = 1, 2, \dots, l \tag{2}$$

and

$$\sum_{j=1}^{l} p_j x_j^i = \sum_{j=1}^{l} p_j \omega_j^i. \tag{3}$$

The demand function $\xi^i(p)$ of the i-th agent is derived from (2) and (3) as a function of p. As is easily seen, ξ^i satisfies

$$\xi^i(\alpha p) = \xi^i(p) \quad \text{for any } \alpha > 0 \text{ and } p, \tag{4}$$
$$\text{(positive homogeneity of degree zero)}$$

and

29 As in other pioneers of modern utility theory, Walras wrote a utility function in additive form:

$$u^i(x^i) = u_1^i(x_1^i) + u_2^i(x_2^i) + \cdots + u_l^i(x_l^i).$$

However, we do not have to worry about it here.

30 An independent set of equations in (2) is given by (for instance)

$$\frac{D_j u^i(x^i)}{D_1 u^i(x^i)} = \frac{p_j}{p_1}, \quad j = 2, 3, \dots, l. \tag{2'}$$

If these $(l-1)$ equations are valid, other equations follow from them. $D_j u^i(x^i)$ denotes the j-th partial derivative of u^i at x^i; i.e. $D_j u^i = \partial u^i / \partial x_j^i$.

$$\sum_{j=1}^{l} p_j \xi_j^i(p) = \sum_{j=1}^{l} p_j \omega_j^i \quad \text{for any } p. \tag{5}$$

The market is in a *general equilibrium* under the price p^* if the excess demand is cleared; that is

$$\zeta(p) \equiv \sum_{i=1}^{m} \xi^i(p) - \sum_{i=1}^{m} \omega^i = 0 \tag{6}$$

for some $p = p^*$. p^* is called an *equilibrium price*. $\zeta(p) = 0$ consists of l equations $\zeta_j(p) = 0$ $(j = 1, 2, \dots, l)$. Since we obtain the Walras law:

$$\sum_{j=1}^{l} p_j \zeta_j(p) \equiv 0 \quad \text{for all } p \tag{7}$$

by adding (5) over all i, not all the l equations in (6) are independent.[31] Thus $p = (p_1, p_2, \dots, p_l)$ can not be solved by (6). However, taking account of

$$\begin{aligned} &\xi^i(p) = \xi^i(1, q_2, \dots, q_l), \\ &\qquad (q_j = p_j/p_1 \quad , \quad j = 2, 3, \dots, l) \end{aligned} \tag{8}$$

by the homogeneity (4), we can solve $q = (q_2, q_3, \dots, q_l)$ by means of $(l - 1)$ independent equation (6).[32] q_j's are prices of commodities in terms of the first commodity. The first commodity playing a role as the standard of value is called a *numéraire*.

This is the simplest framework of general equilibrium completed by Walras in his *Eléments d'économie politique pure*.

3.2 Fisher–Wicksell System

I. Fisher and K. Wicksell presented an equilibrium system which is essentially the same as Walras'.[33] The Fisher–Wicksell system consists of (2), (3) together with

$$\sum_{i=1}^{m} x_j^i = \sum_{i=1}^{m} \omega_j^i, \quad j = 1, 2, \dots, l. \tag{9}$$

---

[31] Assuming (7), we can deduce any one of the equations in (6) from the others.

[32] The existence of equilibrium prices is not guaranteed by this elementary method of counting the number of variables and that of equations. The exposition of the problem in full detail is given in Debreu [24] Chap. 5.

[33] Wicksell was quite familiar with Walras' *Elémens*. On the other hand, Fisher, a Ph.D student of J. W. Gibbs, did not know it. Wicksell [132] ss. 50–56, Fisher [31] pp. 52–54.

Both the number of equations and that of variables (p_j, x_j^i) are the same $(= ml + l)$. But the number of independent equations is (at most) no more than $ml + l - 1$. In fact, we obtain Walras' law again by adding both sides of (3) over all $i = 1, 2, \ldots, m$. Then one of the equations consisting of (9) can be deduced from the others. So we can determine only relative prices, say $p_2/p_1,\ p_3/p_1, \ldots, p_l/p_1$. The essential structure of the Fisher–Wicksell system is the same as Walras' system.

In addition, it deserves a special notice that Fisher presented an ingeneous mechanical/hydraulic apparatus to compute equilibrium prices and allocations.[34] He is "the" precursor of the systematic theory of designing processes for computing fixed points or equilibria, which has been developed after the 1970s.

3.3 Jevons' Exchange Equation

Jevons also explored an equilibrium system called the "exchange equation", which deserves more careful examination.

For the sake of simplicity, we assume that there are three agents $(i = 1, 2, 3)$ and three commodities A, B, C. The utility function of each agent i is expressed in an additive form:

$$U^i(a^i, b^i, c^i) = U_A^i(a^i) + U_B^i(b^i) + U_C^i(c^i),$$

where a^i, b^i and c^i are quantities of his demand for A, B and C, respectively. $MU_A^i(a^i)$, $MU_B^i(b^i)$ and $MU_C^i(c^i)$ denote his marginal utilities of A, B, and C. The initial endowment ω^i $(i = 1, 2, 3)$ is assumed to be given by

$$\omega^1 = (a^*, 0, 0), \quad \omega^2 = (0, b^*, 0), \quad \omega^3 = (0, 0, c^*),$$

i.e. each agent has only one commodity as his initial endowment.

Suppose that any exchange is performed pairwise; that is, the 1st agent and the 2nd agent exchange A and B, the 2nd agent and the 3rd agent exchange B and C, and finally the 3rd agent and the 1st agent exchange C and A, mutually. The quantities of the commodity A supplied by the 1st agent to the 2nd and the 3rd agent are denoted by $a(2)$ and $a(3)$. $b(1)$, $b(3)$; $c(1)$ and $c(2)$ should be interpreted similarly (cf. Fig. 2). Then the following relations must hold true at an equilibrium in the sense of Jevons. As explained above, the notation MU_A^1 means the marginal utility of the 1st agent with respect to the commodity A. Others should be interpreted similarly.

[34] Brainard–Scarf [11].

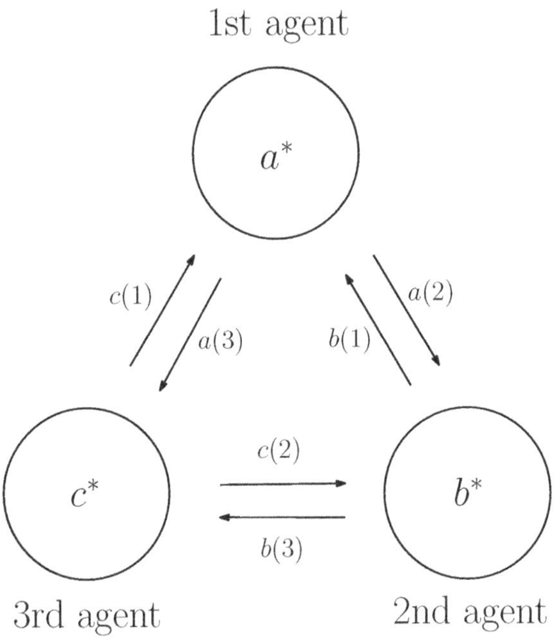

Fig. 2 Jevons' bilateral exchanges

$$\begin{aligned}
\frac{MU_A^1(a^* - a(2) - a(3))}{MU_B^1(b(1))} &= \frac{b(1)}{a(2)} = \frac{MU_A^2(a(2))}{MU_B^2(b^* - b(1) - b(3))}, \\
\frac{MU_A^1(a^* - a(2) - a(3))}{MU_C^1(c(1))} &= \frac{c(1)}{a(3)} = \frac{MU_A^3(a(3))}{MU_C^3(c^* - c(1) - c(2))}, \\
\frac{MU_B^2(b^* - b(1) - b(3))}{MU_C^2(c(2))} &= \frac{c(2)}{b(3)} = \frac{MU_B^3(b(3))}{MU_C^3(c^* - c(1) - c(2))}.
\end{aligned} \tag{10}$$

This system is called the exchange equation.

However, a solution of Jevons' exchange equation has a serious defect. We should recall Cournot's insight into the non-arbitrage condition (cf. p. 5). If the exchange equation is satisfied by, for instance,

$$p(B, A) = \frac{a(2)}{b(1)} = \frac{1}{2}, \quad p(C, B) = \frac{b(3)}{c(2)} = \frac{1}{3}, \quad p(C, A) = \frac{a(3)}{c(1)} = \frac{1}{4},$$

this solution is not stable because such a "distortion" of prices admits some indirect exchanges (arbitrages) which give further benefits to agents.

As far as I know, K. Wicksell first pointed out this defect of the Jevonian concept of equilibrium in his *Über Wert, Kapital und Rente* (1893):

Jevons scheint aber nicht bemerkt zu haben, daß der durch seine Gleichungen bezeichnete Gleichgewichtszustand die Möglichkeit des Zwischenhandels sowie der Geld- und Kreditgechäfte prinzipiell ausschließt, und daß, wenn jene zugelassen werden, das Gleichgewicht sofort aufs neue gestört würde. Er erinnert daran, daß dasselbe Paar von Gütern auf demselben Markte nur ein Tauschverhältnis haben könne, sagt aber nichts davon, daß es bei völlig freiem Verkehr unter drei Gütern nur zwei unabhängige Tauschverhältnisse (und allgemein unter n Gütern nur deren $n-1$) geben kann; er behandelt jene Tauschverhältnisse in der That, als ob sie alle drei unabhängig wären.[35]

Walras also presented a general equilibrium system in the case of pairwise exchanges. For the sake of simplicity, suppose that there are three agents and three goods. The utility function and the endowment of the i-th agent is given as above $(i = 1, 2, 3)$.

In the pairwise exchange economy, the behavior of each agent is described by the following problems ($x_A^1, x_B^2, \ldots,$ denote the demand for A of the 1st agent, the demand for B of the 2nd agent, ...).

$$\begin{aligned}
\text{agent 1:}\quad & \text{Maximize}\quad U^1(x_A^1, x_B^1, x_C^1) \\
& \text{subject to} \\
& \quad p(B, A)x_B^1 + p(C, A)x_C^1 = a(2) + a(3). \\
\text{agent 2:}\quad & \text{Maximize}\quad U^2(x_A^2, x_B^2, x_C^2) \\
& \text{subject to} \\
& \quad p(A, B)x_A^2 + p(C, B)x_C^2 = b(1) + b(3). \\
\text{agent 3:}\quad & \text{Maximize}\quad U^3(x_A^3, x_B^3, x_C^3) \\
& \text{subject to} \\
& \quad p(A, C)x_A^3 + p(B, C)x_B^3 = c(1) + c(2).
\end{aligned}$$

By solving these problems, we derive (pairwise) demand functions. $\xi_{B,A}^1$ (resp. $\xi_{C,A}^1, \ldots$) is the optimal demand of the 1st agent for B in exchange for A (resp. the optimal demand of the 1st agent for C in exchange for $A, \ldots$)

$$\text{agent 1}\begin{cases} \xi_{B,A}^1 = \xi_{B,A}^1(p(B, A), p(C, A)), \\ \xi_{C,A}^1 = \xi_{C,A}^1(p(B, A), p(C, A)), \end{cases}$$

$$\text{agent 2}\begin{cases} \xi_{A,B}^2 = \xi_{A,B}^2(p(A, B), p(C, B)), \\ \xi_{C,B}^2 = \xi_{C,B}^2(p(A, B), p(C, B)), \end{cases}$$

$$\text{agent 3}\begin{cases} \xi_{A,C}^2 = \xi_{A,C}^2(p(A, C), p(B, C)), \\ \xi_{B,C}^2 = \xi_{B,C}^2(p(A, C), p(B, C)). \end{cases}$$

---

[35] Wicksell [132] s. 55.

An equilibrium of the pairwise-exchange economy requires two sets of conditions, I and II.

I. Pairwise equalities of demands and supplies:

$$\begin{aligned} \xi^1_{B,A} &= \xi^2_{A,B}\, p(A,B), & \xi^1_{C,A} &= \xi^3_{A,C}\, p(A,C), \\ \xi^2_{A,B} &= \xi^1_{B,A}\, p(B,A), & \xi^2_{C,B} &= \xi^3_{B,C}\, p(B,C) \\ \xi^3_{A,C} &= \xi^1_{C,A}\, p(C,A), & \xi^3_{B,C} &= \xi^2_{C,B}\, p(C,B). \end{aligned} \tag{11}$$

II. No-arbitrage admitted:[36]

$$\begin{aligned} p(A,B) &= \frac{1}{p(B,A)}, & p(C,B) &= \frac{p(C,A)}{p(B,A)}, \\ p(A,C) &= \frac{1}{p(C,A)}, & p(B,C) &= \frac{p(B,A)}{p(C,A)}. \end{aligned} \tag{12}$$

We now choose, for instance, the 1st commodity as the numéraire and write $p(B,A) = p_B$, $p(C,A) = p_C$. Then it follows from (11) and (12) that

$$\begin{aligned} \xi^2_{A,B} + \xi^3_{A,C} &= p_B \xi^1_{B,A} + p_C \xi^1_{C,A}, \\ \xi^1_{B,A} + \xi^3_{B,C} &= \frac{1}{p_B}\xi^2_{A,B} + \frac{p_C}{p_B}\xi^2_{C,B}, \\ \xi^1_{C,A} + \xi^2_{C,B} &= \frac{1}{p_C}\xi^3_{A,C} + \frac{p_B}{p_C}\xi^3_{B,C}. \end{aligned} \tag{13}$$

However, the first equality can be derived from the other two.[37] Thus we obtain two independent equations to solve two independent variables, p_B and p_C.

It is quite obvious how to generalize this simple system to l-commodities model. In the case there are only two commodities, the problem of arbitrage does not occur.

Walras seemed to learn the significance of the non-arbitrage condition from Cournot. On the other hand, Jevons had not read Cournot's book when he wrote the 1st edition of *Theory of Political Economy* (cf. p. 8). This is, at least, one of the reasons why Jevons did not take notice of the arbitrage problem.[38] The following is Jevons' appraisal of Cournot.

---

[36] Clearly, $B(C,B) = 1/p(B,C)$is implied.

[37] Multiply the second equality by p_B and the third by p_C. Then adding them, we obtain

$$\begin{aligned} p_B\xi^1_{B,A} + p_B\xi^3_{B,C} &= \xi^2_{A,B} + p_C\xi^2_{C,B}, \\ p_C\xi^1_{C,A} + p_C\xi^2_{C,B} &= \xi^3_{A,C} + p_B\xi^3_{B,C}, \end{aligned}$$

from which the first equality follows.

[38] Many years ago, this problem of Jevons' exchange equation was discussed and examined in detail in my seminar for graduate students. Nakano [82] resulted from the seminar and is an excellent report on the topic.

> My impression is that ... ... the third chapter, treating of the conditions of the foreign exchanges, is highly ingenious if not particularly useful; but that by far the most important part of the book commences with the forth chapter upon the "Loi du débit." The remainder of the book, in fact, contains a wonderful analysis of the laws of supply and demand, and of the relations of prices, production, consumption, expenses and profits. Cournot starts from the assumption that the débit or demand for a commodity is a function of the price, or $D = F(p)$; and then, after laying down empirically a few conditions of this function, he proceeds to work out with surprising power the consequences which follow from those conditions.[39]

3.4 Marshall's General Equilibrium Model

The theoretical framework of A. Marshall's *Principles of Economics* (1890) is basically an equilibrium analysis which shares common features with Walras' system. While Walras tried to grasp the mutual interdependence of an entire economy at one stroke, Marshall's attitude was more practical and he focused his attention on some restricted aspects relevant to the problem in question.

> The forces to be dealt with are however so numerous, that it is best to take a few at a time; and to work out a number of partial solutions as auxiliaries to our main study. Thus we begin by isolating the primary relations of supply, demand and price in regard to a particular commodity.[40]

However, we have to keep in mind that Marshall understood the significance of the general interdependence so well that he invented many supplementary tools to reinforce his partial analysis based upon the "*ceteris paribus*" method (cross elasticity, for instance). Actually, Marshall drew up a general equilibrium model with production at a very early stage of his economic research. His system developed in Mathematical Appendix XXI of Marshall [67] illuminates how well he understood the general interdependence theoretically.

Suppose that there are l consumption goods ($j = 1, 2, \ldots, l$) and n productive factors ($k = 1, 2, \ldots, n$). Walras expresses the quantities of demand and supply as functions of all prices. On the contrary, Marshall regards prices as functions of quantities. According to Marshall, the concepts of "demand price" and "supply price" are defined as follows, respectively[41]:

> ... ... there is a demand price for each amount of the commodity, that is, a price at which each particular amount of the commodity can find purchasers in a day or week or year.

> ... ... [a supply price—Maruyama] is the price the expectation of which will just suffice to maintain the existing aggregate amount of production.

The demand price p_j^d of the j-th commodity is assumed to depend only upon the quantity x_j of the demand for it:

---

[39] Jevons [54] pp. xxx–xxxi.

[40] Marshall [67] Preface to the eighth edition, p. xiv.

[41] Marshall [67] pp. 342–343.

$$p_j^d = f_j^d(x_j), \quad j = 1, 2, \ldots, l.$$

The supply price p_j^s is also a function of x_j alone:

$$p_j^s = f_j^s(x_j), \quad j = 1, 2, \ldots, l.$$

The most efficient production coefficient of the k-th productive factor to produce the quantity x_j of the j-th commodity is denoted by $a_k(x_j)$ $(j = 1, 2, \ldots, l;\ k = 1, 2, \ldots, n)$. Given the prices q_k $(k = 1, 2, \ldots, n)$ of productive factors, the supply price p_j^s is more concretely written as

$$p_j^s = a_1(x_j)q_1 + a_2(x_j)q_2 + \cdots + a_n(x_j)q_n, \quad j = 1, 2, \ldots, l. \tag{1}$$

Conversely, when the prices p_j's of commodities are given, the demand prices q_k^d's for productive factors, which each producer is at most able to pay, must satisfy

$$p_j = a_1(x_j)q_1^d + a_2(x_j)q_2^d + \cdots + a_n(x_j)q_n^d, \quad j = 1, 2, \ldots, l. \tag{2}$$

The demand y_k for the k-th productive factor for each x_j is

$$\begin{aligned} y_k &= a_k(x_1)x_1 + a_k(x_2)x_2 + \cdots + a_k(x_l)x_l \\ &= \sum_{j=1}^{l} a_k(x_j)x_j \quad , \quad k = 1, 2, \ldots, n, \end{aligned} \tag{3}$$

The supply price q_k^s's of the k-th productive factor is expressed as a function of its quantity y_k of supply:

$$q_k^s = g_k^s(y_k), \quad k = 1, 2, \ldots, n. \tag{4}$$

The equality of demand and supply in the factor markets is expressed as

$$\sum_{j=1}^{l} a_k(x_j)x_j = y_k, \quad k = 1, 2, \ldots, n. \tag{5}$$

Substituting $q_k^d = q_k^s = g_k^s(y_k)$ $(k = 1, 2, \ldots, n)$ into (1), we obtain the equilibrium condition for commodity markets in the form

$$f_j^d(x_j) = \sum_{j=1}^{n} a_k(x_j)g_k^s(y_k)\ (= f_j^s(x_j)), \quad j = 1, 2, \ldots, l. \tag{6}$$

Thus the Marshallian equilibrium must satisfy (5) and (6). The number of unknown variables is $(l + n)$ and there is the same number of equations.

In Marshall's model, the demand or supply price of each commodity is assumed to be a function of the quantity of the commodity in question alone. Hence this model, in a sense, still remains in the realm of the partial-equilibrium analysis. Someone may also point out the defect that the relation between the demand for consumption goods and the income resulting from the supply of factors is not clearly explained. The determinants of $a_k(x_j)$ are left unexplained.

Although Marshall's system does not reach the full clarity of Walras', it correctly grasps the mutual interdependence among all the sectors by simultaneous equations. There is no essential difference between Marshall and Walras.

3.5 ♣ *Mr. Kawamata on Characterizations of Competitive Equilibrium*

Consider an economy described in Sect. 3.1. An allocation $x = (x^1, x^2, \ldots, x^m)$ (list of activities of all agents) is called a Walras allocation if it is in a general equilibrium under some price.

We now compare a pair of feasible[42] allocations, $x = (x^1, x^2, \ldots, x^m)$ and $y = (y^1, y^2, \ldots, y^m)$ from another viewpoint. y is said to be superior to x if (i) $u^i(x^i) \leqq u^i(y^i)$ for all i and (ii) $u^{i'}(x^{i'}) < u^{i'}(y^{i'})$for some i'. A feasible allocation x is called Pareo-optimal if there is no feasible allocation which is superior to x. This concept was formulated by V. Pareto in [90][43] and became a basic criterion by which to judge the efficiency of allocations.

A Walras allocation is, in a sense, a solution of a non-cooperative game in a competitive market. On the other hand, F. Y. Edgeworth [29] considered a cooperative game, in which agents are permitted to form coalitions and to cooperate to improve their economic situations. The core is the set of all feasible allocations which can be interpreted as solutions or a "final settlements" of this game.

We thus obtain the sets of feasible allocations which satisfy three different requirements:

(W) the set of all the Walras allocations,

(P) the set of all the Pareto-optimal allocations, and

(C) the core.

Roughly speaking, (W), (P) and (C) are coincide under certain assumptions. This deep insight illuminates marvelous roles of competitive price mechanism.

Another sort of allocation was invented by the Austrian School, in particular by F. von Wieser [121], in the context of the theory of "Zurechnung" (imputation).

Kawamata rigorously compares these concepts of equilbrium and tries to seek for their respective origins in the prehistory of the Marginal Revolution.[44]

---

[42] An allocation $x = (x^1, x^2, \ldots, x^m)$ is feasible if $\sum_{i=1}^{m} x^i = \sum_{i=1}^{m} \omega^i$.

[43] Pareto [90] pp. 617–618 (Appendix 89).

[44] For the basic literature, see the reference of Kawamata's chapter.

3.6 ♣ Mr. Cornet on Exchange Economies with Linear Consumers

The Walras–Cassel theory of production paved the way to the evolution of various linear economic models. The pioneering works by A. Wald and J. von Neumann, W. Leontieff's input-output analysis and so on belong to this genealogy. It also seems remarkable that the new fresh development of convex analysis in mathematics was deeply resonant with "linear economics".

On the other hand, Cornet examines the workings of a simple exchange economy with linear utility functions. Converting the equilibrium problem to a concave programming problem, he shows the existence of equilibrium and the convexity of equilibrium set (even uniqueness). It embodies some aspects in the new trends of linear economics, the origin of which can be traced back to the Marginal Revolution.

4 Utility and Demand

Every economist who played an active role in the Marginal Revolution based his theory upon a subjective element—utility or human wants. In Sect. 2, we briefly observed the long history of continuous succession of utility theory over two thousand years since Aristotle. We now review the development of utility theory during the Marginal Revolution.[45]

4.1 From Utility to Demand

Every author in the utility tradition emphasized the close relationship between utility and demand. However, even Davanzati and Galiani could not show how to connect these two concepts rationally.

Menger wrote in his *Grundsätze*:

> Wo immer Menschen wohnen, und welche Stufe der Culturentwickelung sie auch immer einnehmen, überall können wir beobachten, wie die wirthschaftenden Individuen die Bedeutung der Befriedigung ihrer verschiedenen Bedürfnisse im Allgemeinen und jene der Einzelnen zur mehr oder minder vollständigen Befriedigung derselben führenden Acte insbesondere gegen einander abwägen, und sich schliesslich von dem Resultate dieser Prüfung in der auf die möglichst vollständige Befriedigung ihrer Bedürfnisse gerichteten Thätigkeit (Wirthschaft) bestimmen lassen. Ja, es ist dies Abwägen der verschiedenen Bedeutung der Bedürfnisse, die Wahl zwischen jenen, weche, unbefriedigt bleiben, und jenen, welche, je nach den verfügbaren Mitteln, zur Befriedigung gelangen, und die Bestimmung des Grades, bis zu welchem diese letzteren ihre Befriedigung finden sollen, jener Theil der ökonomischen Thätigkeit der Menschen, welcher ihre Geister mehr als irgent ein anderer erfüllt,

[45] Stigler [110] is the most reliable report on this topic. See slso Howey [41], Kauder [58] and Schumpeter [103] pp. 1053–1073.

> auf ihre ökonomischen Bestrebungen den weittragendsten Einfluss nimmit, und von jedem wirthschaftenden Subjecte fast ununterbrochen geübt wird.[46]

However, we can not find any explicit exposition here to connect utility and demand. Jevons arrived at a sound understanding about the condition of utility maximization:

> The keystone of the whole Theory of Exchange, and of the principal problems of Economics, lies in this proposition — *The ratio of exchange of any two commodities will be the reciprocal of the ratio of the final degrees of utility of the quantities of commodity available for consumption after the exchange is completed.*[47]

On the other hand, he clearly observed that demand and supply depend upon prices.

> The ordinary laws of supply and demand treat entirely of quantities of commodity demanded or supplied, and express the manner in which the quantities vary in connection with the price. In consequence of this fact the laws *are* mathematical.[48]

Very regrettably, he did not give any theory to derive the demand function through utility maximization. Taking account of his passage as above, we can not help feeling it somewhat curious that Jevons did not grasp the concept of the demand function. This defect is due to his unfamiliarity with Cournot's book when he wrote the first edition of *Theory* (cf. p. 8).

It is definitely Walras who completely solved the problem to derive the demand function from utility through its maximization and to connect the two concepts rationally. Walras' solution of this problem was first published in the summer in 1873,[49] one year before the release of *Eléments*. His works prior to 1873 left no hint as to the approach to his goal.[50] How did he find a correct way? We can learn a historical episode from W. Jaffé's masterpiece [50], which clarifies this problem. The most important material is A. P. Piccard's reply to Walras' inquiry (Walras' letter is not archived)[51] concerning how to derive the demand function from utility maximization. Walras grasped the "problem" in the correct manner, but he was not able to solve it mathematically by himself. It is thought to be in autumn or winter of 1872, that Piccard's reply was sent to Walras. Piccard's exposition is restricted to the elementary case of two-commodities. It is Walras who generalized it by himself to the case

---

46 Menger [77] ss. 94–95.

47 Jevons [54] p. 95.

48 Jevons [54] pp. 3–4.

49 Walras [127].

50 Walras' works on economics prior to 1873 are: "Application de mathématiques à l'économie politique",
1ère tentative, 1860,
2e tentative, 1869–1870.
"Système des phenomènes économiques", Program of proposed courses of lectures at Geneve January 2, 1872. Jaffé [50] Vol. I., pp. 216–221, pp. 293–296.

51 Jaffé [50] Vol. I, pp. 309–311. (undated). Piccard was professor of mechanics at the Academy of Lausanne.

of multi-commodities. Walras' article in 1873 was thus completed with the indispensable assistance of Piccard. On October 25, 1873, Walras wrote to Piccard to acknowledge his contribution.

> J'ai communiqué à l'Académie des sciences morales et politiques, à Paris, dans ses séances des 16 et 23 août dernier, le principe de la *Théorie mathématique de l'échange* que je vous ai soumise et que vous avez pris la peine d'examiner dans le courant de l'hiver dernier. Cette communication doit être reproduite en janvier prochain dans le recueil des Séances et Travaux de l'Académie.
>
> Une application simple et facile de calcul différentiel et intégral, que vous avez aperçue et énoncée vous-même, si vous vous en souvenez, dès que je vous ai posé la question, motre que c'est *que le rapport des raretés soit égal au prix*.[52]

Like most of his contemporaries, Marshall expressed utility functions in additive form:

$$U_1(x_1) + U_2(x_2) + \cdots + U_l(x_l).$$

He also imposes an additional restriction that the marginal utility of the numéraire (say, the first commodity) is always constant. Then the condition of utility maximization is

$$\frac{dU_j/dx_j}{p_j} = \frac{dU_1}{dx_1} \equiv \alpha \text{ (constant)}, \quad j = 2, 3, \ldots, l.$$

$$\text{i.e. } p_j = \frac{1}{\alpha} \cdot \frac{dU_j(x_j)}{dx_j}, \quad j = 2, 3, \ldots, l.$$

We immediately obtain the demand function of the j-th commodity by multiplying its marginal utility by $1/\alpha$.[53]

4.2 Ordinal or Cardinal

Let X be a subset of $\mathbb{R}^l$, which is interpreted as the set of all possible consumptions $x = (x_1, x_2, \ldots, x_l)$ for an agent. His preference is assumed to be represented by a complete preordering $\precsim$ on X.[54] Mathematically, a real-valued function $u : X \to \mathbb{R}$ is called a utility function which represents $\precsim$ if it satisfies

[52] Jaffé [50] Vol. I, pp. 345–347. Letter ♯ 239.

[53] Marshall [67] pp. 94–97. Such a simple derivation of the demand curve becomes possible by accepting the Marshallian additional assumption. But this assumption guarantees some very strong results. For example, he effectively made use of this concept in his theory of the price adjustment process from any disequilibrium position to an equilibrium. His non-tâtonnement process converges to the same equilibrium price irrespective of the adjustment paths. This result is obtained by the Mashallian assumption on utility; cf. Walker [123].

[54] A binary relation $\precsim$ on X is called a preordering if it satisfies (i) $x \precsim x$ (reflexivity) and (ii) $x \precsim y, y \precsim z \Rightarrow x \precsim z$ (transitivity) for any x, y and $z \in X$. A preordering is said to be complete if either $x \precsim y$ or $y \precsim x$ holds good for any $x, y \in X$. See Debreu [24] Chap. 4 for the existence of utility function.

$$x \precsim y \iff u(x) \leqq u(y), \quad x, y \in X.$$

Let Φ be a family of all monotonically increasing real-valued functions of $\mathbb{R}$. Then the composition $\varphi \circ u$ is also a utility function representing the same $\precsim$ for any $\varphi \in \Phi$. Hence if there exists at least a utility function, then infinitely many utility representations are admitted for the same preordering $\precsim$. An economist is called an ordinalist if he regards any proposition as valid provided that it does not depend upon the choice of $\varphi \in \Phi$. His standpoint is called ordinalism.

For instance, the following statements are not admissible to ordinalists because their validity depends upon the choice of $\varphi \in \Phi$.[55]

a. A marginal utility is decreasing.
b. $u(y) - u(x) > u(z) - u(x), \quad x, y, z \in X.$

Let $\Phi_a \subset \Phi$ be a family of all the affine functions $\varphi : \mathbb{R} \to \mathbb{R}$ of the form $\varphi(t) = at + b$ with $a > 0$. If anyone regards a proposition as valid provided that it does not depend upon the choice of $\varphi \in \Phi_a$, his standpoint is called cardinalism. From the viewpoint of cardinalism, the statements a and b are easily seen to be admissible.

For a cardinalist, utility is a "measurable concept" like length, volume, weight, ... But for an ordinalist, utility is not measurable (with some origin and unit) but only admits mutual comparison. In other words, ordinalists admit utility as no more than a "comparative concept" and reject it as a "measurable concept".[56]

Walras derived the demand function by maximizing utility. It is an important observation that resulting demand function is determined independently of the choice of utility function which represents the same preference. Thus Walrasian demand theory must be admitted from the ordinalist viewpoint.

This is exactly what Walras achieved in *Eléments*. However, he suffered from some criticisms which did not justify the use of non-measurable concepts like utility in science. For instance, H. Laurent, a mathematician, wrote to Walras (May 13, 1900):

> Vous supposez qu'il existe une mesure à l'utilité. Ne pourriez-vous pas vous passer de ce postulatum en présentant les choses ainsi, ce qui choquera un peu moins les mathématiciens purs?[57]

---

[55] Assume that $l = 1$ and so $X \subset \mathbb{R}$ for the sake of simplicity. $\varphi(t) \in \Phi$ is specified as $\varphi(t) = t^3$.

a. $v(x) = \varphi \circ u(x) = u(x)^3$ is also a utility function. The marginal utility of v is given by $v'(x) = \varphi'(u(x))\, u'(x) = 3u(x)^2\, u'(x)$. However, $v''(x) = 6u(x)\,(u'(x))^2 + 3u(x)^2 u''(x)$ may not be negative even if $u''(x) < 0$.
b. If $u(x) = 1$, $u(y) = 4$, $u(z) = 6$, for instance, then $u(y) - u(x) > u(z) - u(x)$. But $v(y) - v(x) < v(z) - v(x)$.

Thus the statements a and b contradict the ordinalism because their validity depends upon how to choose $\varphi \in \Phi$.

[56] For the philosophical observations on measurement, see Hempel [37] Chap. III and Carnap [14] Part II (especially Chaps. 5–6).

[57] Jaffé [50] Vol. III, p. 116. Letter ♯ 1452.

In his reply (May 22, 1900). Walras tried to explain his view again—utility is certainly not a measurable concept, but his theory does not require its measurability.

> Ce rapport commun correspond à votre *facteur d'intégrabilité* μ. Vous pouvez, si vous le jugez à propos, faire figurer ce facteur dans la définition des fonctions de rareté. Mais j'appelle toute votre attention sur le fait que le problème économique, une fois posé comme il l'a été plus haut, se résout sans que μ soit supposé ni *mesuré*, ni même *mesurable*. Pour ma part, je me passe complètement de l'étalon d' utilité et de rareté.[58]

Walras, still being anxious about the validity of his method, wrote a letter of inquiry to H. Poncaré.[59] Poincaré's reply (September 30, 1901) explained the crucial points so clearly that a historian once called it the "Magna Carta" of ordinalism in utility theory.[60] The important points exposed in Poincaré's short letter are as follows.[61]

1. Clear distinction between the comparative concept and the quantitative concept.

    > Voici comment je la justifierais. La satisfaction peut-elle se mesurer? Je puis dire que telle satisfaction est plus grande que telle autre, puisque je préfère l'une à l'autre. Mais je ne puis dire que telle satisfaction est deux fois ou trois fois plus grande que telle autre. Cela n'a aucun sens par soimême et ne pourrait en acquérir un que par une convention arbitraire.
    >
    > La satisfuction est donc une grandeur, mais non une grandeur mesurable.

2. Any monotone transformation of a utility function is also a utility function.

    > Maintenant, une grandeur non-mesurable sera-t-elle par cela seul exclue de toute spéculation mathématique? Nullement. La température par exemple (au moins jusqu'à l'avènement de la thermodynamique qui a donné un sens au mot de température absolue) était une grandeur non-mesurable. C'est arbitrairement qu'on la définissait et la mesurait par la dilatation du mercure. On aurait pu tout aussi légitimement la définir par la dilatation de tout autre corps et la mesurer par une fonction *quelconque* de cette dilatation pourvu que *cette fonction fût constamment croissante*. De même ici vous pouvez définir la satisfaction par une fonction arbitraire pourvu que cette fonction croisse toujours en même temps que la satisfaction qu'elle représente.

3. Theoretical consequences derived from a utility function should be independent of its arbitrary choice.

---

58 Jaffé [50] Vol. III, p. 119. Letter ♯ 1454.

59 Jaffé [50] Vol. III, pp. 161–162, September 26, 1901. Letter ♯ 1495. Poincaré was professor of University of Paris.

60 Schumpeter points out Walras' French character shared with Poincaré:

> Marie Esprit Léon Walras (1834–1910) was a Frenchman and not only by virtue of his birthplace. The style of his reasoning and the nature of his achievement are characteristically French in the same sense in which Racine's plays and J.H. Poncaré's mathematics are characteristically French.

Schumpeter [103] p. 828.

61 Jaffé [50] Vol. III, pp. 164–165. Letter ♯ 1496.

> Dans vos prémisses vont donc figurer un certain nombre de fonctions arbitraires; mais une fois ces prémisses posées, vous avez le droit d'en tirer des conséquences par le calcul; si, dans ces conséquences, les fonctions arbitraires figurent encore, ces conséquences ne seront pas fausses, mais elles seront dénuées de tout intérêt parce qu'elles seront subordonnées aux conventions arbitraires faites au début. Vous devez donc vous efforcer d'*éliminer* ces fonctions arbitraires, et c'est ce que vous faites.

The last words "c'est ce que vous faites", which assured the validity of Walras' method, completely satisfied Walras. He included Poincaré's letter in his last article "Economique et mecanique".

4.3 Forms of Utility Function

Almost all the earlier contributors (including Jevons, Walras, Marshall, ...) express utility in the additive form:

$$u(x) = U_1(x_1) + U_2(x_2) + \cdots + U_l(x_l).$$

F.Y. Edgeworth and I. Fisher proposed a utility function of general form $u(x) = u(x_1, x_2, \ldots, x_l)$ first.[62] R. Auspitz and R. Lieben started to study substitutive or complementary relations between commodities by means of generalized utility functions.[63] According to their classification, the i-th commodity and the j-th commodity are substitutive (resp. complementary) with each other if the second derivative $D^2_{x_i,x_j} u = \partial^2 u/\partial x_i \partial x_j$ is negative (resp. positive). However, this classification is not approved from the ordinalist viewpoint because the sign of the second derivative depends upon the choice of $\varphi \in \Phi$.

The definition permitted by ordinalism was discovered much later by Hicks.[64]

4.4 Comparative Statics

Consider a system of equations

$$f_i(x_1, x_2, \ldots, x_n; \theta_1, \theta_2, \ldots, \theta_m) = 0, \quad i = 1, 2, \ldots, n,$$

with parameter $\theta = (\theta_1, \theta_2, \ldots, \theta_m)$. A solution $x^* = (x_1^*, x_2^*, \ldots, x_n^*)$ of the system, of course, depends upon the parameter θ. The study of laws of change of $x^*(\theta)$ in response to a parametric shift of θ is called comparative statics.[65]

---

62 Edgeworth [29] p. 20, p. 26, Fisher [31] p. 82.

63 Auspitz-Lieben [5] s. 482.

64 Hicks [39] Chap. III.

65 This terminology should be used when the system does not involve time element. If the system involves time, "comparative dynamics" seems to be a better naming.

In the modern theory of consumers' behavior, the condition of utility maximization is expressed by a system of equations:

$$D_j u(x) - \lambda p_j = 0 \quad , \quad j = 1, 2, \ldots, l,$$

$$\sum_{j=1}^{l} p_j x_j - \sum_{j=1}^{l} p_j \omega_j = 0.$$

Jevons and Walras successfully discovered the condition which characterizes the optimal consumption when prices and initial endowments are given. However, they did not answer the problem of comparative statics.

In this case, we regard $p = (p_1, p_2, \ldots, p_l)$ as parameters, putting aside $\omega = (\omega_1, \omega_2, \ldots, \omega_l)$. λ is the Lagrange multiplier. The demand function $\xi(p) = (\xi_1(p), \xi_2(p), \ldots, \xi_l(p))$ (together with $\lambda(p)$) is the solution of this system for each given value of parameters $(p_1, p_2, \ldots, p_l)$. How does $\xi_j(p)$ vary when some p_k changes? This is a typical problem of comparative statics.

It was Pareto who showed a correct approach to this basic problem.[66] His theorem is summarized as follows. The relation

$$D_i \xi_i(p) = \frac{\partial \xi_i}{\partial p_i}(p) < 0$$

holds true if we assume that:

(i) utility function $u(x)$ is additive: $u(x) = U_1(x_1) + U_2(x_2) + \ldots + U_l(x_l)$,
(ii) $D_j u(x) = \partial u / \partial x_j(x) = U_j'(x_j) > 0$ for all j,
(iii) $D_j^2 u(x) = \partial^2 u / \partial x_j^2(x) = U_j''(x_j) < 0$ for all j, and
(iv) $\xi_i(p) - \omega_i \geqq 0$.

Pareto proved the theorem for $i = 1$ as a representative case.

Pareto's proof. The conditions which assure the maximization are

$$\sum_{j=1}^{l} p_j(\xi_j(p) - \omega_j) = 0, \tag{1}$$

$$\frac{U_1'(\xi_1(p))}{p_1} = \frac{U_2'(\xi_2(p))}{p_2} = \cdots = \frac{U_l'(\xi_l(p))}{p_l} = \lambda(p). \tag{2}$$

By (2), we obtain[67]

---

[66] Pareto [86] pp. 119–126 in [93]. $D_i \xi_i(p)$ denotes the partial derivative of ξ_i with respect to the i-th variable at p. $D_j^2 u(x)$ is the second partial derivative of u with respect to the j-th variable at x.

[67] $U_j', U_j'', \ldots$ are evaluated at $\xi_j(p)$. For the sake of simplicity, I omitted it.

$$D_1\lambda(p) = \frac{1}{p_1}U_1^{''}D_1\xi_1(p) - \frac{1}{p_1^2}U_1^{'},$$
$$D_1\lambda(p) = \frac{1}{p_j}U_j^{''}D_1\xi_j(p) \quad , \quad j = 2, 3, \ldots, l. \tag{3}$$

It follows that

$$D_1\xi_1(p) = p_1 D_1\lambda(p)\frac{1}{U_1^{''}} + \frac{1}{p_1}\frac{U_1^{'}}{U_1^{''}},$$
$$D_1\xi_j(p) = p_j D_1\lambda(p)\frac{1}{U_j^{''}} \quad , \quad j = 2, 3, \ldots, l. \tag{4}$$

Multiplying both sides of the j-th equation of (4) by p_j $(j = 1, 2, \ldots, l)$ and adding them all, we obtain

$$\sum_{j=1}^{l} p_j D_1\xi_j(p) = \left(\sum_{j=1}^{l} \frac{p_j^2}{U_j^{''}}\right) D_1\lambda(p) + \frac{U_1^{'}}{U_1^{''}}. \tag{5}$$

If we write

$$S_1 = \sum_{j=1}^{l} p_j D_1\xi_j(p) \quad \text{and} \quad T = \sum_{j=1}^{l} \frac{p_j^2}{U_j^{''}},$$

then the formula (5) can be simplified as

$$S_1 = T D_1\lambda(p) + \frac{U_1^{'}}{U_1^{''}}. \tag{5'}$$

Differentiating (1) with respect to p_1, we also get

$$[\,\xi_1(p) - \omega_1\,] + S_1 = 0. \tag{6}$$

Combining (5$'$) and (6), we can deduce the relation

$$D_1\lambda(p) = \frac{1}{T}\left(S_1 - \frac{U_1^{'}}{U_1^{''}}\right) = \frac{1}{T}\left[-(\xi_1 - \omega_1) - \frac{U_1^{'}}{U_1^{''}}\right] \tag{7}$$
$$= -\frac{1}{T}\left[(\xi_1 - \omega_1) + \frac{U_1^{'}}{U_1^{''}}\right].$$

Thus we arrived at the desired result

$$
\begin{aligned}
D_1\xi_1(p) &= p_1 \frac{1}{TU_1''}\left[-(\xi_1-\omega_1)-\frac{U_1'}{U_1''}\right]+\frac{U_1'}{p_1U_1''} \\
&\qquad\qquad\text{by (4) and (7)} \\
&= \frac{1}{TU_1''}\left[-p_1(\xi_1-\omega_1)-\frac{p_1U_1'}{U_1''}+\frac{TU_1'}{p_1}\right] \\
&= \frac{1}{TU_1''}\left[-p_1(\xi_1-\omega_1)+\frac{U_1'}{p_1}\left(\frac{-p_1^2}{U_1''}+T\right)\right] \\
&= \frac{1}{TU_1''}\left[-p_1(\xi_1-\omega_1)+\frac{U_1'}{p_1}\left(\sum_{j=2}^{l}\frac{p_j^2}{U_j''}\right)\right]. \qquad (8)
\end{aligned}
$$

It is quite easy to confirm $D_1\xi_1(p) < 0$ under the assumptions (i)–(iv).

This completes the proof of the "law of demand".

However, we have to keep in mind several insufficiencies in Pareto's result:

1. Utility function is restricted to additive form. His way of proof can not be generalized to the case of a general utility function.
2. Assumption (iii) is not admissible from the viewpoint of ordinalism.
3. When the i-th commodity is supplied (i.e. $\xi_i(p) - \omega_i < 0$), the sign $D_i\xi_i(p)$ is undetermined.

In the French edition of *Manuale*,[68] Pareto evaluated $D_i\xi_j(p)$ in the case of a general utility function $u(x_1, x_2, \ldots, x_l)$. However, he could not successfully determine the sign of, say, $D_1\xi_1(p)$ except the special case of an additive utility function.

Pareto and his Italian colleagues[69] formed a mathematical economics group based on Walras' general equilibrium theory. They are sometimes called the Lausanne School in Italy. Ample technical results accumulated by them attracted the deep interest of a young Ukrainian mathematician, E. Slutsky.[70] The complete solution of the comparative statics problem in consumer theory was provided by Slutsky in 1915. His article was written in Italian and published in *Giornale degli economisti*. This episode implicitly tells us the indispensable historical role played by the Lausanne School in Italy.[71] In this context, I should mention the name of V. Volterra, who was a leading figure of the scientific world in Italy. It is a pity that I have to leave this theme for another occasion.[72]

---

68 Pareto [90] French ed., pp. 579–585.

69 For instance, Demaria, Amoroso, Ricci, de Pietri-Tonelli, and so on.

70 Slutsky [105].

71 Matsuura [70], Schumpeter [103] pp. 855–861 and de Pietri-Tonelli Bousquet [26].

72 Volterra's interest in economics is shown in Volterra [119]. Consult Goodstein [35], and Ingrao–Israel [47] Chaps. 5–6. See also Maruyama [69] pp. 231–243.

4.5 ♣ Mr. Grandmont on New Comparative Statics

After around 1970, the scope of comparetive statics became much wider. A system of basic elements of an economy which determine its equilibrium is called an economic structure. For instance, a set of agents and a combination of preference and initial endowment of each agent constitute an economic structure. If we designate the set of possible preferences by $\mathscr{P}$ and the set of possible endowments by Ω, then $\mathscr{E} = (\mathscr{P} \times \Omega)^m$ is the set of all economic structures (the set of m agents being fixed). We assume the set $E(e)$ of the general equilibrium prices for any economic structure $e \in \mathscr{E}$ is not empty. (The set $E(e)$ is not necessarily a singleton.) Then the analysis of the set-valued mapping $e \mapsto\!\!\!\rightarrow E(e)$ is, in a sense, a subject of comparative statics. The problem is—how does the set $E(e)$ change in response to a shift of e? Debreu [25] formulated a framework for comparative statics of this kind and studied the generic "regularity" of $E(\cdot)$.

When we assume that the set A of agents constitutes a measure space (with measure μ), we may regard a measurable mapping $e : A \rightarrow \mathscr{P} \times \Omega$ as an economic structure. Or alternatively, the distribution ν of μ via e may be interpreted as an economic structure. Comparative statics in this setting is to analyze the mapping $\mu \mapsto\!\!\!\rightarrow E(\mu)$ (or $\nu \mapsto\!\!\!\rightarrow E(\nu)$).[73]

Grandmond's essay in this volume is a typical example in this direction. An economic structure is defined by a probability distribution ν on $\mathscr{P}$. For the sake of simplicity, the income of each agent is assumed to be the same constant b. A particular group of commodities is specified. $W_\nu(p, b)$ denotes the aggregate expenditure function for this group of commodities when the economic structure (distribution on $\mathscr{P}$) is ν. $S_\nu(p, b) = W_\nu(p, b)/b$ is the budget share for the group. Grandmond's problem is, how does the elasticity of S_ν with respected to b change in response to a shift of the economic structure? His work is nothing other than analysis of the mapping

$$\nu \mapsto \frac{b\frac{\partial S_\nu}{\partial b}(p, b)}{S_\nu(p, b)}$$

in a more elementary setting.

[73] Of course, certain topological or measurable structures must be defined on $\mathscr{P}$, Ω and the space of measures on $\mathscr{P} \times \Omega$. An economy is regarded as relatively heterogeneous (resp. homogeneous) if the variance of ν is large (resp. small).

Suppose that ν is a distribution on a space of expected rate of interest. It expresses the dispersion of expectations among agents. Keynes claimed that the larger the variation of ν, the bigger the effect of one unit of monetary policy. Such an insight can also be formulated in a similar setting.

4.6 Integrability

In the Walrasian theory of consumers' behavior, demand function is generated by utility function. While demand function is an observable concept, utility is considered to be unobservable. Many naive empiricists criticized the Walrasian demand theory incorporating such an empirically unobservable concept as utility.

Suppose now that a demand function (or inverse demand function) is observed. Is there any utility function which generates the given demand function through maximization? If the answer is positive, we may regard the demand function as being derived from it hypothetically even if it is not observed directly. The Walrasian attitude may be justified in this way.

Assume that the demand function $\xi(p)$ admits the inverse function $P(x) = (P_1(x), P_2(x), \ldots, P_l(x))$. $P(x)$ is called the inverse demand function. The problem is—are there any functions $\varphi(x)$ and $\lambda(x) \neq 0$ which satisfy

$$D_j\varphi(x) = \lambda(x)P_j(x), \quad j = 1, 2, \ldots, l \tag{1}$$

when $P(x)$ is given? If the answer is positive, the partial differential equation (1) is said to be integrable. In this case, we may regard φ to be a utility which generates $P(x)$.

We define some notations:

$$\begin{aligned} a_{ij} &= D_jP_i - D_iP_j, \quad i, j = 1, 2, \ldots, l, \\ A_{ijk} &= a_{jk}P_i + a_{ki}P_j + a_{ij}P_k, \quad i, j, k = 1, 2, \ldots, l. \end{aligned}$$

The following theorem due to Frobenius [32] characterizes the integrability of the Eq. (1).

(1) *is integrable if and only if*

$$A_{1ij} = 0, \quad i, j = 2, 3, \ldots, l \quad i \neq j, \tag{2}$$

$$\text{i.e.} \quad P_1\left(D_jP_i - D_iP_j\right) + P_i\left(D_1P_j - D_jP_1\right) + P_j\left(D_iP_1 - D_1P_i\right) = 0, \quad i, j = 2, 3, \ldots, l, \quad i \neq j, \tag{2'}$$

provided that $P_1, P_2, \ldots, P_l$ *are sufficiently smooth.*

It is easy to show that (2) implies $A_{ijk} = 0$ for all distinct i, j, k.
For the sake of simplicity, we consider the problem:

$$\begin{aligned} &\text{Maximize} \quad u(x) \\ &\text{subject to} \quad \sum_{j=1}^{l} p_jx_j = w, \end{aligned}$$

where w(purchasing power) appears instead of ω.

Since demand function is homogeneous of degree zero, we may assume $P_1(x) = 1$.

Since $P_1 = 1$, $D_i P_1 = D_j P_1 = 0$, Frobenius' integrability condition (2′) is simplified as

$$D_j P_i - P_j D_1 P_i = D_i P_j - P_i D_1 P_j, \quad i, j = 2, 3, \ldots, l, \quad i \neq j. \tag{3}$$

Denoting Slutsky's substitution term by K_{ij},[74] Samuelson [98] much later proved the equivalance:

$$\begin{aligned} &K_{ij} \text{ is symmetric}^{75} \\ &\Longleftrightarrow \left(D_j P_i - P_j D_1 P_i\right)^{-1} \quad \text{is symmetric} \\ &\Longleftrightarrow D_j P_i - P_j D_1 P_i \quad \text{is symmetric.} \end{aligned} \tag{4}$$

Therefore the Eq. (1) is integrable if and only if K_{ij} is symmetric.

Summing up,

(1) is integrable $\Leftrightarrow$ (2) $\Leftrightarrow$ (3) $\Leftrightarrow$ (4).

It is noticeable that G. B. Antonelli [3] already observed that the condition (3) assured the existance of utility function which generates the given demand (inverse demand) function.

Antonelli considered an equation

$$D_j\varphi - P_j(x) D_1\varphi = 0, \quad j = 2, 3, \ldots, l, \tag{1′}$$

which is equivalent to (1) if we replace $\lambda(x)$ by $D_1\varphi(x)$ and assume $P_1(x) = 1$. If we define a differential operator X_j $(j = 2, 3, \ldots, l)$ by $X_j\varphi = D_j\varphi - P_j(x)D_1\varphi$, then (1′) is expressed as $X_j\varphi = 0$ $(j = 2, 3, \ldots, l)$. This is a system of linear homogeneous partial differential equations of the first order. It is called a Jacobian system if $(X_i X_j)\varphi = (X_j X_i)\varphi$ for all $i, j = 2, 3, \ldots, l$. Antonelli knew the theorem:

If (1′) is a Jacobian system, then it has a solution.

It is easy to show that the condition (3) implies that (1′) is a Jacobian system.[76] Hence there exists a solution of (1′) in this case.

---

[74] If demand is a function $\xi(p, w)$ of price p and purchasing power w, K_{ij} is given by

$$K_{ij} = \frac{\partial \xi_j}{\partial p_i} + \xi_i \frac{\partial \xi_j}{\partial w}.$$

$w = \sum_{j=1}^{l} p_j \omega_j$ in our framework.

[75] That is, $K_{ij} = K_{ji}$.

[76] Actually, we need the condition $D_1\varphi \neq 0$. Antonelli overlooked it. The method of solving a system of linear partial differential equations of the first order was developed by, e.g. Clebsch [18]

He had recourse to relatively advanced knowledge about the theory of partial differential equations. Although his article is full of vague reasonings and strange expressions, his mathematical maturity seemed to be much higher than Walras. Antonelli's work was published half a century before Samuelson.

A few years passed since Antonelli's brochure had appeared. Pereto started to publish a series of articles titled "Considerazioni",[77] in which he gave an exposition of the meaning and the significance of integrability problem in utility theory. However, the evalutions of later critics have not been very favorable for Pareto.[78] Indeed, it is difficult for anyone to disprove some of their objections. It is true that Pareto did not show explicitly the mathematical condition which guarantees the integrability. And neither could he distinguish rigorously the two different systems of differential equations:

$$\begin{aligned} &\text{a. } D_j\varphi = P_j, \quad j = 1, 2, \ldots, l, \\ &\text{b. } D_j\varphi = \lambda P_j, \quad j = 1, 2, \ldots, l, \\ &\qquad \text{where } \lambda \text{ is an integrating factor.}^{79} \end{aligned}$$

We can not deny that Pareto's confusions bewildered many readers.[80]

I have to add one more remark here. Even if the integrability condition is satisfied and some utility function u is derived, the given demand function may not be recovered by maximizing u. In such a case, the solution of the problem should be regarded as very incomplete.[81] Neither Antonelli nor Pareto noticed this difficulty.

4.7 ♣ Mr. Mogi on Jevons as a Social Reformer

The hegemony of the Ricardo–Mill school suddenly declined in England after the 1850s.[82] In the latter half of the 19th century, various reactions occurred against the

---

just before Antonelli's work. See Dieudonné [28] for the history of this discipline of mathematics. For a modern treatment, consult Carathéodory [13].

77 Pareto [86]. This article was already quoted in the discussion of comparative statics.

78 Samuelson [98] pp. 355–356, Stigler [110] pp. 121–126.

79 As Volterra [119] highly appreciated, Pareto endeavored to discover a utility function from empirically observable data. Suda [116] proposed an interpretation to vindicate Pareto in which he made use of type a (resp. type b) when he looked for a utility in the cardinal (resp. ordinary) sense properly.

80 As we read through Walras' correspondence (Jaffé [50]), we find two correspodents who discussed the integrability problem with Walras. One is L. Perozzo, a statistician. In his long letter to Walras dated February 24, 1890 (Jaffé [50] Vol. III, pp. 394–398, Letter ♯ 962). Perozzo explained Antonelli's idea based upon the Jacobian system. Another one is H. Laurent, a mathematician. He asked Walras how to interpret the integrating factor. Walras gave a reply that the integrating factor could have an economic meaning as the marginal utility of numéraire. See Laurent's letters dated May 13, 24, 1900 (Jaffé [50] Vol. III, pp. 116–117, Letter ♯ 1452, and p. 121, Letter ♯ 1455) as well as Walras' letter dated May 15, 1900 (Jaffé [50] Vol. III, pp. 117–120, Letter ♯ 1453).

81 For a complete solution, see Hurwicz–Uzawa [43].

82 Hutchison [45] Chap. 3.

orthodox Ricardo–Mill views. For instance, some sympathies for romanticism and historicism motivated resistance against the selfish and physical growth-oriented view of men and society. Jevons' economic theory was invented as a destructive challenge against the Classical School based upon the labor (or cost) theory of value.

As is well-known, the area of Jevons' activities was so vast that Mogi calls him a Victorian polymath. Mogi tries to paint a portrait of Jevons as a social reformer. In contrast with the physical economic growth mind set of established society, Jevons, a utility theorist, advocated social reforms to increase subjective fulfilment.

Bentham's utilitarianism consists of several aspects:

[A] Hedonic calculus as a driving force which regulates men's behavior.

[B] Hedonism as an individual moral discipline.

[C] A social standard of legislation—"Maximum happiness of the greatest number".

It is well known that Jevons organized his economic theory of exchange based upon aspect [A] of Bentham's utilitarianism. On the other hand, Mogi successfully proves that Jevons' idea of social reforms depends upon aspect [C]. Thus it is clarified that Jevons was a Benthamite both individually and socially. Among Mogi's new findings, it also deserves a special mention that Jevons and Bentham share a definitely negative view on the classical dogma of laissez-faire.[83]

But it is an irony that Jevons shared the Benthamite standpoint with his enemies, Ricardo and Mill.

4.8 ♣ Ms. Ingrao on Lasting Imprint of Utility

Many economists joining the Marginal Revolution, Walras and Pareto in particular, endeavored to forge the old concept of utility as a purely scientific tool to analyze agents' choice behaviors, scraping off its philosophical and metaphorical nuances. Walras' serious correspondence with Poincaré and Pareto's invention of the "Pareto efficiency" criterion reflected their attitude towards "scientification" of the utility concept.

However, as Mogi also illustrates, various social problems were often discussed and assessed in terms of moral nuances of utility. Even in economic theory, the shadow of utility or utilitarianism never disappeared. Economic thoughts of Edgeworth, A. C. Pigou and H. Sidgewick are definitely based upon Benthamite utilitarianism. (Pigou's *Economics of Welfare*, 1920, was published more than ten years after Pareto had invented the concept of Pareto optimality.) It is not hard for us to find out lasting "imprints" of the utility concept here and there outside economic theory. These were partially Romantic reactions against the modern tendencies.

Observing them all carefully, Ingrao concludes that "the pretense to break up ideas of utility from any philosophical background, notably the utilitarian one, did not finally succeed" (p. 155).

[83] Consult Hutchison [46] for Bentham's attitude concerning roles of nations.

Indeed we can not neglect the revival of the cardinal utility as well as the interpersonal comparison in some field of game theory, nonetheless they were once discarded. The possibility of measurement of utility, considered negatively in the old days, has been reexamined approvingly in the development of experimental psychology.

As I wrote in Sect. 2, the tradition of utility theory has a quite long history. Ingrao restricts here the scope of her research mainly to the imprint of Benthamite utilitarianism.

5 Production and Distribution

5.1 Two Origins of Modern Theory—Ricardo and von Wieser

The Ricardian theory of rent was a dominant orthodoxy during the period of the Classical School in England. According to Ricardo, an economy is composed of three classes, that is, laborers, capitalists and land-owners.[84] Ricardo's rent theory gives a law which regulates the distribution of the produce to the three classes. We now provide a rough outline of Ricardo's idea for the sake of comparison with the modern theory. Suppose that there are several ranks of land classified by their qualities. The cost of production on the most inferior (marginal) land is divided into wages and profit. Wages are determined (in the long-run) at the "subsistence level" by Malthus' law and the excess of the total produce over wages is absorbed by capitalists as profit. The price of the produce must be equal to the unit cost on the marginal land. But the unit cost on superior (infra-marginal) land is less than the price. The difference between them is the rent accruing to the owners of superior land. The rent of the marginal land is zero. The distinct principles are prepared for explaining the distribution to the three classes. This is possible because there are only three productive factors. If there are many kinds of productive factors, we must admit that it is impossible to prepare an independent principle for each of the factors. A certain "uniform" principle seems to be required for a general theory of distribution. But suppose that there remains surplus (positive or negative) of the produce after all the remunerations are paid out according to the "uniform principle". Then this surplus is left unexplained by the principle. So the principle obtains the "full explanatory power" only if the whole produce is exhausted with no surplus after distributions.

P. H. Wicksteed criticized this aspect of the Ricardian rent theory in his *Essay on the Coordinations of the Laws of Distribution* (1894).[85]

> In investigating the laws of distribution it has been usual to take each of the great factors of production such as Land, Capital and Labour, severally, to enquire into the special circumstances under which that factor co-operates in production, the special considerations which act upon the persons that have control of it, and the special nature of the service that

[84] D. Ricardo, *On the Principles of Political Economy, and Taxation* (John Murray, London), 1817.

[85] Stigler [109] is the classical reference for the historical study of the marginal productivity theory. Yasui [139] is also an indispensable work on this topic. Yasui's systematic studies on the entire aspects of Walras' economic theory are collected in Yasui [140].

> it renders, and from all these considerations to deduce a special law regulating the share of the product that will fall in distribution to that particular factor.
>
> Now as long as this method is pursued it seems impossible to co-ordinate the laws of distribution and ascertain whether or not the shares which the theory assigns to the several factors cover the product and are covered by it. For in order that this may be possible it seems essential that all the laws should be expressed in common terms. As long as the law of rent, for example, is based on the objective standared of fertility of land, while the law of interest is based on the subjective standard of estimate of the future as compared with the present, it is difficult even to conceive any calculus by which the share of land and the share of capital could be added together and an investigation then instituted as to whether the residual share will coincide with what the theory assigns as the share of wages.[86]

Wicksteed found the germ of his new distribution theory in a careful and critical study of Ricardo's orthodox theory. Another source of modern distribution theory is found in the Austrian School. Their idea is sometimes called the "Zurechnung" (imputation) theory. Menger wrote:

> Es ist demnach auch nicht der Werth der Güter niederer Ordnung in der Gegenwart, wornach sich der Werth der entsprechenden Güter höherer Ordnung richtet, sondern vielmehr unter allen Umständen der voraussichtliche Werth des Productes welcher das massgebende Princip des Werthes der bezüglichen Güter höherer Ordnung ist.[87]

The value of a productive factor is evaluated by the "loss principle"—that is, the value of a productive factor is evaluated by the loss of value of its produce when one unit of the factor is removed from the production process. F. von Wieser presented the problem in a more explicit form.[88]

We assume, following von Wieser, that there are three products and three productive factors. a_{ij} $(i, j = 1, 2, 3)$ denotes the production coefficient of the j-th productive factor for the i-th product. von Wieser specified a_{ij} as follows:

$$\begin{aligned} &\text{1st product} \quad (a_{11}, a_{12}, a_{13}) = (1, 1, 0), \\ &\text{2nd product} \quad (a_{21}, a_{22}, a_{23}) = (2, 0, 3), \\ &\text{3rd product} \quad (a_{31}, a_{32}, a_{33}) = (0, 4, 5). \end{aligned}$$

The prices of the three products are also specified as $(p_1, p_2, p_3) = (100, 290, 590)$. Then the following relations hold goods under the competitive situation:

$$q_1 + q_2 = 100, \quad 2q_1 + 3q_3 = 290, \quad 4q_2 + 5q_3 = 590,$$

Solving the above simultaneous equations, we obtain $q_1 = 40$, $q_2 = 60$, $q_3 = 70$. These values are interpreted as productive contributions of the factors imputed to them.

---

86 Wicksteed [135] p. 7.

87 Menger [77] s. 126.

88 Wieser [121]. The mathematical structure of "Zurechnung" can be interpreted as the duality problem of linear programming. See Uzawa [118] and Maruyama [69] pp. 186–191.

K. Wicksell criticized von Wieser's idea in his *Über Wert, Kapital und Rente* (1893) and made a big stride towards a systematic theory of distribution by means of the concept of marginal productivity.

> Allein es ist klar, daß man durch ein solches Verfahren, und mag man die Zahl der Gleichungen noch so sehr vergrößern, überhaupt nichts mehr erfahren wird, als man schon im voraus wußte, nämlich daß bei freier Konkurrenz das Entgelt oder der Ertragsanteil eines und desselben „Produktionsmittels" in allen Geschäften annäherungsweise derselbe sein muß. ... ... Wenn also Wieser mit seinem „produktiven Beitrage" nur das thatsächlich ausfallende Entgelt der verschiedenen produktiven Faktoren ... ... bezeichnen wollte, so hat er einen wahren, aber selbstverständlichen Satz ausgesprochen; hat er etwas anderes gemeint, so muß seine „Lösung" a priori als falsch erklärt werden.[89]

5.2 Marginal Productivity

We now start by giving a brief exposition of the modern marginal productivity theory.

A producer tries to choose an optimal production plan (a combination of inputs and outputs). When the quantities of inputs are determined under the given prices, the remuneration to each productive factor is determined (price times quantity employed). Thus the theory of production and that of distribution are the head and the tail of a coin.

In the following, we assume that there is only one kind of output for a producer.

The technological relation between inputs y_j $(j = 1, 2, \ldots, n)$ and output x is expressed by a production function

$$x = f(y_1, y_2, \ldots, y_n).$$

The j-th partial derivative $D_j f(y_1, y_2, \ldots, y_n)$ is called the marginal productivity of the j-th input. p and q_j denote the price of the output and of the j-th input, respectively $(j = 1, 2, \ldots, n)$.

A producer is assumed to choose a combination of output x and inputs $y_j (j = 1, 2, \ldots, n)$ which maximizes his profit subject to technological constraint.

$$\begin{aligned} &\text{Maximize} \quad px - \sum_{j=1}^{n} q_j y_j \\ &\text{subject to} \quad x = f(y_1, y_2, \ldots, y_n), \end{aligned}$$

where p and $(q_1, q_2, \ldots, q_n)$ are given.

The marginal productivity theory consists of two statements:

(I) The producer maximizes his profit when the value of the marginal productivity of each input is equal to its factor price:

$$q_j = p \cdot D_j f(y_1, y_2, \ldots, y_n), \quad j = 1, 2, \ldots, n,$$

[89] Wicksell [132] s. XII.

given p and q_j.

(II) Assume that either A or B is satisfied:

A. f is positively homogeneous of degree one. i.e. $f(\alpha y_1, \alpha y_2, \ldots, \alpha y_n) = \alpha f(y_1, y_2, \ldots, y_n)$ for any $\alpha > 0$.

B. $p = \text{average cost} = \frac{1}{x}\sum_{j=1}^{n} q_j y_j$.

Then all the products are just exhausted: i.e.

$$px = \sum_{j=1}^{n} p \cdot D_j f(y_1, y_2, \ldots, y_n) y_j$$
$$\Longleftrightarrow x = \sum_{j=1}^{n} D_j f(y_1, y_2, \ldots, y_n) y_j$$

if the inputs of all factors are determined according to the condition (I).

The marginal productivity theory was developed primarily in the 1890s, two decades later than the marginal utility theory.

5.3 *Wicksell–Wicksteed–Barone*[90]

In the 1st and 2nd editions of *Eléments*, Walras' general equilibrium theory with production is formulated with constant coefficients of production. Hence the first two editions do not contain any clue to the marginal productivity theory.[91] However, the problem of variable coefficients became an important topic in the correspondence between Walras and Parero since Pareto published his "Il massimo di utilità dato della libera concorrenza" in 1874.[92] Pareto received a complimentary copy of Wicksell's *Über Wert, Kapital und Rente* and suggested to E. Barone, in his letter of March 5, 1894, that he write a review article.

Barone, after a careful scrutiny, discovered the undeveloped marginal productivity theory embedded in Wicksell's book. Barone's "Sopra un libro del Wicksell"[93] was published in 1895, in which he obtained the statement (I) (in the preceding section) without any vagueness.

> Infatti la condizione che il prezzo di costo π sia minimo dà luogo appunto a tante relazioni logiche — a tante equazioni, se meglio cosi piace — quante sono le variabili da determinare,

---

90 I am much Indebted to Jaffé [49] for the materials of this section.

91 But Walras pointed put, in §247 of the 2nd ed. of *Eléments*, that production coefficients should have been determined so as to minimize cost when they were variable. Nevertheless, he mentioned neither the concept of a production function nor that of marginal productivity.

92 Pareto [87]..

93 Barone [6]..

cioè i vari coefficienti di fabbricazione e la durata del processo produttivo: perchè la condizione del minimo costo per ogni unità del prodotto richiede che ciascun fattore si impieghi nella produzione fino a quel punto, oltre il quale un ulteriore incremento di esso darebbe un incremento di prodotto non più sufficiente a rimunerare lo incremento del fattore.[94]

His exposition is not so explicit about the statement (II). In a letter to Walras dated September 20, 1894, Barone wrote about the successful completion of a "new theorem of maximum net product".

A présent je suis occupé à des questions de distribution du produit entre les services producteurs. En reprenant et développant des observations que Thünen, Jevons, Marshall, Clark, Weekstead [should read Wicksteed — Maruyama] ont déjà faites, il me semble d'avoir réussi à formuler une théorie assez simple de la distribution en général et du profit de l'entrepreneur en particulier sur la base d'un nouveau théorème *du produit net maximum pour l'entrepreneur.*

... ... la quantité P du produit (A) est, pour un entrepreneur donné, une certaine fonction des quantités de services qui concourent à la fabbrication

$$P = \phi(K, L, T \ldots).$$

....

Mathématiquement, puisque:

$$dP = \frac{\partial P}{\partial K} dK + \frac{\partial P}{\partial L} dL + \frac{\partial P}{\partial T} dT + \cdots$$

l'entrepreneur obtiendra le produit net maximum lorsqu'il fera

$$\frac{\partial P}{\partial K} = \frac{p_K}{p_a}; \frac{\partial P}{\partial L} = \frac{p_L}{p_a}; \frac{\partial P}{\partial T} = \frac{p_T}{p_a}; \ldots,$$

savoir lorsque *les degrés finals de productivitié des services seront proportionnels aux prix.*

He also stated that all the product is absorbed into the remunerations of factors.[95] Taking account of this letter, it seems very possible for us to suppose that: (i) Barone's idea of "Sopra" was almost completed by September 20, and (ii) the statement (II) had also been obtained in Barone's mind if not explicit in "Sopra".

It was in the summer in 1894 that Wicksteed's *Coordinations* were received by some members of the Lausanne Circle. Walras received it on July 14. Pareto stayed in Paris during the summer, and wrote to Walras (in October) that he had not read it yet.[96] So

---

[94] Barone [6] p. 536, [7], Vol. I, pp. 138–139.

[95] Jaffé [50] Vol. II, pp. 619–621, September 20, 1894. Letter ♯ 1191. In "Studi sulla distribuzione" (1896), both of the statements (I) and (II) are much more clarified. See, in particular its "Note mathematiche".

In tali condizioni la legge delle produttività marginali si estende anche alla rimunerazione dello imprenditore e *dopo aver rimunerato tutti i fattori* (l'opera dell'imprenditore compresa) *in ragione delle produttività marginali* (da ciascuna delle quali va sottratto l'interesse sulla quota del capitale d'anticipazioni) *non resta alcun residuo non distribuito* (Barone [7] Vol. I, p. 196.)

[96] Jaffé [50] Vol. II, p. 622, October 7, 1894. Letter ♯ 1193.

we may also suppose that Barone had arrived at the marginal productivity theory by himself even before he looked at *Coordinations*.

Wicksteed successfully established (I) and (II) of the marginal productivity theory under the assumption A on p. 39. A. W. Flux's short review article[97] served to clarify its mathematical structure. Walras immediately read *Coordinations* and wrote "Note sur la réfutation de la théorie anglaise du fermage de M. Wicksteed" (dated September, 1894). After careful discussions with Pareto and Barone for a year, Walras sent his manuscript to J. Berney (Dean of the Faculty of Law, the University of Lausanne) for publication, on September 30, 1895.[98]

On the other hand, Barone also studied *Coordinations* and wrote a review article "Sopra un recente libro del Wicksteed", in which he brushed up the marginal productivity theory and established an exhaustion theorem without the homogeneity assumption A. This manuscript was sent to F. Y. Edgeworth, the editor of *Economic Journal*. However, Edgeworth rejected it for publication.

Walras borrowed Barone's manuscript from Pareto and made a French translation for his own use. Walras wrote to Barone on October 23, 1895:

> De retour en ville, je suis allé, dimanche, voir M.Pareto qui m'a remis votre propre note *Sopra un recente libro del Wicksteed*, destinée à l'*Economic Journal*, que j'ai lue la plume à la main en en faisant une traduction française.[99]

Walras filled in a memorandum[100] "Envoyé à Edgeworth pour l'Economic Journal" at the end of his translation manuscript, which tells us that he did not know yet about Edgeworth's rejection. Walras seemed to be informed of the event for the first time by Barone's reply (dated October 26, 1895). Thus Walras understood the situation. Barone's letter ran as follows:

> Je vous remercie bien de votre très aimable lettre. Mon ami, M. Pareto, vous aura, peut-être, déjà dit pourquoi ma note ne sera plus publiée dans *l'Economic Journal*. M. Edgeworth, après m'avoir prié d'examiner le travail de M. Wicksteed, et de développer dans un article les observations que j'avais faites dans une de mes lettres dirigée à M. Edgeworth même; après m'avoir écrit que ma note était, selon lui, un nouveau rayon de lumière sur une question très importante, un beau jour m'a fait savoir qu'il lui était impossible de la publier. Ma note a, évidemment, choqué ces messieurs-là, qui n'ont pas compris ni votre théorie, ni le rôle de votre entrepreneur; c'est évident; ma note n'a pas eu le bonheur d'être agrée par le pontife des économistes anglais, M, Marshall, et le pauvre M. Edgeworth s'est trouvé dans l'embarras![101]

Walras made up a summary of Barone's review article as the "Postscript" to "Note sur la réfutation" and sent it to Berney on December 2.[102]

---

97 *Economic Journal*, **4** (1894), 308–313.

98 See Walras' letter to Barone dated October 23, 1895. Jaffé [50] Vol. II, p. 643. Letter ♯ 1215.

99 Jaffé [50] Vol. II, p. 693. Letter ♯ 1215.

100 Jaffé [50] Vol. II, p. 648.

101 Jaffé [50] Vol. II, p. 648. Letter ♯ 1216.

102 Jaffé [50] Vol. II, p. 666, December 2, 1895. Letter ♯ 1229.

Walras' "Note sur la réfutation" was published in 1896 together with the "Post-Scriptum".[103] Regrettably, we can not scrutinize Barone's original manuscript because it was lost. But Walras' translation[104] was deposited in the Lausanne Archives. Comparing it with the "Post-scriptum", we can confirm that the contents of the latter turn out to be completely due to Barone. It was republished as Appendice III in the 3rd edition of *Eléments*, and finally, it was moved to the text in the 4th edition.

The completed form of the marginal productivity theory as we now see was presented by Wicksell in his article "Om gränsproduktiviteten såsom grundval för den nationalekonomiska fördelningen"[105] and Vol. I of *Lectures on Political Economy*. We are impressed by Wicksell's work in the history of this topic. Indeed the history was started by Wicksell and completed by him.[106]

5.4 Walras Versus Marshall

Walras got so angry at Edgeworth's rejection of Barone's article, as shown in his correspondence, that he could not restrain himself from expressing it publicly.

When Walras wrote the "Post-Scriptum" to "Note sur la réfutation", he intended to add one more piece titled "Petite Préface". Barone thoughtfully advised him not to publish it since it contained passages that were too provocative. Walras accepted Barone's opinion and deleted the "Petite Préface" from the manuscript. The deleted part is deposited in the Laussanne Archives.

> Dans le courant de, ce mois [Octobre 1895], M. Pareto m'a communiqué à son tour une note de M. Barone destinée à l'*Economic Journal*, qui m'a paru compléter l'étude de la question et dont j'ai fait le résumé qu'on trouvera à la fin de la mienne.
>
> Maintenant, j'apprends que M. Edgeworth, éditeur de l'*Economic Journal*, qui, dans ses discussions avec nous, s'est toujours abstenu, avec un soin significatif, de nous répondre dans son propre recueil, et qui a trouvé dans ceux de nos amis une hospitalité sans limites pour des explications plus prolixes que probantes, refuse à M. Barone l'insertion de cette *Note* de quatre ou cinq pages où l'auteur éclaire complètement, au moyen d'une démonstration sans réplique, une théorie dont i'importance est évidente; et je m'empresse de protester contre ce procédé en même temps que je publie ma propre *Note* et son *Post-Scriptum*.[107]

---

[103] *Recueil publié par la Faculté de Deroit de l'Université de Laussanne à l'occasion de l'exposition nationale suisse* (Viret-Gentou, Geneva) 1896.

[104] Jaffé [50] Vol. II, p. 644–648.

[105] English translation is contained in Wicksell [134].

[106] We now add one more episode. Walras sent a letter of inquiry to H. Amstein (mathematician). Regrettably the date is not specified. In the case of variable coefficients of production, they are determined by minimizing cost. Walras asked Amstein how to solve this problem. Amstein replied to Walras on January 6, 1877. His solution included the first application of the Lagrange multiplier method in economics. But Walras did not understand what Amstein did and forgot it. If Walras's mathematical ability had been a little bit better, the completion of the marginal productivity theory would have been realized in 1877. cf. Jaffé [50] Vol. I, pp. 517–518. Letter ♯ 364.

[107] Jaffé [50] Vol. II, p. 660.

In the 1880s,s, the communications between Marshall and Walras were relatively friendly as shown by their correspondence. However, Marshall's letter to Walras on September 19, 1889 deserves special attention because it illuminates Marshall's attitude towards mathematical economics.

> I have to thank you very heartily for your new edition of Elements d'Econ. Pol. I have not myself retired from the conclusion that I think I communicated to you some time ago, viz that the right place for mathematics in a treatise on Economics is the back-ground. But I think it is most desirable that different seekers after truth should take different routes; I rejoice much that the pure mathematical route is being developed by your great ability and energy.[108]

Walras kept silent for a while. But after a few years, he could not endure his situation and began to write harshly about Marshall in his correspondence.

> ... il est pitoyable de voir Marshall, *ce grand éléphant blanc de l'économie politique*, et Edgeworth par ⟨ jalousie, inintelligence et mesquinerie restaurer et s'obstiner ⟩ impuissance et jalousie s'évertuer à remettre en honneur la théorie de Ricardo et de Mill sur le prix des produits.[109]
>
> J'aurai plaisir à connaître votre opinion sur ce point si vous étudiez un jour ou l'autre la courbe d'offrre de Marshall comme vous étudiez aujourd'hui sa courbe de demande. Mais je ne discuterai pas la question avec lui. Il a écrit (dans la Préface de son livre, à ce qu'il me semble) que "les traductions de raisonnements économiques en formules mathématiques étaient peut-être très utiles à ceux qui les faisaient, mais ne servaient à rien aux autres personnes", et je vous avoue, que, depuis lors, je l'ai classé parmi les adversaires de l'école mathématique et non parmi les adeptes dont vous êtes, à mes yeux, un des plus éminents.[110]

The conflict aroused by Edgeworth's rejection definitely deteriorated the situation and deepened the gap between Walras' circle and Marshall's. This is one of the most regrettable episodes in the history of economics.[111]

5.5 ♣ *Mr. Kurz on the Heterogeneity Inside the Vienna Circle*

The term "Marginal Revolution" in economics may suggest implicitly some novelty and homogeneity of the Menger–Jevons–Walras achievements. Kurz definitely refuses to regard the trio as the real pioneers of the marginal analysis in economics because there were precursors who elaborated both of the marginal utility theory and the marginal productivity theory in the first half of the 19th century. Furthermore, Kurz points out several important heterogeneities among the trio.[112]

However, it is still more impressive among Kurz's observation that there were sharp discrepancies between Menger and his disciples—von Wieser and von Böhm-Bawerk.

---

[108] Jaffé [50] Vol. II, pp. 613–615, August 26, 1894. Letter ♯ 1188.

[109] From Walras to Pareto, Jaffé [50] Vol. II, pp. 486–487, March 12, 1892. Letter ♯ 1051.

[110] From Walras to Barone, Jaffé [50] Vol. II, pp. 613–615, August 26, 1894. Letter ♯ 1188.

[111] See Jaffé [49] for a detailed description of the conflict.

[112] See also Jaffé [51].

Although von Wieser's Zurechnung theory brought about a deep insight leading to the marginal productivity theory, as I already noted, its basic ideas had been created by Menger.

But it is well-known that according to Menger's illustration, the total of renumerations paid out to productive factors may exceed the output. The arithmetic solvability of Menger's problem also seemed dubious. von Wieser criticized Menger's defects of this sort in his Zurechnung theory. In addition, von Böhm-Bawerk attacked Menger's capital theory,[113] which is the weaker point in his *Grundsätze*. As a consequence, Kurz "can not avoid concluding that "Menger's theory of income distribution is beset by serious difficulties. The criticisms his pupils and then colleagues levelled at it drew a permanent wedge between him and them" (p. 115).

6 Some Remarks on Methodology

This final section is a memorandum on the methodological attitudes of a few representative economists in the era. I start by investigating Menger's *a priori* essentialism, particulary from the viewpoint of Aristotelian philosophy. Then, I proceed to discuss the methodology of the Lausanne School just briefly.[114]

6.1 Menger's Exact Law

Menger's philosophical foundations of economics is very far from the Jevons–Pareto stance based upon empiricism. Basic characteristic elements of his methodology, called analytic-synthetic (or analytic-composite) method, are summarized as follows.[115]

(1) Any concrete complicated phenomenon is generated by some simpler elements. A scientist has to start by discovering the simplest elements or the ultimate general causes of the phenomenon in question. Another name for the simplest elements is the essence (das Wesen) of the phenomenon.

(2) The simplest elements can be discovered by getting back from the phenomenon toward its origin. Then conversely, we have to reconstitute a process of occurrence of the phenomenon from the simplest elements stated in (1). The former procedure is called the analytic method and the latter the synthetic or compos-

---

[113] I do not go into details concerning capital theory. Reading the description of the Menger–Böhm-Bawerk conflict, I am reminded of the modern controversy between Joan Robinson and Robert Solow.

[114] Regrettably I do not mention Jevons' philosophy of science, which deserves scrutiny. See Black [8, 9] and Mays [73].

[115] Important materials concerning this topic are found in Menger [77, 78] and his correspondence with Walras collected in Jaffé [50].

ite method. The combination of these two procedures forms a basic framework which Menger proposed as the method of economic theory.[116]

(3) A chain of the causal relations leading from the simplest elements to the goal concrete phenomenon forms a theory which explains the phenomenon. The step-by-step causal relations composing the chain are called the exact laws.

> Sind die Quantitäten von Gütern welche im Verkehre gegen einander ausgetauscht werden (Quantitäten welche nach Zeit und Ort wechseln !) willkürliche, oder folgen sie bestimmten *Gesetzen ?* Dies ist die Frage. Habe ich Sie richtig verstanden, wenn ich annehme, dass das Streben nach diesen *Gesetzen* das hauptsächliche Ziel Ihrer Forschung sei? Ist dies aber der Fall, so ist zugleich klar, dass dies Ziel Ihrer Forschung keineswegs auf specifisch mathematischem Wege erreichbar ist. Es ist vielmehr nothwendig, dass wir auf die einfachsten Elemente der hier in Rede stehenden, zumeist sehr complicirten Erscheinungen zurückgehen — also auf *analytischem Wege* die letzten constitutiven Factoren der Preiserscheinungen bestimmen, hierauf das der Natur dieser Elemente entsprechende *Mass* an dieselben legen und mit Festhaltung dieses letztern die *Gesetze* festzustellen unternehmen, nach welchen sich aus jenen einfachen Elementen die complicirten Erscheinungen des menschlichen Verkehres entwickeln. Die Methode welche wir bei diesen Untersuchungen anzuwenden haben, wird demnach die *analytisch-synthetische,* oder *analytisch-compositive,* keineswegs aber schlechthin die *mathematische* sein. Die Mathematik wird uns bei dem obigen Erkenntnissprocesse wichtige Dienste leisten, wie überall, wo es sich unter anderem auch um die Messung von Grössen handelt, indess dieselbe wird doch nur den Charakter einer *Hilfswissenschaft* für uns haben.[117]

(4) The exact laws are valid of necessity without any exception. It is a logical contradiction to verify their validity by any empirical procedures.

> Sie sucht die einfachsten Elemente alles Realen zu ergründen, Elemente, welche, eben weil sie die einfachsten sind, streng typisch gedacht werden müssen. Sie strebt nach der Feststellung dieser Elemente auf dem Wege einer nur zum Theile empirisch-realistischen Analyse, d.i. ohne Rücksicht darauf, ob dieselben in der Wirklichkeit als selbständige Erscheinungen vorhanden, ja selbst ohne Rücksicht darauf. ob sie in ihrer vollen Reinheit überhaupt selbständig darstellbar sind.[118]

> Nichts ist so sicher, als dass die Ergebnisse der exacten Richtung der theoretischen Forschung, mit dem Massstabe des Realismus gemessen, gleichwie auf allen übrigen Gebieten der Erscheinungswelt, so auch auf jenem der Volkswirthschaft als unzureichend und unempirisch erscheinen. Dies ist indess selbstverständlich, indem die Ergebnisse der exacten Forschung, und zwar auf allen Gebieten der Erscheinungswelt, nur unter bestimmten Voraussetzungen wahr sind, unter Voraussetzungen, welche in der Wirklichkeit nicht immer zutreffen. Die Prüfung der exacten Theorie der Volkswirthschaft an der vollen Empirie ist eben ein methodischer Widersinn, eine Verkennung der Grundlagen und Voraussetzungen der exacten Forschung, zugleich aber auch eine solche der besondern Zweche, welchen die exacten Wissenschaften dienen. Die reine Theorie der Volkswirthschaft an der Erfahrung in ihrer vollen Wirklichkeit erproben zu

---

[116] Menger's terminologies, analytic and synthetic, have completely different meanings from the same words used by Kant. We have to be careful enough to avoid possible confusion. We will give later brief expositions of these terminologies in the sense of Kant.

[117] Letter from Menger to Walras (in February 1884). Jaffé [50] Vol. II, p. 4. Letter ♯ 602.

[118] Menger [78] s. 41.

wollen, ist ein Vorgang, analog jenem eines Mathematikers, welcher die Grundsätze der Geometrie durch Messungen realer Objecte berichtigen wollte, ohne zu bedenken, dass diese letzteren ja mit den Grössen, welche die reine Geometrie supponirt, nicht identisch sind, auch jede Messung nothwendig Elemente der Ungenauigkeit in sich schliesst. Der Realismus in der theoretischen Forschung ist gegenüber der exacten Richtung der letztern nicht etwas höheres, sondern etwas verschiedenes.[119]

(5) What are, then, the simplest elements of economic phenomena? In *Grundsätze*, Menger gave a unique answer to this question.[120]

Die Preise, oder mit andern Worten, die im Tausche zur Erscheinung gelangenden Güterquantitäten, so sehr sie sich auch unseren Sinnen aufdrängen und desshalb den gewöhnlichsten Gegenstand der wissenschaftlichen Beobachtung bilden, sind doch nichts weniger als das Wesentliche der ökonomischen Erscheinung des Tausches. Diesses liegt vielmehr in der durch den Tausch herbeigeführten besseren Vorsorge für die Befriedigung der Bedürfnisse der beiden Tauschenden. Die wirthschaftenden Menschen haben das Bestreben, ihre ökonomische Lage nach Möglichkeit zu verbessern. Zu diesem Zwecke setzen sie ihre wirthschaftliche Thätigkeit überhaupt in Bewegung und zu diesem Zwecke tauschen sie auch die Güter aus, wo immer hiedurch derselbe erreicht werden kann. Die Preise sind hiebei aber lediglich accidentielle Erscheinungen, Symptome des ökonomischen Ausgleiches zwischen den menschlichen Wirthschaften.[121]

(6) The simplest elements discovered by analytic-synthetic method are not constructs of human mind. They have their existence or reality which is independent of man's mind.

Alle diese Factoren sind in letzter Linie unabhängig von der menschlichen Willkür, durch die jeweilige Sachlage gegeben: der Ausgangspunkt und der Zielpunkt aller Wirthschaft (Bedarf und verfügbare Güterquantität einerseits und die erreichbare Vollständigkeit der Deckung des Güterbedarfs andererseits) sind in letzter Linie den wirthschaftenden Menschen gegeben, ...[122]

6.2 Menger and Kant

Taking account of (1)–(5), we are naturally inclined to categorize the knowledge obtained by Menger's method as synthetic-*a priori* in the sense of Kant (cf. footnote 116).

Kant classified all the statements based upon two criteria: the logical classification and the epistemological one. By the logical criterion, all the statements are divided into analytic and synthetic ones. A statement is analytic if it "involves nothing more than the

[119] Menger [78] s. 54.

[120] Menger [77] s. 45. In *Untersuchungen*, he added two more elements (Menger [78] s. 45).
Die ursprünglichsten Factoren der menschlichen Wirthschaft sind die Bedürfnisse, die den Menschen unmittelbar von der Natur dargebotenen Güter (sowohl die bezüglichen Genussals Productionsmittel) und das Streben nach möglichst vollständiger Befriedigung der Bedürfnisse (nach möglichst vollständiger Deckung des Güterbedarfes).

[121] Menger [77] s. 172.

[122] Menger [78] s. 45.

meaning relations of the terms". It is synthetic if it "goes beyond the assigned meaning of the terms", that is, it "tells something about the nature of the world". On the other hand, according to the epistemological criterion, all the statements are classified as *a priori* or *a posteriori*. A statement is *a priori* if it "is independent of experience". It is *a posteriori* if its assertion "cannot be justified without reference to experience".[123]

Now do the demarcations in the two criteria coincide? It is clear that any analytic statement is *a priori*. However, is there any synthetic as well as *a priori* statement? Kant admitted room for synthetic-*a priori* statements. But empiricists deny it and claim that the demarcations coincide.

As far as we take account of (1)–(5), it seems quite convincing for us to interpret Menger's methodology in the context of Kant.[124] As a consequence, the exact laws might be regarded as being in the category of synthetic *a priori*.

6.3 Menger and Aristotle

However, according to Menger's thought (6), the simplest elements exist outside the human mind. This ontological attitude of Menger clearly contradicts to Kant, who regarded "Ding an sich" as unknowable. We now quote Russell's concise exposition:

> According to Kant, the outer world causes only the matter of sensation, but our own mental apparatus orders this matter in space and time, and supplies the concepts by means of which we understand experience. Things in themselves, which are the causes of our sensations, are unknowable; they are not in space or time, they are not substances, nor can they be described by any of those other general concepts which Kant calls 'categories'.[125]

Some philosophers and historians of economics suggested a way to interpret Menger as a successor of Aristotle rather than Kant.[126] In view of ample evidence supporting their views, I also agree with them to a large extent. Now what are those aspects of Aristotle's philosophy which are relevant to Menger's methodology?

(i) By a simple comparison, we can find a lot of similarities between the works of the two. In order to exemplify it. I quote just one passage from Aristotle's *Politics*.

---

[123] Carnap [14] pp. 177–178. B. Russell's explanation is as follows.

> An 'analytic' proposition is one in which the predicate is part of the subject; ... ... A 'synthetic' proposition is one that is not analytic. ... ... An 'empirical' proposition is one which we cannot know except by the help of sense-perception, either our own or that of someone else whose testimony we accept. ... ... An 'a priori' proposition, on the other hand, is one which, though it may be *elicited* by experience, is seen, when known, to have a basis other than experience.

(Russell [97] pp. 679–680.).

[124] I also tried to look for a philosophical root of Menger's methodology in Kant's *Reinen Vernunft*. cf. Maruyama [69] pp. 254–268.

[125] Russell [97] p. 680.

[126] Alter [1, 2], Kauder [57], Smith [106]..

> As in other departments of science, so in politics, the compound should always be resolved into the simple elements or least parts of the whole. We must therefore look at the elements of which the state is composed, in order that we may see in what the different kinds of rule differ from one another, and whether any scientific result can be attained about each one of them.[127]

The similarity with Menger's exposition of analytic-synthetic method is indisputable.

(ii) Matter and form. We know the concept of money. A bullion of gold is transformed into money, watch, pen, ... Gold is a material of money, watch, pen, ... According to Aristotle's terminology, gold is the matter of money (watch, pen, ...). And this concept is its form. People have to know the concept of money (watch, pen ...) before they produce money (watch, pen, ...). Being led by the form of money (watch, pen, ...), gold is transformed into money (watch, pen, ...) which has a real existence. Thus the substance[128] of money is realized. Gold as matter can potentially be transformed into various things. Hence matter is no more than a potentiality of a form. "It is in virtue of the form that the matter is some definite thing, and this is the substance of the thing."[129]

The essence of a thing means "those of its properties which it cannot change without losing its identity."[130]

The form of a thing is its essence and primary substance which exists independently of the matter.[131]

Matter, being led by a form, becomes a substance. Then this substance turns into new matter, which, being led by a new form, becomes a new substance, and so on ... . Such a chain of alternating interactions between matter and form—that is, of causal relations, can be identified with what Menger sought for by the analytic-synthetic method. Each causal relation composing this chain corresponds to an exact law.

Menger's ontological view of essence seems to be understandable in a manner compatible with the matter–form theory in Aristotle's metaphysics.

(iii) Aristotle's *Analytics*.

The validity of an exact law is regarded as being independent of experience. That is, an exact law carries *a priori* knowledge in the sense of Kant.

However, Menger's *a priorism* may come from Aristotles' logic exposed in his *Analytics* rather than from Kant.

Compare a couple of deductions as follows:[132]

---

[127] Aristotle [4] Vol. II, *Politics*, 1252 a 18–22.

[128] Here a substance means what is signified by a proper name. It guarantees the real existence. On the other hand, a universal means what is signified by an adjective or class-name. A universal can not exist by itself, but only in particular things. cf. Russell [97] p. 176; Aristotle [4] Vol. II, *Metaphysics*, Book VI (E). Book VII (Z), pp. 1619–1644.

[129] Russell [97] p. 178.

[130] Russell [97] p. 211.

[131] Russell [97] p. 178. If this interpretation is correct, we may have to be confronted with a cynical contradiction. That is, Aristotle's criticism of Plato's doctrine of ideas may attack himself, in his turn.

[132] Aristotle [4] Vol. II, *Posterior Analytics*. 78 a 30–35. Basic syllogism is discussed in *Aprior Analytics*.

[A] deduction of the reason

Things near do not twinkle.
Planets are near

Planets do not twinkle.

[B] deduction of the fact

What does not twinkle is near.
Planets do not twinkle.

Planets are near.

The second presupposition in each inference ("Planets are near" in [A] and "Planets do not twinkle" in [B]) is called the middle term. Aristotle seemed to regard the middle term as the cause of the conclusion.[133]

The conclusion of [A] ("Planets do not twinkle") is more or less well-known to us by repeated experience and is difficult to deny. It may be regarded as a true (or almost true) fact. The validity of the middle term ("Planets are near") is not trivial. It can be verified only by making use of some scientific procedures. Hence [A] is a systematic explanation of the well-known fact by means of a middle term (cause) which is relatively difficult to know. In order to find the cause of an superficial phenomenon, we first have to look backwards from a known fact to deeper unknown world. And then the explanation based on the cause comes next. [A] is, so to say, a "backward inference" in this sense. [A] is necessarily true because its conclusion is a known undeniable fact.

On the other hand, [B] is a "forward inference" since it has a structure to predict an unknown fact from relatively well-known knowledge. Whether [B] is true or false depends upon the validity of the conclusion.[134]

We now recall Menger's analytic-synthetic method. According to his methodology, we have to start from some given (known, observed) phenomenon to discover its essence, and then reconstitute a chain of causal relations leading to the given phenomenon from its essence.

Menger's way of thinking seems to correspond exactly to Aristotle's deduction of the reason. If Menger's exact law is supported by the deduction of the type [A], it must be *a* priori. This explains the root of Menger's *a priorism*.

---

[133] The reason for Aristotle's view on the relation between "middle term" and "cause" is not clear to me. cf. Oide [85] pp. 103–109.

[134] Of course, it is type [B] which gives a basic framework of modern empirical science. However, Aristotle said nothing about the prediction of unknown facts, while he devoted many pages to type [A]. cf. Oide [85], pp. 91–120.

6.4 Walras and Pareto

We now turn to the methodology of the Lausanne School. Walras distinguished pure economics from political economy. Although he regarded political economy as both an experimental and rational science, he characterized pure economics exclusively as a rational one. He simply wrote in 3me Leçon of *Eléments*:

> La méthode mathématique n'est pas la méthode *experimentale*, c'est la méthode *rationnelle*. Les sciences naturelles proprement dites se bornent-elles à décrire purement et simplement la nature et ne sortent-elles pas dè l'expérience? Je laisse aux naturalistes le soin de répondre à cette question. Ce qui est sûr, c'est que les sciences physico-mathématiques, comme les sciences mathématiques proprement dites, sortent de l'expérience dès qu'elles lui ont emprunté leurs types. Elles abstraient de ces types réels des types idéaux qu'elles définissent; et, sur la base de ces définitions, elles bâtissent *à priori* tout l'échafaudage de leurs théorèmes et de leurs démonstrations. Elles rentrent, après cela, dans l'expérience non pour confirmer, mais pour appliquer leurs conclusions.[135]

I quoted (pp. 45–46) a passage from Menger's letter to Walras. However, Walras did not seem to understand what Menger meant. For instance, Walras presupposed a well-organized competitive market, and then analyzed the exchange phenomena in this framework. On the other hand, Menger considered it an error of the rational method to start with such an arbitrary axiom (like the assumption of competition) without seeking for the ultimate and general cause by the analytic-synthetic method. According to Menger, the Walrasian market itself has to be explained by exact laws starting from its essence.

Thus the methodology of Walras' pure economics is completely devoid of any empirical aspect.[136]

Pareto's attitude is quite far from Menger and Walras. His methodology is, in a sense, characterized as textbook-empirical nominalist. While Menger is a believer in essentialism ontology, Pareto declaired himself as the most nominalist of nominalists.

> Io sono il più *nominalista* dei *nominalisti*. Per me non ci sono oggettivamente se non casi concreti; le classificazioni di questi sono opera nostra, e quindi arbitrarie, eccettochè stabiliamo, sempre arbirariamente, il fine di una classificazione; nei qual caso essa si deduce logicamente da quel fine.[137]

According to Pareto, the correct scientific method is a combination of the rational method and the empirical method.

> Le progrès, dans toutes les sciences naturelles, s'est accompli par l'union de ces deux méthodes; c'est-à-dire qu'on a fait certaines hypothèses, dont on a déduit les conséquences au

---

[135] Walras [128] ed. def. p. 29.

[136] I did not go into the "normative bias" of Walras' pure economics. See Jaffé [53] and Walker [125]..

[137] Pareto, "Sul principio economico", *Giornale degli economisti*, **22** (1901), 131–138, reprinted in Pareto [93] 448–455. See p. 448 for the quotation.

Marchionatti–Gambino [66] and Marchionatti [65] are important works about the compairison between Walras and Pareto from the methodological viewpoint.

> moyen de la méthode rationnelle. Ces conséquences ont été comparées aux vérités découvertes par la méthode empirique. Si les unes et les autres se sont trouvées d'accord, les hypothèses ont été acceptées, sinon elles ont été rejetées. Puisque telle est la voie que les hommes ont dû suivre pour apprendre tout ce qu'ils savent des phénomènes naturels, il est extrêmement probable que telle sera la voie que devra suivre aussi la sociologie pour devenir une vraie science. C'est précisément de la sorte que nous voyons actuellement se constituer les nouvelles doctrines économiques.[138]

Thus the procedure of scientific research is composed of several steps:

1. Setting up some hypothesis.
2. Deduction of certain consequences from the hypothesis.
3. Comparison of the consequences with empirical facts, according to which the hypothesis is accepted or rejected.
4. In case of rejection, the hypothesis must be replaced by new one.

$\Rightarrow$ return to 1.

In view of the experimental property of science, no theory can be expected to be absolutely true. What we can expect is no more than successive approximations.

The realm of pure economics is restricted to a specific world where agents exclusively act towards economic benefit extremely rationally. But Pareto was also interested in the working of society in the case where people are irrational or their behaviors are influenced by the subconscious. Pareto committed the investigations into such problems to his sociology.[139] But, in his *Trattato di sociologia generale*, he applied the same empiricist methodology in sociology as in pure economics. What was the result? H. S. Hughes, historian of social thought, criticized Pareto's attitude as follows.

> And indeed what we find is that the bulk of the two-thousand-page *Trattato* consists of illustrative examples. More specifically, we find a complicated methodology only tenuously connected with an entertaining catalogue of *faits divers*. It is these letter that give the *Trattato* its verve and perennial appeal.[140]

Pareto's ambitious venture to surmount the Marginal Revolution did not seem successful.[141]

* The mark † in the Referneces denotes the sources referred to in the text when different editions of the same book are listed together.

---

[138] Pareto, "Comment se pose le problème de l'economie pure", Mémorie présenté, en décembre 1898, à la Société Stella. *Cahier Vilfredo Pareto*, **1** (1963), 121–130. Reprinted in Pareto [92] 102–109. See p. 102 for the quotation.

[139] We are immediately reminded of M. Weber and G. Freud who shared their interests with Pareto. However, Weber's sociology and Freud's psychoanalysis were based upon peculiar methodologies quite different from Pareto's.

[140] Hughes [42] p. 265.

[141] See also Matsuura [71] and Tarascio [117].

References

1. Alter, M.: What do we know about Menger. In: Caldwell (ed.) [12] (1990), pp. 313–348
2. Alter, M.: Carl Menger and the Origin of Austrian Economics. Westview Press, Boulder/San Francisco/Oxford (1990)
3. Antonelli, G.B.: Sulla teoria matematica della economia politica. Nella Topografia del Folchetto, Pisa (1886). Reprinted in Giornale degli economisti, **10**, nuova serie, 223–263 (1951). †
4. Aristotle: The Complete Works of Aristotle, 2 Vol's. In: Barnes, J. (ed.) Princeton University Press, Princeton (1984). Among the works contained in these two volumes, the following are referred to in the text: Prior Analytics, Posterior Analytics, Vol. I; Metaphysics, Nicomachean Ethics, Politics, Vol. II
5. Auspitz, R., Lieben, R.: Untersuchungen über die Theorie des Preises. Dunker & Humblot, Leipzig (1889)
6. Barone, E.: Sopra un libro del Wicksell. Giornale degli Econ. **11**(2nd ser.), 524–539 (1895). Reprinted in Barone [7]. †
7. Barone, E.: Le opere economiche, volume primo. Nicola Zanichelli Editore, Bologna (1936)
8. Black, R.D.C.: Jevons, Bentham and De Morgan. Economica **39**, 119–134 (1972)
9. Boos, M.: W.S. Jevons and the foundation of modern economics. Hist. Polit. Econ. **4**, 364–378 (1972)
10. Boos, M.: Die Wissenschafts-Theorie Carl Mengers. Hermann Böhlaus Nochf, Graz/Wien (1986)
11. Brainard, W.C., Scarf, H.E.: How to compute equilibrium prices in 1891. In: Dimand, R.W., Geanakoplos, J. (eds.) Celebrating Irving Fisher; The Legacy of a Great Economist. American Journal of Economics and Sociology, Malden/Oxford/Victoria, pp. 57–83 (2005)
12. Caldwell, B.J. (ed.): Carl Menger and His Legacy in Economics. Duke University Press, Durham/London (1990)
13. Carathéodory, C.: Calculus of Variations and Partial Differential Equations of the First Order, Vol. I. Trans. by R.B. Dean and J. Brandstatter. Holden-Day, San Francisco (1965)
14. Carnap, R.: Philosophical Foundations of Physics. Basic Books, New York/London (1966)
15. Chipman, J.S.: An empirical implication of Auspitz-Lieben-Edgeworth-Pareto complementarity. J. Econ. Theory **14**, 228–231 (1977)
16. Chipman, J.S.: German Utility Theory. Routledge, Oxon/New York (2014)
17. Chipman, J.S., Hurwicz, L., Richter, M.K., Sonnenschain, H.F. (eds.): Preferences, Utility and Demand. Harcount Brace Jovanovich, New York/Chicago/San Francisco/At lanta (1971)
18. Clebsch, A.: Ueber die simultane Integration linearer partielle Differential-Gleichungen. J. Reine Angew. Math. **65**, 257–268 (1866)
19. Cournot, A.A.: Recherches sur les principes mathématiques de la théorie des richesse. chez L. Hachette, Paris (1838). Œuvres complètes de Cournot, édité par G. Jorland, tome VIII. J. Vrin, Paris (1980). †
20. Cready, J.: Edgeworth and the Development of Neoclassical Economics. Blackwell, Oxford (1986)
21. Cready, J.: Some recent interpretations of Mathematical Psychics. Hist. Polit. Econ. **12**, 267–276 (1980)
22. Davanzati, B.: Scisma d'inghilterra con altre Operette del Sig. Bernaldo Davanzati. Nella nuova stamperia. Maffi e Landi, Fiorenza (1638)
23. Davanzati, B.: A Discourse upon Coins, By Signor Bernaldo Davanzati. A Gentleman of Florence; Being Publickly Spoken in the Academy there, Anno 1588. Trans. by J. Toland. Awsbam and Churchill, London (1696). Original version is contained in [22]
24. Debreu, G.: Theory of Value. Wiley, New York (1959)
25. Debreu, G.: Economies with a finite set of equilibria. Econometrica **38**, 387–392 (1970)
26. de Pietri-Tonelli, A., Bousquet, G.H.: Vilfredo Pareto-Neoclassical Synthesis of Economics and Sociology. Trans. by J. Bomford. Macmillan, Hampshire/London (1994)

27. de Roover, R.: San Bernardino of Siena and San Antonino of Florence; the Two Great Economic Thinkers of the Middle Ages. No. 19 of the Kress Library of Business and Economics. Harvard Graduate School of Buisiness Administration, Boston (1967)
28. Dieudonné, J.: Abrégé d'histoire des mathématiques, 2 Vol's. Hermann, Paris (1978)
29. Edgeworth, F.Y.: Mathematical Psychics. Kegan Paul, London (1881)
30. Fabian, R., Simons, P.M.: The second Austrian school of value theory. In: Grassel, W., Smith, B. (eds.) Austrian Economics, pp. 37–101. Croom Helm, London/Sydney (1986)
31. Fisher, I.: Mathematical Investigations in the Theory of Value and Prices. Transactons of the Connecticut Academy, Vol. IX (1892). Reprinted by Yale University Press, New Haven (1925). †
32. Frobenius, G.: Ueber die Determinante mehrerer Funktionen einer Variablen. J. Reine Angew Math. **77**, 245–257 (1874)
33. Frobenius, G.: Ueber das Pfaffsche Problem. J. Reine Angew Math. **82**, 230–315 (1877)
34. Galiani, F.: Della moneta. Presso Giuseppe Raimondi, Napoli (1750). Edizione seconda. Stamperia Simoniana, Napoli (1780). †
35. Goodstein, J.R.: The rise and fall of Vito Volterra's world. J. Hist. Ideas **45**, 607–617 (1984)
36. Harsanyi, J.C.: Cardinal welfare, individual ethics, and interpersonal comparison of utility. J. Polit. Econ. **63**, 309–321 (1955)
37. Hempel, C.G.: Fundamentals of Concept Formation in Empirical Sciences. University of Chicago Press, Chicago/London (1952)
38. Hicks, J.R.: Léon Walras. Econometrica **2**, 338–348 (1934). Reprinted in Hicks [40]. †
39. Hicks, J.R.: Value and Capital, 2nd edn. Clarendon, Oxford (1946)
40. Hicks, J.R.: Classics and Moderns. Blackwell, Oxford (1983)
41. Howey, R.S.: The Rise of the Marginal Utility School 1870–1889. Columbia University Press, New York/Oxford (1989) †. Originally published by University of Kansas Press, Lawrence (1960)
42. Hughes, H.S.: Consciousness and Society. Revised Vintage Books, New York (1977)
43. Hurwicz, L., Uzawa, H.: On the integrability of demand functions. In: Chipman et al. (eds.) [17]
44. Hutchison, T.W.: A Review of Economic Doctrines 1870–1929. Oxford University Press, Clarendon (1953)
45. Hutchison, T.W.: Revolution and Progress in Economic Knowledge. Blackwell, Oxford (1981)
46. Hutchison, T.W.: The Politics and Philosophy of Economics. Blackwell, Oxford (1981)
47. Ingrao, B., Israel, G.: The Invisible Hand; Economic Equilibrium in the History of Science. MIT Press, Cambridge/London (1990)
48. Israel, G., Nurzia, L.: Fundamental trends and conflicts in Italian mathematics between the two world wars. Arch. Int. d'hist. Sci. **39**, 111–143 (1989)
49. Jaffé, W.: New light on an old quarrel. Cahier Vilfredo Pareto **3**, 61–102 (1964). Reprinted in Walker (ed.) [123]
50. Jaffé, W. (ed.): Correspondence of Léon Walras and Related Papers, 3 Vol's. North-Holland, Amsterdam (1965)
51. Jaffé, W.: Menger, Jevons and Walras de-homogenized. Econ. Inq. **14**, 511–524 (1976). Reprinted in Walker (ed.) [123]
52. Jaffé, W.: The Walras-Poincaré correspondence on the cardinal measurability of utility. Can. J. Econ. **10**, 300–307 (1977). Reprinted in Walker (ed.) [123]
53. Jaffé, W.: The normative bias of the Walrasian model: Walras versus Gossen. Q. J. Econ. **91**, 371–387 (1977). Reprinted in Walker (ed.) [123]
54. Jevons, W.S.: The Theory of Political Economy. Macmillan, London (1871). 4th ed, Macmillan, London (1911). †
55. Jevons, W.S.: The Principles of Science. Macmillan, London, (1874). 3rd ed, Macmillan, Lndon (1879). †
56. Jevons, W.S.: Investigations in Currency and Finance. Macmillan, London (1884)
57. Kauder, E.: Intellectual and political roots of the older Austrian school. Zeitschrift für Nationalökonomie **17**, 411–425 (1958)

58. Kauder, E.: History of Marginal Utility Theory. Princeton University Press, Princeton (1965)
59. Lagrange, J.L.: Mecanique Analytique. Desaint, Paris (1788)
60. Langholm, O.: Price and Value in the Arislotelian Tradition. Universitetsforlaget, Oslo (1979)
61. Langholm, O.: The Legacy of Scholasticism in Economic Thought. Cambridge University Press, Cambridge/New York/Melbourne (1998)
62. Lowry, S.T.: Aristotle's mathematical analysis of exchange. Hist. Polit. Econ. **1**, 44–66 (1969)
63. Lowry, S.T.: The Archaeology of Economic Ideas. Duke University Press, Durham (1975)
64. MacLennan, B.: Jevons' philosophy of science. Manch. Sch. Econ. Soc. Stud. **40**, 53–71 (1972)
65. Marchionatti, R.: The methodological foundations of pure and applied economics in Pareto. Revue Eur. Sci. Soc. **37**, 277–294 (1999)
66. Marchionatti, R., Gambino, E.: Pareto and political economy as a science: methodological revolution and analytical advances in economic theory in the 1890s. J. Polit. Econ. **105**, 1322–1348 (1997)
67. Marshall, A.: Principles of Economics. Macmillan, London (1890). 9th ed, Macmillan, London (1961). †
68. Martin, T.: Probabilités et critique philosophique selon Cournot. Vrin, Paris (1996)
69. Maruyama, T.: A Picture of Walras. Keisoshobo, Tokyo (2008). (in Japanese)
70. Matsuura, T.: The economics of the Lausanne school in Italy. Mita J. Econ. **61**, 929–950 (1968). (in Japanese)
71. Matsuura, T.: Walras and Pareto. Mita J. Econ. **64**, 1039–1061 (1971). (in Japanese)
72. Mayer, A.: Ueber unbeschränkt integrable Systeme von linearen totalen Differentialgleichungen. Math. Ann. **5**, 448–470 (1872)
73. Mays, W.: Jevons's conception of scientific method. Manch. Sch. Econ. Soc. Stud. **30**, 223–249 (1962)
74. Mclean, I., Hewitt, F. (eds.): Condorcet and Modernity. Cambridge University Press, Cambridge (2004)
75. McNulty, J.: Economic theory and the meaning of competition. Q. J. Econ. **82**, 639–656 (1968)
76. Menard, C.: Three forms of resistance to statistics: Say, Cournot, Walras. Hist. Polit. Econ. **12**, 524–541 (1980)
77. Menger, C.: Grundaätze der Volkswirtschaftslehre. Wilhelm Braumüller, Wien (1871). Gesammelte Werke, Hrsg. von F.A. Hayek, Band I J.C.B. Mohr, Tübingen (1968). †
78. Menger, C.: Untersuchungen über die Methode der Socialwissenschaften und der Politischen Oekonomie insbesondere. Duncker & Humblot, Leipzig (1883). Gesammelte Werke, Hrsg. von F.A. Hayek, Band II J.C.B. Mohr, Tübingen (1969). †
79. Monroe, A.E.: Early Economic Thought. Harvard University Press, Cambridge (1924)
80. Mosselmans, B.: William Stanley Jevons and the Cutting Edge of Economics. Routledge, London/New York (2007)
81. Mutoh, I.: Léon Walras and the tradition of 'mathématique sociale'. Mita J. Econ. **86**, 169–204 (1993). (in Japanese)
82. Nakano, S.: A reconsideration of Jevons's theory of exchange I. Mita J. Econ. **82**, 333–352 (1989). (in Japanese)
83. Oakley, A.: The Foundations of Austrian Economics from Menger to Mises. Elgar, Cheltenham (1997)
84. O'Brien, D.P.: Marshall's Work in Relation to Classical Economics. In: Whitaker (ed.) [130]
85. Oide, A.: Genealogy of Knowledge Revolution. Iwanami, Tokyo (2004). (in Japanese)
86. Pareto, V.: Considerazioni sui principii fondamentali dell' economia politica pura, I–V, Giorunale degli Econ. **4**, 389–420, 485–512 (1892); **5**, 119–157 (1892); **6**, 1–37 (1893); **7**, 279–321 (1893). Also in Pareto [93]. †
87. Pareto, V.: Il massimo di utilità dato dalla libera concorrenza. Giornale degli Econ. **9**(2nd ser.), 48–66 (1894). Also in Pareto [93]
88. Pareto, V.: Cours d'économie politique. Rouge, Lausanne (1896–1897). Œuvres complètes, tome I Droz, Genève (1964). †

89. Pareto, V.: Manuale di economia politica. Società editrice libraria, Milano (1906)
90. Pareto, V.: Manuel d' économie politique. Giard & Brière, Paris (1909). Œuvres complètes, tome VII Droz, Genève (1966). †
91. Pareto, V.: Trattato di sociologia generale. Barbera, Firenze (1916)
92. Pareto, V.: Marxisme et économie pure Œuvres complètes, tome IX. Droz, Genève (1966)
93. Pareto, V.: Ecrits d'économie politique pure. In: Busino, G. (ed.) Œuvres complètes, tome XXVI Droz, Genève (1982)
94. Pareto, V.: Lettere a Maffeo Pantaleoni, 3 Vol's a cura di G. De Rosa Œuvres complètes, tome 28, 1 Droz, Genève (1984)
95. Robbins, L.: The place of Jevons in the history of economic thought. Manch. Econ. Soc. Stud. **7**, 1–17 (1936)
96. Robertson, R.M.: Jevons and his Precursors. Econometrica **19**, 229–249 (1951)
97. Russell, B.: History of Western Philosophy, 2nd edn. Allen & Unwin, London (1961)
98. Samuelson, P.A.: Foundations of Economic Analysis. Harvard University Press, Cambridge (1947)
99. Samuelson, P.A.: The problem of integrability in utility theory. Economica **17**, 355–385 (1950)
100. Schabas, M.: A World Ruted by Number. Princeton University Press, Princeton (1990)
101. Schneewind, J.B.: Sidgwick's Ethics and Victorian Moral Philosophy. Oxford University Press, Oxford (1977)
102. Schneider, E.: Vilfredo Pareto—The economist in the light of his letters to Maffeo Pantaleoni. Banca Nazionale del Lavoro, Q. Rev. **14**, 247–295 (1961)
103. Schumpeter, J.A.: History of Economic Analysis. Oxford University Press, New York (1954)
104. Schwalbe, U.: Irving fisher's mathematical investigations in the theory of value and prices. In: Loeh, H.-E., Monissen, H.G. (eds.) The Economics of Irving Fisher, pp. 281–303. Elgar, Cheltenham (1999)
105. Slutsky, E.: Sulla teoria del bilancio del consumatore. Giornale degli Econ. **51**(3), 1–26 (1915)
106. Smith, B.: Aristotle, Menger, Mises: an essay in the metaphysics of economics. In: Caldwell (ed.) [12], pp. 263–288
107. St. Augustine: Concerning the City of God against the Pagans. Trans. by H. Bettenson. Penguin Books, London (2003)
108. Steiner, p.: Pareto contre Walras: le problème de l'économie sociale. Economie et sociétés, Série Œconomia, Histoire de la pensée économique **PE n° 20–21**, 53–73 (1994)
109. Stigler, G.J.: Production and Distribution Theories. Macmillan, New York (1941)
110. Stigler, G.J.: The development of utility theory. J. Polit. Econ. **58**, 307–327, 373–396 (1950). Reprinted in Stigler [112]. †
111. Stigler, G.J.: Perfect competition, historically contemplated. J. Polit. Econ. **65**, 234–267 (1957). Reprinted in Stigler [112]. †
112. Stigler, G.J.: Essays in the History of Economics. Chicago University Press, Chicago (1965)
113. Stigler, G.J.: Place of Marshall's *Principles* in the development of economics. In: Whitaker (ed.) [131]
114. Streissler, E.W.: To what extent was the Austrian school marginalist? Hist. Polit. Econ. **4**, 426–441 (1972)
115. Streissler, E.W.: The influence of German economics on the work of Menger and Marshall. In: Caldwell (ed.) [12], pp. 31–68
116. Suda, S.: Vilfred Pareto and the integrability problem of demand functions. Mita J. Econ. **99**, 637–655 (2007). (in Japanese)
117. Tarascio, V.: Pareto's Methodological Approach to Economics. The University of North Carolina Press, Chapel Hill (1968)
118. Uzawa, H.: A note on the Menger-Wieser theory of imputation. Zeitschrift für Nationalökonomie **18**, 318–334 (1958)
119. Volterra, V.: L'economia, matematica ed il nuovo manuale del prof. Pareto. Giornale degli Econ. **32**, 296–301 (1906)
120. von Böhm–Bawerk, E.: Grundzüge der Theorie des wirtschaftlichen Güterwerts. Jahrbücher für Nationalökonomie und Statistik, N.F. **13**, 1–82, 477–541 (1886)

121. von Wieser, F.: Über den Ursprung und die Hauptgesetze des wirtschaftlichen Werthes. Alfred Hölder, Wien (1884)
122. von Wieser, F.: Der Natürliche Werth. Alfred Hölder, Wien (1889)
123. Walker, D.A.: Marshall's theory of competitive exchange. Can. J. Econ. **2**, 590–598 (1969)
124. Walker, D.A. (ed.): William Jaffé's Essays on Walras. Cambridge University Press, Cambridge/NewYork/Melbourne (1983)
125. Walker, D.A.: Is Walras's theory of general equilibrium normative scheme? Hist. Polit. Econ. **16**, 445–469 (1984)
126. Walras, A.A.: De la nature de la richesse et de l'origine de la valeur. Librarie Flix Alcan, Paris (1838)
127. Walras, L.: Principe d'une théorie mathématique de l'échange", Mémorie lu à l'académie des sciences morales et politique, à Paris (séances des 16 et 23 août) (1873). Reprinted in Walras [129]
128. Walras, L.: Éléments d'économie politique pure, 2 Vol's. Corbaz, Lausanne (1874–1877). Édition définitive. Pichon, Paris/Rouge, Lausanne (1926). †
129. Walras, L.: Théorie mathématique de la richesse sociale. Guillaumin, Paris (1877). Reprint edition (Otto zeller, Osnabrück) (1964). †
130. Whitaker, J.K.: Some neglected aspects of Alfred Marshall's economic and social thought. Hist. Polit. Econ. **9**, 162–197 (1977)
131. Whitaker, J.K. (ed.): Centenary Essays on Alfred Marshall. Cambridge University Press, Cambridge/New York/Melbourne (1990)
132. Wicksell, K.: Über Wert, Kapital und Rente. Neudruck der Ausgabe, Jena (1893)
133. Wicksell, K.: Lectures on Political Economy. Trans. by E. Classen, Vol. I. Routledge, London (1934)
134. Wicksell, K.: Selected Papers of Economic Theory. In: Lindahl, E. (ed.) Allen & Unwin, London (1958)
135. Wicksteed, P.H.: Essay on the Coordination of the Laws of Distribution. Macmillan, London (1894)
136. Williams, D.: Condorcet and Modernity. Cambridge University Press, Cambridge (2004)
137. Wold, H.: A synthesis of pure demand analysis, Parts I, II and III. Skandinavisk Aktuarietidskrift **26**, 85–118, 220–263 (1943); **27**, 69–120 (1944)
138. Yasui, T.: Pure economics and the theory of price. J. Econ. **3**, 1242–1299 (1933). Reprinted in Yasui [140] † (in Japanese)
139. Yasui, T.: Theory of imputation and marginal productivity. J. Econ. **4**, 452–513 (1934). Reprinted in Yasui [140] † (in Japanese)
140. Yasui, T.: Collected Works of Takuma Yasui, vol. I. Sobunsha, Tokyo (1970). (in Japanese)

Toward the Marginal Revolution: The Foundational Status of Antoine-Augustin Cournot

Claude Menard

Abstract Cournot's *Researches into the Mathematical Principles of the Theory of Wealth* (1838) was a major breakthrough towards modern economics. It was so both in form and content. With respect to form, it was the first rigorous and extensive mathematical model using advanced calculus to explore key economic issues. With respect to the content, it initiated the representation of demand and supply as well-shaped functions, from which an equilibrium could be determined. Cournot went further, exploring how different equilibria correspond to different market structures. In that process, through the examination of the case of duopoly he introduced the concept of a non-cooperative solution based on the "reaction functions" of the competitors, providing a first insight of what would be revisited over a century later as a Nash-equilibrium. Besides a reminder of these major results, this chapter examines personal as well as contextual factors that could explain this radical transformation in economic analysis. It also discusses why this book was ignored for decades before being rediscovered by the founders of the marginal revolution and fully integrated into modern economics. To sum up, this chapter demonstrates that Cournot was a radical innovator whose contribution remains refreshing reading.

1 Introduction

The mathematical revolution in economics can be precisely dated: it started with Cournot's *Researches into the Mathematical Principles of the Theory of Wealth* [8].[1] Although there was a substantial literature on the role of demand and supply long before Cournot [1] as well as some fuzzy usage of mathematics applied to economic issues (e.g., [6, 37]), Cournot's book represents a major breakthrough.

[1] From now on, all references to *Researches* are to the English translation by Nathaniel Bacon, with a preface by Irving Fisher (who initiated the project of this translation).

C. Menard (✉)
Centre d'Economie de La Sorbonne, University of Paris-Pantheon-Sorbonne, 106 Bd de L'Hôpital, Cedex 13, 75634 Paris, France
e-mail: claude.menard@univ-partis1.fr

T. Maruyama (ed.), *Marginal Revolution in Economics*, Monographs in Mathematical Economics 6, https://doi.org/10.1007/978-981-99-4342-5_2

The radical transformation it introduced comes from the approach to demand and supply as well-shaped functions, with their relation determining the equilibrium of the system of exchange. Cournot went even further in showing how different conditions determining variations in demand and supply would lead to different equilibria, defining different market structures. In exploring the consequences of this theoretical setting with the powerful mathematical tools available in his time, Cournot provided the foundation of modern economics and the scaffolding of what we now identify as micro-economics.

The very representation of what was called at the time "political economy" was to be substantially modified by this innovative approach. Indeed, although *Researches* was initially almost totally ignored, the book percolated through what can be called the French school of "social engineering" (Dupuit, Walras, Bertrand, etc.) as well as through the strong influence of Marshall and Edgeworth in the UK, Fisher, Moore and others in the USA, thus shaping the development of mathematical economics. Indeed, all these authors referred to and discussed Cournot.

His influence also comes from the very type of mathematics that he introduced at the core of economics. Indeed, in establishing an analogy between the structure of exchanges and a physical system of dots and forces, Cournot justified the introduction of mathematical calculus, the most powerful tool sustaining the development of classical mechanics as well as astronomy at the juncture between the 18th and the 19th century. In doing so his contribution not only transformed the perspective on economic issues, but also the way to deal with them. In connecting market structures and pricing he introduced the idea of market power. In identifying wealth to the monetary value of traded goods, he introduced a representation of the economic system as a set of dots (prices) submitted to forces (demand and supply), which provided the foundation for the analogy between economics and mechanics. Thus, this book opened room for the "marginalist revolution".

However, the destiny of *Researches* remains somehow paradoxical. Notwithstanding this major influence on the founders of mathematical economics and more specifically of the marginalist revolution, the extension of Cournot's contribution to shaping modern economics remains largely ignored, with the possible exception of the now ritual references to Cournot's duopoly and the so-called "Cournot-Nash Equilibrium". Although Cournot later published other books in economics (e.g., [12, 13], hoping to attract the attention of a wider audience, it is ultimately the more technical 1838 book that represents a landmark in economics. This chapter focuses exclusively on *Researches* and intends to pay due to its role in revolutionizing economic thought. To do so, I shall present different facets of the book and will stick as much as possible to its initial notation to give the gist of how innovative it was, thus avoiding a "translation" into a modern mathematical formulation that would dissipate this sense of novelty.

The chapter proceeds through the following steps. Section 2 reviews some contextual elements, both personal and societal, that made this theoretical revolution possible. Section 3 highlights the key role of the mathematical definition of "the law of demand" as one of Cournot's main contributions to modern economics. Section 4 summarizes how Cournot built a rigorous theory of market structures and pays special attention to the infamous case of duopoly, often viewed as a first step towards the analysis of players' strategies and, ultimately, the development of game theory. Section 5 discusses a paradox of this contribution, which opted for using calculus rather than probabilities as key tool of investigation, and the many obstacles it had to overcome in order to permeate economic reasoning, which it did after a long wait. Section 6 concludes by emphasizing how *Researches* represents a fundamental episode in the history of the so-called marginal revolution.

2 Background: 1838 in Context

To understand what shaped the conditions of emergence of *Researches* and to better capture how innovative the book was, it is necessary to have at least a quick look at some aspect of the itinerary of Cournot as well as to contextual elements that contributed to create a fertile ground for this revolutionary approach to economics.

2.1 Personal Factors

It is of course impossible (and meaningless) to try to identify all personal elements that could explain how Cournot became … Cournot! However, some elements relevant for understanding the origin of *Researches* can be pointed out.

Cournot was born on August 26, 1801, in Gray, a small town in Haute-Saone, on the south-east of Burgundy (France).[2] He therefore grew up in the challenging years that followed the French Revolution and spent the first period of his life under the Napoleonic regime. Notwithstanding the surrounding turmoil, he benefited from the reforms of the educational system implemented in the spirit of the "Enlightenment", particularly the development of a public educational system implemented on the entire territory of France. Thus, Cournot attended local schools. He was already a brilliant student, particularly in mathematics, and successfully passed through the selective examination in sciences to be admitted (in 1821) to what later became known as the prestigious "Ecole Normale Supérieure," another creation of the French Revolution dedicated to the training of highly qualified teachers and researchers. The scientific environment surrounding a student in Paris at that time was quite exciting and totally dominated by the prestigious scientific figures of the late 18th to early nineteenth century, particularly Laplace and Lagrange. For Cournot their influence percolated through the teaching (and support) of his mathematical mentors, Lacroix (a specialist of calculus) and Poisson (mostly known for his contributions

[2] He passed away in March 1877 in Paris. This is to say that in his lifetime he went through four revolutionary periods: the Napoleonic regime in the aftermath of the French revolution of 1789, the revolutions of 1830 and 1848; and the invasion of Prussia followed by the 'Commune de Paris' in 1870–1871.

to probabilities). He shared ideas (and likely debated the political situation) with close friends also involved in scientific activities (e.g., Hachette, who later became his publisher; Saigey, etc.). However, Cournot's training was threatened when the government closed the school in 1822 for political motives (it reopened a few years later): this was the time of the conservative "*Restauration*" regime that was imposed on France after the defeat of Napoleon and the school was considered a stronghold of (condemnable) liberal ideas.

Cournot nevertheless pursued his program in mathematics at the Sorbonne, where he graduated one year later. He then worked for 10 years as personal secretary to field Marechal de Gouvion Saint-Cyr (a "hero" of the Napoleonic wars), helping him to write his memoirs and tutoring his son while simultaneously attending the Law school (in which there was some introduction to economics, a discipline without a specific program at that time; he graduated in 1827) and preparing a Ph.D. in Mathematics (he defended his thesis in 1829). At the time (and this remained true in France until the early 1970s), getting a Ph. D required two theses: a main thesis and a secondary one. Significantly, with respect to his later contribution to economics (see below), Cournot did his main thesis in mechanics and his secondary one in astronomy, both being then considered as "royal disciplines" in sciences. So, throughout all this politically and socially agitated period,[3] he pursued his formation and became close to a young generation of mathematicians and physicist (besides Hachette and Saigey he was close to Dirichlet who studied in Paris in 1822–1826). This is to say that during all his formative period Cournot evolved in a highly stimulating intellectual environment. He also largely benefited from the support of his mentor, Poisson, who played a decisive role in Cournot's career, helping him to get appointed as professor at the University of Lyon in 1834, then as rector of the Academy of Grenoble and other administrative responsibilities in 1835 and after.

2.2 *Leaning Towards Economic and Social Issues*

An important occupation between the end of Cournot's term with the family de Gouvion Saint-Cyr and his appointment in Lyon was the translations of two renowned English textbooks in physics and astronomy, Kater and Ladner [21] and Herschel [18], to which he added appendices of his own (Chapter 23 of the former, an "Addition" in the latter). These appendices already suggested his interest for issues of economics and social sciences (more on this below). His teaching on the *Theory of Functions* at the Chair of Mathematics and Mechanics in Lyon, with a very diversified audience, reinforced this conviction and led him to work on the preliminaries of *Researches* while under the pressure of his friend Hachette (who also published the 1838 book) to transform his mathematical lessons into a textbook, ultimately

[3] Partially related to the emergence of a large urban-working class population in relation to the development of the industry.

published in 1841 and that became quite influential.[4] This was the time when he wrote *Researches*.

Actually, his interest in economics and social issues may well go back to his years at the Law School and the discussions he likely had with his friends and colleagues at the Sorbonne in a time of social instability and political turmoil. These concerns also show through his scientific activities. In the late 1820s he already raised the question of the possible extension of principles developed in "rational mechanics" *à la Lagrange* to issues such as social problems, as can be perceived through his short notes to the *Bulletin des sciences mathématiques, physiques et chimiques*, better known as "Bulletin de Ferusac," to which Cournot collaborated regularly between 1826 and 1829. However, sparse indications in the additions he made to his two translations of 1834 provide better clues to his increasing interest in that direction. His addition to Herschel contains an important methodological note (p. 504) in which he regrets that probabilities ("le calcul des chances") have been applied almost exclusively to problems of games, purely speculative problems, problems of mediocre practical interest, or problems of social statistics ("statistique sociale"), issues that he considered by far too complex for this type of mathematical investigation, while statistics and probabilities can be fully relevant for astronomy through the building of series supporting theory.[5] The chapter added to his translation of Kater and Lardner (Chapter 23, on "De la mesure des forces et du travail des machines") includes remarks and developments that support the idea that calculus might be a much better tool than statistics and probabilities for analyzing social issues. In #405 of that addition, Cournot introduces a fundamental analogy between mechanics and economy through the analysis of how to obtain a certain output from machines as well as from the effort of laborers. This analogy is then developed through several paragraphs (up to #421), leading to a comparison between the conditions of equilibrium of mechanical forces and of human (labor) forces.

This strategy of relying on inspiration provided by mechanics (and the associated tool of calculus) rather than on probabilities is confirmed by Cournot's remarks in the *Exposition on the theory of chances*[6] that he published only in 1843 but that he started writing in the mid-1830s at a time he was interacting with Poisson.[7]

---

[4] This textbook (as well as his main dissertation and a few short papers published in the late 1820s) shows that Cournot was well informed and knowledgeable about the progress of mathematics of his time. However, his appointment in Lyon also meant addressing an audience including notables from that city with little familiarity with mathematics but expecting insights that could be helpful in the running of their businesses. One must be aware that universities at that time also played a role of social venues to build networks. This suggests that Cournot had to find illustrations from the business world, something he mentions in his *Souvenirs* ([13]; published posthumously in 1913).

[5] "Jusqu'à présent on n'a guère appliqué le calcul des chances qu'à des problèmes sur les jeux, problèmes purement spéculatifs ou d'un médiocre intérêt pratique, et à des faits de statistique sociale dont les causes se dérobent par leur complication à toute investigation mathématique, et pour lesquels nous n'avons d'autres données que celles de l'expérience" (Addition, #7, p. 504).

[6] John Maynard Keynes highly regarded and was influenced by the *Exposition* [22].

[7] His friend Hachette, who was becoming a major scientific publisher, continuously pushed Cournot – distracted by his increasing administrative responsibilities – to wrap up his scientific

From 1835 on, Cournot became increasingly involved in administrative responsibilities at a high level in the French educational system and later published mostly in philosophy, with the major exception of his *Exposition on the theory of chances* [10] which he likely wrote in the previous decade, stimulated by the encouragement of Poisson. Although this significant contribution to probabilities (and their diffusion as part of the research program in mathematics in the 19th century and after) goes beyond the main focus of our chapter, it contains remarks that help us understand why Cournot opted for calculus rather than probabilities in his *Researches*, an intriguing aspect to be discussed further in Sect. 5 of this chapter.[8]

All the indications above draw a configuration of Cournot's interest for the extension of mathematics, mainly the theory of functions (calculus), to issues relating to the "question sociale," a growing concern among intellectuals of that period. It is through the analogies he saw between mechanics and astronomy on the one hand and socio-economic problems on the other hand that Cournot connected his intellectual interest with the social expectations and problems of his time. In that respect he was very much in the continuity of the values and concerns of the philosophy of the Enlightenment as carried through the French Revolution by the great mathematicians (Condorcet, Laplace, Lagrange, Poisson and several others) who provided Cournot his scientific inspiration and his interest in societal issues.

2.3 Cournot and the Economics of His Time

But what about economics? Reading *Researches*, it is striking how few references there are to the existing economic literature, something for which he was later criticized (de Fontenay, 1963). Smith, Ricardo, Say (the leading figures in economics of that time) are mentioned mostly *en passant* and almost exclusively in the Preface and in the last chapter (on international trade) of *Researches*. The classic contribution of Smith is mentioned only eight times; Say, the most influential French economist at the time, is graced with four mentions; and there is only one reference to Ricardo. Although paying lip service to the value of their empirical analyses, the Preface primarily mentions them to point out the opposition to the usage of mathematics in what was then identified as political economy. The other references are in Chapter XII, the last one, on the impact of open international trade on national income, in which Cournot refers to Smith and Say only to challenge some of their conclusions about the advantages of international free trade (what we now refer to as "globalization"). He also mentions "the doctrine of Quesnay's school" (in the French original

---

books while apparently considering that *Researches* should not be a priority. He nevertheless secured its publication.

8 Among his later publications, two books help understanding Cournot's approach to economics: *Principes de la Théorie des Richesses* [11]; *Revue Sommaire des Doctrines Economiques* [12]. Cournot suggested that they can be read, particularly the one from 1863, as *Researches* revisited ...without mathematics (see *Souvenirs*, [13]). They are actually quite different in many aspects from the rigorous approach that prevailed in his *Researches*.

expressed even more negatively as "la doctrine de la secte de Quesnay") in his chapter on taxation (#38). And the only usage of mathematics referenced in his book is to a book from Canard [6] which he considered "so radically at fault, and the application of them so erroneous, that the approval of a distinguished body of men was unable to preserve the work from oblivion" (Preface).

So, very little in Cournot's contribution of 1838 can be directly connected to the existing economic literature. *Researches* at most picked up some elements feeding the initial intuitions, e.g., the relation between prices and demand. Otherwise, there is such a distance between this book and the existing authorities of that period in economics and it was so innovative that it is no surprise if it was so largely ignored by his contemporaries [26, Chap. 4; 35].

Actually, the main influences in shaping Cournot's approach to economics may come from an intellectual environment much more difficult to capture, that is: the general "ambience" surrounding the generation of scientists trained in the 1820s and expressing interest in economic and social issues.[9] Through the involvement of their scientific mentors in the French Revolution, they inherited from social thinkers (e.g., Condorcet and the so-called "Ideologues" Destutt de Tracy and Cabanis) concerns about the need to deal with social and political problems by building a scientific approach (the "mathématique sociale" of the Marquis de Condorcet) that could provide the foundations of "social engineering." The underlying idea is that this was the way to follow in order to proceed to a "rational reform" of the society without falling into the turmoil experienced after 1789; it fed what conservatives viewed as a "liberal orientation" threatening the restoration of the monarchy and of old privileges, leading to the decision from 1822 to close the school that would formally become the Ecole Normale Supérieure over two decades later. What social engineering meant was that the society should be understood as a system sharing similarities to mechanical systems, susceptible of an analysis with the mathematical tools available to physicists and engineers, and making it possible to smoothly reform societies, the reduction of "social disturbances" having analogies with the reduction of frictions in a mechanical system. In that respect, Cournot's appendices to his translation of Kater and Ladner and of Herschel already suggest that he was leaning in that direction and participating in this general movement of ideas.

Although the other facet of Cournot's intellectual activity, his philosophical publications, remain highly considered, at least in France (a complete edition in thirteen volumes of his contributions and part of his correspondence was published in France between 1973 and 2010), this short bio-sketch has mainly emphasized those elements that provided the background in which to find the building blocks of *Researches,* indisputably his major economic contribution and maybe his main contribution overall.

---

[9] Among this generation, besides Cournot, we find Jules Dupuit (1804–1866) and Adolphe Quételet (1796–1874) among others.

3 Foundation of the 1838 Model: The Mathematical Law of Demand

In his *Researches*, Cournot intended to explore two main issues, one that modern economics would identify as part of the core of micro-economics (how can markets lead to an equilibrium that maximizes buyers' gains and suppliers' profits), and the other one as part of macro-economics (how can generalization of exchanges though open and competitive markets lead to optimizing national income). What follows focuses primarily on the first issue, which corresponds to the key contribution of Cournot. Modelling the second problem, notwithstanding some interesting indications he provided in that direction, would have required tools that had to await the much later development of national income accounts and aggregation techniques (Kuznets et al., with national accounting starting in the mid-1930s). However, in considering how to maximize national revenue, Cournot had intuitions that brought him close to the general equilibrium theory, largely associated to the name of Walras [16].

To build what became the foundational micro-economic model, now fully integrated in our conception of economics, Cournot had to overcome two major obstacles: (1) how to depart from the classical (and quasi-metaphysical) debates about the origin of value (Labor? Utility? Scarcity?) as distinct from market prices and to rather focus on how values are actually embedded in and captured through prices determined in the actual process of exchange; and (2) how to move beyond the often-verbose debates about "economic laws" and the rare and mostly erroneous efforts to introduce rudimentary forms of calculation in economics (e.g., Canard, Whewell). With Cournot's mathematical background, the solution to these problems required elaborating assumptions that would allow the rigorous usage of the most recent developments in the mathematics of his time, mainly calculus. As stated upfront in *Researches* (Preface: p. 3): "…most authors who have devoted themselves to political economy … imagined that the use of symbols and formulas could only lead to numerical calculations … But those skilled in mathematical analysis know that its object is not simply to calculate numbers, but that it is also employed to find the relations between magnitudes which cannot be expressed in numbers and between functions whose laws is not capable of algebraic expression." This is key to understanding how Cournot built the cornerstone of his model.

3.1 The Central Intuition

Chapter 4 of *Researches* ("Of the law of demand") is absolutely crucial in that respect as it introduces key elements of what became mathematical economics. The central intuition is that on markets of specific goods (what we would qualify a "partial equilibrium" approach), agents operate as rational maximizers. "We shall invoke but a single axiom, or, if you prefer, make but a single hypothesis, i.e., that each one seeks to derive the greatest possible value from his goods or his labour" (*Researches*, #20[10]). These maximizing agents interact through markets and the associated price mechanism: rational buyers (resp. sellers) adjust their demand (resp. supply) according to the indications that prices provide in the exchange process, allowing the system to reach an equilibrium (the adequation of demand and supply) through "*tâtonnement*," as illustrated by the stock exchange (#22). It is the interaction of buyers and sellers through the price mechanism that structures economic choices, leading to market equilibrium. "The laws which govern the movement of values" make the way markets work analogous to the movement of points in space: economic systems share similarities with physical systems, with prices determined through the interacting forces encapsulated in the functions of demand and supply. This brings economics away from the prevailing and, according to Cournot, sterile debate about the origin of value.[11]

The introduction of these two functions, and particularly the law of demand, therefore provided the building blocks of modern mathematical economics, opening the avenue to the rigorous usage of calculus.

3.2 Towards a Rigorous Concept of Equilibrium in Economics

Once we reposition this reasoning in the context of what economics was in the early 19th century, making these steps was not obvious at all. First it required the explicit introduction of a behavioral assumption that goes beyond the Smithian idea of agents motivated by their self-interest and that became central in micro-economics: agents are rational maximizers. Second, it required a shift in the concept of markets. Indeed, Cournot's *Researches* departs upfront from the classical approach to markets, no more viewed as the result of considerations on the origin of values but as a starting point, the very central mechanism through which economic systems operate, providing the very foundation of economic theory understood as the science of traded goods and services. In Cournot's model, the market is the abstract space in which free trade prevails so that homogeneous prices can be levelled rapidly and promptly.[12] Trading requires the organization of transactions between two sets of partners, those willing to sell (the supply side) and those willing to buy (the demand side), which is exactly what markets make possible. And parties trade up to the point where they reach an

[10] From now on and if not otherwise specified all references to paragraph numbers are those used in *Researches*.

[11] "The abstract idea of *wealth* or of *value in exchange*, a definite idea, and consequently susceptible of a rigorous treatment in combinations, must be carefully distinguished from the accessory ideas of utility, scarcity, and suitability to the needs and enjoyment of mankind, which the world *wealth* still suggests in common speech" (#3).

[12] Cournot's exact expression states that a market designates "the entire territory of which the parts are so united by the relations of unrestricted commerce that prices there take the same level throughout, with ease and rapidity" (#23).

equilibrium because they are maximizers. For a theoretician trained in Lagrange's mechanics, as Cournot was, maximization refers to calculus, which usage imposed defining the supply and demand functions more rigorously.

On the supply side, producers intend to maximize the net value of what they sell, their "profit," which is a function of the price of the good (or service) traded. To reach that goal they must satisfy consumers. And they cannot target individual consumers, whose demand might be volatile and difficult to predict (#21): they need to reach a set of consumers large enough to establish some regularity in the demand for the good (or service) offered. Hence, the profit of the supplier(s) depends on the demand function, which becomes a very central piece of the model ... and the major problem to solve: how to switch from the then prevailing idea that to understand (and predict) demand one had to look at either individual behavior and/or at numbers provided by statistics to an abstract conception of demand as a function of prices that should guide the collection of data rather than the other way around!

Indeed, Cournot was highly skeptical about the possibility of building a demand function through the accumulation of statistics on individual choices, because the variability of individual demand for a specific homogenous good is hardly tractable: "[S]ince so many causes capable of neither enumeration nor measurement affect the law of demand, it is plain that we should no more expect this law to be expressible by an algebraic formula ..." (#21). This skepticism regarding the possibility to build a demand function on "statistics, or what is called social arithmetic" (#21; see also his Preface)[13] does not deter Cournot, who circumvents the difficulty by looking in a different direction, relying on the combination of lessons from calculus, in which unknown functions can still have properties that under certain conditions (e.g., continuity) can be explored (e.g., study of variations); and from probabilities, thanks to the law of large numbers.

This last point is crucial. Cournot borrowed the law from his mentor, Poisson.[14] This now well-known law (often mentioned as LLN or "Poisson's Law") states that if a number of events or observations with properties randomly distributed is large enough, individual variations can be ignored and predictions can be made regarding the outcome (e.g., the average is close to the expected value). In the case under discussion, Cournot inferred from this law that when the number of randomly distributed consumers of a homogeneous good is large enough, the distribution of their demand can be considered as a continuous function of prices. Thus, a contribution from probabilities provided Cournot the key assumption that sustains the analogy between economics and physics: the abstract shape of the demand function and its "continuity," which is a necessary condition for its differentiation. If attention was focused on individual demand, small changes in individual preferences could potentially cause large variations or even discontinuity in individual demand (e.g.,

[13] For the opposition of Cournot to statistics, see Menard [28]. What follows was partially developed in ([26, pp. 23 sq.]; 1981) but has been largely improved through exchanges with Professor Maruyama. All errors are mine, of course.

[14] An interesting indication of their relationship is provided in Poisson's letter to Cournot from January 26, 1836, quoted in [11, pages VI–VII].

substituting a different good). Calculus could not be applied to such a non-smooth function. The assumption of a large number of consumers is supported by observations, for example by the possibility to estimate demand over a yearly period, which allows cancelling irregularities and making demand a well-behaved function tractable with the method of calculus.[15] Even if unpredictable "moral" factors may impact individual demand and even if an exact empirical estimation of the aggregated demand cannot be reached, general properties of the function can nevertheless be explored. As stated by Cournot in this crucial passage:

> But even if this object [numerical applications] were unattainable (on account of the difficulty of obtaining observations of sufficient number and accuracy, and also on account of the progressive variations which the law of demand must undergo in a country which has not yet reached a practically stationary condition), it would be nevertheless not improper to introduce the unknown law of demand into analytical combination, by means of an indeterminate symbol ... Unknown functions may none the less possess properties or general characteristics which are known; as, for instance, to be indefinitely increasing or decreasing, or periodical, or only real between certain limits (#21).

3.3 *The Resulting Law of Demand*

It is therefore in a sense paradoxical that the law of large numbers allowed Cournot to move away from the complex and uncertain world of statistics and even probabilities and to rely entirely on the well-established calculus branch of mathematics. It provided the rationale needed to remain at the abstract level of the "law of demand," with demand a function of prices[16]:

$$D = \mathrm{F}(p).$$

"The wider the market extends, and the more the combinations of needs, of fortunes, or even of caprices, are varied among customers, the closer the function F (p) will come to varying with p in a continuous manner" (#22).[17] The introduction of this well-defined function is followed by a short discussion of the impact of marginal variations of p, which brings Cournot very close to the concept of elasticity and, above all, clearly makes him the founder of the marginalist revolution in economics. Moving forward, he deduces from the demand thus defined "the function p F (p),

---

[15] Two other conditions are actually needed: a mathematical one (the sufficient condition requires a certain shape of the demand curve, an issue discussed in #24), and an economic one (there should be no radical changes in the markets for inputs or in the production system since such changes would alter the conditions shaping demand).

[16] To make the idea more concrete to his readers, Cournot suggested that p could denote "the annual average price" for a certain homogenous good (#23).

[17] Cournot is well aware that some goods may not fulfill the conditions justifying continuity (e.g., luxury goods) but he considers that they are either exceptions or that these "exceptional cases" will tend to be "suppressed" by "the extension of commerce" (conclusion of #22).

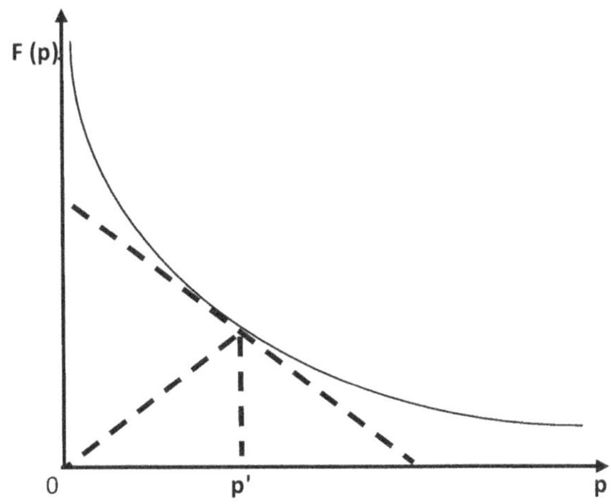

Fig. 1 Cournot's law of demand[18]

which expresses the total value of the quantity annually sold", which is also continuous, its differentiation (condition of first order: Cournot then explicitly refers to Lagrange) leading to the maximum value of p, determined by the equation

$$\mathrm{F}\,(p) + p\mathrm{F}'(p) = 0$$

This relation determines the value at which the marginal revenue is equal to zero (which also implicitly means that elasticity is equal to one). Assuming a Gaussian distribution of consumers, Cournot then draw the now classic representation of the "law of demand" (#24), likely for pedagogical purposes (Fig. 1).

4 Market Structures

There is more coming out of the $p\,\mathrm{F}\,(p)$ function, which has a bell shape (described with precision although not drawn in #24–25). First, it leads naturally to a discussion of first-order and second-order conditions to determine the maximum value and to a subtle discussion of why the well-shaped concavity condition on the demand function allows one to determine a unique maximum while otherwise multiple maxima (or minima) would be possible. Because it corresponds to the expected total value which depends on variations in demand (and therefore price), it is this function that suppliers should consider while trying to maximize their profit.

[18] Notation slightly modified. Cournot does not have the concept of budget constraint; the maximum value p' is determined through a purely mathematical reasoning.

4.1 The Challenge

Adjusting supply to different levels of demand, themselves a function of price, should lead to a continuous supply function as an implicit function of price. However, it is not that simple. On the demand side, the law of large numbers allowed discarding the behavior of individual buyers, which could have challenged the properties of a well-defined function.[19] On the supply side, assuming a large number of suppliers is more problematic if one takes into consideration what is going on in the "real" world. Here, there is a shift in Cournot's reasoning. He rather focuses on the provision of goods by varying number of suppliers, depending on the market under review, the resulting equilibrium coming out of the progressive adjustment of the supplier(s) to variation in prices (and therefore demand) by "*tâtonnement*" (#26).[20] Therefore, the construction of the supply function is a thought experiment. It also led to a major innovation: a theory of market structures.

Indeed, Cournot was aware of the arbitrariness of assuming a large number of suppliers, whatever the market under consideration. Many markets are dominated by one or a handful of producers. Cournot then adopted an approach different to the one that led to the expression of the "law of demand." Abandoning the law of large numbers, he opted to "proceed from the simple to the complex," from absolute monopoly to extended competition. In doing so, he introduced two major contributions to modern economics: (1) he provided the first mathematical model of market structures, with a rigorous approach to the impact of (respectively) monopoly, duopoly, oligopoly, and unlimited (pure) competition on equilibrium price, on its way adding remarkable insights on the impact of collusion on prices: (2) he introduced a pre-game theoretic reasoning with the infamous development of the "reaction functions" in the context of a duopoly, leading to what can be called the "Cournot–Nash equilibrium." Without entering into the details of what occupies five chapters in *Researches*, what follows only intends to give the gist of these very fundamental contributions.

4.2 Market Structures

Because he did not consider that extended competition prevails on markets, Cournot did not take "pure competition" (which is a poor designation and that Cournot more appropriately calls "unlimited competition"—see his Chapter VIII) as the starting

[19] The idea of a utility function clearly did not occur to Cournot, who considered the very idea of "utility" in economics as "accessory" (#3).

[20] The term "*tâtonnement*" became dear to Walras, but took a very different meaning in economics (see [19]). In the general equilibrium model that Walras initiated [36], the term "*tâtonnement*" refers to a price-adjustment process leading to equilibrium and (that is the main point) without any transaction out of equilibrium. I am grateful to Professor Maruyama for having reminded me of this important difference between the two models (Walras vs Cournot).

point of his model, contrary to what Walras would do about 40 years later. He rather considered that since many markets are dominated by one or a handful of producers it would be more logical from a mathematical point of view to proceed by considering an increasing number of suppliers. In doing so, he introduced a rigorous concept of different market structures, (1) using the number of competitors as the key to understand how markets actually work; and (2) defining a cost function as central to the maximizing strategy of suppliers, the combination of these two elements leading to the determination of different equilibria. Adopting this apparently simple approach to market structures through the number of competitors allowed Cournot to define rigorously the very concept of competition, previously the source of much confusion ("Every one has a vague idea of the effects of competition." #43) and to capture its variations.

Since these results are now an integral part of our heritage as economists, what follows provides only an overview of the results, with an emphasis on how Cournot got there.

Monopoly

The case of monopoly, his starting point, illustrates well, and is a contribution that remains at the core of introductory microeconomic textbooks. At the time, "monopoly" was a blurred concept, most of the time identified with public control over a resource. Cournot's approach is straightforward: a monopoly corresponds to the control of a resource by a single agent, so that he/she is the price-maker, although as a maximizer he/she is operating under the constraint of the law of demand. The goal of the owner of the resource is therefore to reach through "tâtonnements"(#26) the price corresponding to the maximum value of p F (p), which is given by the equation

$$\mathrm{F}(p) + p\mathrm{F}'(p) = 0.$$

Cournot goes even further. Besides identifying the demand for his good (or service), the supplier must also consider his costs, which vary with that demand. Cournot thus introduce the cost function as a composite of two functions, φ and F:

$$C = \varphi(D) = \varphi(\mathrm{F}(p))$$

with the first condition (the maximization of the net revenue $p\,\mathrm{F}(p) - \varphi(\mathrm{F}(p))$ leading to the now well-known condition that an equilibrium is reached when *marginal revenue equals marginal cost.*

From Oligopolies to "Unlimited" Competition

Cournot then goes on, extending the reasoning to cases when there is 2, 3, up to an unlimited number of competitors. I will come back to the case of duopoly since it is the one with which the name of Cournot remains largely identified. To study different degrees of competition, Cournot pushed further the reasoning adopted in the case of a monopoly. As plainly stated (#43): "To make the abstract idea of monopoly comprehensible, we imagined one spring and one proprietor. Let us now imagine two proprietors and two springs of which the qualities are identical, and which, on account of their similar positions, supply the same market in competition." These suppliers must then face the split demand $D = D_1 + D_2$, defining a duopoly (the term is not in Cournot). And so on for 3, 4, or more producers (what we now call oligopolies), in each case suppliers operating "independently"; we are in situations of non-cooperative games (issues of collusion are discussed in Chapter VII). When it comes to a market with a very large number of suppliers, "unlimited" competition prevails. Unlimited competition is rigorously defined as follows: "The effects of competition have reached their limit when each of the partial productions D_k is *inappreciable,*[21] not only with reference to the total production $D = \mathrm{F}(p)$, but also with reference to the derivative $\mathrm{F}'(p)$, so that the partial production D_k could be subtracted from D without any appreciable variation resulting in the price of the commodity" (#50). This is to say that there is no market power anymore, with suppliers becoming price-takers. And the maximization of net receipts (profits) leads to the remarkable result

$$p = \varphi_k'(D_k),$$

price in unlimited competition is equal to marginal cost. This condition allows the definition of the supply function Ω in pure competition:

$$\Omega(p) = \sum_k (\varphi_k')^{-1}(p),$$

so that a competitive market reaches its equilibrium when $\Omega(p) = \mathrm{F}(p)$.

These are remarkable results and obtained very elegantly with the advanced mathematics available to Cournot (and that we still use!). These general results are also followed by very interesting discussions about the impact of taxes and subsidies on prices (e.g., Chapter VI in the case of monopoly, #53 in the case of unlimited competition), of interdependent producers (Chapter IX), of the opening of markets to competition (Chapter X), as well as considerations on interdependent markets (e.g., Chapter III on general exchanges; or Chapter XI, on social income, which brings him very close to a general equilibrium approach[22]).

---

[21] Italics are from Cournot. In French, the term is "insensible", which would be better translated by "unnoticeable" (in modern terminology: "marginal").

[22] For a discussion on how close Cournot came to a general equilibrium model, see Dos Santos Fereira [16: 85 sq]. However, Cournot remained pessimistic about the possibility to build a rigorous model of interdependent markets (and a general equilibrium) because of the complexities involved.

4.3 *Reaction Functions in a Duopoly: A Spectacular Contribution*

Notwithstanding the breakthrough they represented, these results now fully integrated in microeconomics were so much ahead of their time that they remained ignored for decades. Even today, the role of Cournot in shaping modern economics is very rarely acknowledged, with the exception of his analysis of competition between two competitors (a "duopoly") that introduced very explicitly the idea of a non-cooperative equilibrium. Cournot did not use this terminology, of course; it had to wait over a century for John Nash to find its full development and terminology, so that it is now sometimes referred to as a Cournot–Nash equilibrium. One possible explanation of this focus on the case of duopoly, indeed a remarkable contribution, might be from the controversial interpretation provided by Bertrand [2], in which the two parties make decisions focusing on adjustment through quantities rather than through prices, leading to wild competition.

To better capture how innovative was the discussion of this case in *Researches*, what follows summarizes Cournot's analysis (#43 to #45) remaining as close as possible to his terminology.[23] The fundamental assumptions are the following (#43):

(1) The two competitors are profit maximizers.
(2) They produce a homogeneous good ("the qualities are identical"), with (2a) identical conditions of production and (2b) no constraints on quantities, so that production costs are the same and can therefore be put aside of the reasoning.
(3) Producers compete in the same market.
(4) They are positioned symmetrically with respect to consumers so that "the price is necessarily the same for each proprietor" (#43); the price is unique.
(5) "*each of them* [the competing suppliers] *independently* will seek to maximize this income as large as possible" (#43; italics from Cournot): this is a non-cooperative market. This last condition is of course essential. As clearly stated by Cournot: "We say each *independently* [italics from Cournot], and this restriction is very essential, … [otherwise] the results would be entirely different, and would not differ, so far as consumers are concerned, from those obtained in treating of a monopoly."

The issue then becomes: how do these competitors reached a *stable* equilibrium, that is, the price that maximizes their profit while meeting the demand? In other terms, Cournot's goal is to understand how prices are established in this configuration and how suppliers adjust to these specific conditions. To make it clear that he is primarily concerned by price adjustments, he adopts upfront an inverse function of the law of demand (which is possible thanks to the properties associated with this function) with $p = \mathrm{f}(D)$. The $p\,\mathrm{F}(p)$ profit function initially defined as the one to maximize monopolist" income then becomes, for each non-cooperative agent (1) and (2), respectively:

[23] For a more detailed discussion, see Menard [29] and Menard and De Villé [31].

$$D_1 \mathrm{f}\,(D_1 + D_2)$$

and

$$\mathrm{D}_2\mathrm{f}\,(\mathrm{D}_1 + \mathrm{D}_2)$$

The reasoning then goes as follows, leaving no ambiguity about the fact that adjustment is through prices (and not quantities as later argued by Bertrand, in 1883):

> "Proprietor (1) can have no direct influence on the determination of D_2; all he can do, when D_2 has been determined by proprietor (2), is to choose for D_1 the value which is best for him. This he will be able to accomplish by properly adjusting his[24] price, except as proprietor (2), who, seeing himself forced to accept this price and this value of D_1, may adopt a new value for D_2 more favorable to his interests than the preceding one." (#43)

This is to say that the maximizing conditions for non-cooperative players are respectively:

$$\frac{\mathrm{d}[D_1 \mathrm{f}(D_1 + D_2)]}{\mathrm{d}D_1} = 0$$

and

$$\frac{\mathrm{d}[D_2 \mathrm{f}(D_1 + D_2)]}{\mathrm{d}D_2} = 0$$

so that the equilibrium price p will be determined by the system of equations

(1) $\mathrm{f}\,(D_1 + D_2) + D_1\mathrm{f}'(D_1 + D_2) = 0,$
(2) $\mathrm{f}\,(D_1 + \mathrm{D}_2) + D_2\mathrm{f}'(D_1 + D_2) = 0.$

Cournot then discusses how this non-cooperative adaptation between the two competitors will lead to a stable equilibrium, the infamous Cournot–Nash equilibrium, which he illustrates with the also famous "reaction" curves D_1 and D_2 (represented here in a linear form, with no loss of generality with respect to Cournot's reasoning) (Fig. 2):

Because this represented such a breakthrough in economic analysis, it is worth quoting Cournot extensively in his own terminology and notation:

> Let us suppose the curve m_1n_1 to be the plot of equation (1), and the curve m_2n_2 that of equation (2), the variables D_1 and D_2 being represented by rectangular coordinates. If proprietor (1) should adopt for D_1 a value represented by ox_1, proprietor (2) would adopt for D_2 the value oy_1, which, for the supposed value of D_1 would give him the greatest profit. But then, for the same reason, producer (1) ought to adopt for D_1 the value ox_{11}, which gives the maximum profit when D_2 has the value oy_1. This would bring producer (2) to the value oy_{11} for D_2, and so forth; from which it is evident that an equilibrium can only be established where the coordinates ox and oy of the point of intersection i represent the values of D_1 and

[24] The exact expression used by Cournot is "adjusting the price." Bacon's translation slightly differs, emphasizing that each party is price-maker, although conditional on what the other one is doing.

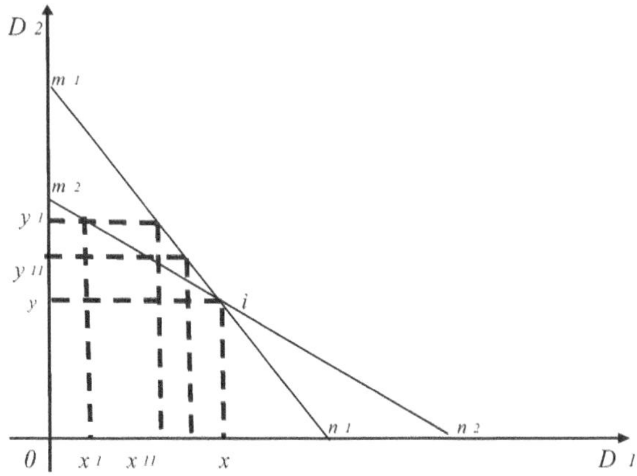

Fig. 2 Non-cooperative adaptation

> D_2. The same construction repeated on a point of the figure on the other side of the point *i* leads to symmetrical results.
>
> The state of equilibrium corresponding to the system of values *ox* and *oy* is therefore *stable*; i.e., if either of the producers, misled as to his true interest, leaves it temporarily, he will be brought back to it by a series of reactions, constantly declining in amplitude, and of which the dotted lines of the figure give a representation by their arrangement in steps. (#43; all italics are from Cournot)

This is quite remarkable not only by its content, but also by the clarity of the presentation, Cournot intending to reach even those readers not familiar with advanced calculus. Unfortunately, it did not work that well!

5 Opening a New World … Long Before It Began Being Explored!

Indeed, notwithstanding the elegance of its demonstrations, Cournot's contribution had to wait decades to be acknowledged and a century before some key results began being fully and more extensively integrated into modern mathematical economics. It may be worth briefly considering what might have been the obstacles to this reception and how it progressively permeated economic theory.

5.1 Major Obstacles to Overcome

In his letter to Walras (September 3, 1873) in which he mentioned having abandoned his mathematical agenda in the early 1840s due to eye disease,[25] Cournot noted how the usage of even very elementary mathematics was an obstacle for most readers ("It is what I tried to do in my book from 1863, which have had even less success than my *plaquette* from 1838").[26] So it was the math, although not only the math!

Mathematics

Notwithstanding some early efforts in using mathematics in economics that were either "so radically at fault" as with Canard [6] (*Researches*, Preface, p. 2) or did not introduce mathematics in a productive way [37],[27] there was nothing comparable in depth and scope to *Researches* before and even decades after Cournot published his essay. At most, some economists would respond to public demand for empirically relevant analysis by using "custom-house abstracts, statistical documents, and government reports" (Preface, p. 1). As argued by Cournot, this is not theory: "I will add that theory should always have some part, small though it may be, in the development of a science" (ibid.). In that respect, *Researches* represented a major leap forward.

As already indicated, Cournot did so by using state-of-the art calculus. It is intriguing that he did not push further the usage of statistics or, even more surprisingly, of probabilities beside the initial and strategic reference to the law of large numbers (see Sects. 3.2 and 3.3), notwithstanding his consideration for the potential role of these new instruments in "penetrating as deeply as possible the knowledge of things in themselves" [11: p. 184]. Indeed, he was familiar with recent developments in both disciplines. In his addendum to Herschel [18] as well as his book from 1843, on which he was already working in the 1830s, he pointed out the potential role of a rigorous usage of statistic to understand … games and, even more relevant, to establish solid foundations to judicial judgements; however, he made no specific references to their potential usage in economics. When it comes to probabilities, his close relation to Poisson and his book from 1843 shows a deep knowledge of the breakthrough from the decades before 1838, with Condorcet [14], Laplace [23] and Poisson [33] pointing out very explicitly their potential role in what we would now identify as the emerging social sciences[28] as well as in providing support to the "social engineering" that should provide guidance to political reforms. However, Cournot

[25] For decades Cournot needed the help of a personal secretary and he was almost blind when he passed away.

[26] "C'est ce que j'ai essayé de faire dans mon volume de 1863, qui a eu encore moins de succès chez tous les économistes que ma *plaquette* de 1838" (italics from Cournot; see Jaffe (ed.), [20, Vol. I, pp. 331–332]).

[27] Cournot does not seem to be aware of Whewell's contributions.

[28] Significantly, although Cournot refers to the "statistiques sociales," I could not find any reference, even in 1843, to the term "mathématique sociale," even when discussing Condorcet.

choose tools provided by Lagrange rather than Laplace or his mentor Poisson as a source of inspiration for *Researches*. Why?

There is no straightforward answer to this question. Cournot's skepticism regarding the usage of statistics in economics or, more generally, in social sciences, can be well documented through several passages of *Researches* (particularly the Preface) and the *Exposition* [11]. Notwithstanding his conviction that extensive homogeneous observations of repeated events can open room to the identification of regularities [11, p. 73 and pp. 181–182] by overcoming variations across specific events and effects of fortuitous causes (addendum to Herschel, p. 514), he remained highly skeptical about the possibility of building a rigorous economic model based on statistical series on social issues, due to the complexity of social phenomena and the dependence of human behavior on contexts ("milieu ambiants"—1843, p. 116).[29] Critical about references to individual preferences and utility considerations, which were to play such an important role in the coming "marginalist revolution," Cournot preferred to stick to the simple and in a sense very "Smithian" behavioral assumption that economic agents are rational maximizers. From a scientific point of view, Cournot shared the epistemological view that "the degree of precision of the result generally depends on the degree of precision of the least precise element" (Addendum to Herschel). In that respect, and notwithstanding what he considered a relevant usage of statistics in disciplines such as demography, he remained highly skeptical about the efforts to build "models of man" based on statistics, e.g., the "average man" promoted by Quételet [34].[30]

When it comes to probabilities, the answer is more complex. Besides the reference to the law of large numbers as key to the definition of the law of demand, Cournot never referred to probabilities as a useful tool in *Researches* or in his other books on economics. Part of the explanation might be his skepticism about the possibility to develop statistics on human behavior to provide rigorous foundations to a probabilistic approach in economics. A complementary explanation might lie in his initial choice to overcome this difficulty by positing the assumption of economic agents as rational maximizers. This reference to maximization quite naturally led Cournot (who did his Ph. D dissertation on astronomy, translated the classic textbook from Herschel on the same topic, and was close in his formative years to a group of intellectuals looking for the development of "social engineering" as a way to reform society while avoiding the turmoil of revolution (see Sect. 2)) to rather turn to calculus (the "theory of functions" in Cournot's vocabulary[31]) as a more promising path to build the abstractions required for a scientific approach to economic phenomena. His analysis of the movement of prices and the positioning of economic agents as sets of moving points analogous to what astronomy does when studying the movement

---

[29] See the following statement in the addendum to Herschel [18, p. 504]: "Jusqu'à présent on n'a guère appliqué le calcul des chances qu'à des problèmes sur les jeux … et à des faits de statistique sociale dont les causes se dérobent par leur complication à toute investigation mathématique et pour lesquels nous n'avons d'autres données que celles de l'expérience.».

[30] For a detailed discussion of the relation of Cournot to statistics, see Menard [28].

[31] See Bru [4] for a comment on the mathematical significance of this terminology at the time.

of planets by postulating spherical and homogeneous globes is significant in that respect.

So, it was that analogies with physics led Cournot opting for this abstract theory that made *Researches* almost undecipherable to economists of his time and of the following decades [30].

Political Economy at Stake

However, this blindness to the novelty of *Researches* involved more than the resistance of the economists of his time to the technicalities of relatively sophisticated usage of mathematics, a situation that persisted for a long time, as shown by the institutional difficulties met by Walras in the development of his career.

As the reaction of de Fontenay [17], a leading intellectual figure in France in the mid-19th century, shows in his comment on the *Principes de la théorie des richesses* [12], the conclusions that Cournot drew from his equations was also viewed as a major ideological challenge. Indeed, his considerations regarding taxation, international trade, and more generally risks associated with "unlimited" (pure) competition and open markets led him to endorse positions opposed to the dominant views of the time, which promoted open competition policies that we would today identify as "globalization" through liberalization. De Fontenay argued that the 1863 book did not differ substantially from *Researches*, which he viewed as leading to the erroneous support of the Ricardian theory of rent, to protectionism, and to misleading ideas regarding the "question sociale" (which refers primarily in the literature of the time to the emergence of a combative working class, a main concern for policy-makers and the "classes bourgeoises"—see [7]). De Fontenay found the roots of these errors and what he viewed as the opacity of the underlying reasoning in the usage of these "*alarming* hieroglyphs"[32] that he considered participating in a "militant" algebra [17: 232 sq.]. This position was shared by almost all leading economists of the time.[33] Nonetheless, Cournot maintained his position in political economy, assigning a positive role to the State in tempering the impact of competition and adopting a moderate protectionist view, positions that he made very explicit in a letter to Walras (September 3, 1873) in which he argued that the conclusions derived from Walras' model about the advantages of "pure competition" would lead to environmental as well as social catastrophes.[34]

Hence the rigorous mathematical reasoning introduced in *Researches* in combination with the positions adopted by Cournot on some key debates of his time

[32] "Pourquoi vous êtes-vous amusé à nous parler en hiéroglyphes *effarouchants* ..." (italics are from de Fontenay).

[33] For a discussion of the strong resistance of the French adepts of liberal policies of the nineteenth century to the usage of mathematics, see Menard [26, pp. 239–242]; Le Van-Lemesle [25], Breton [3], Sigot[35].

[34] "Mais je tremble que vos courbes "d'utilité intensive et extensive" ne vous mènent au pur laissez-faire, c'est-à-dire, en économie nationale, au déboisement du sol, et en économie internationale, à l'étouffement de la plèbe des races par une race privilégiée conformément à la théorie de M. Darwin." (Letter to Walras from September 3, 1873; in Jaffe ed., [20, vol. 1, p. 332].).

in political economy created major obstacles to the perception, not to speak the acknowledgement of how innovative this book was.

5.2 At Last Percolating Economic Theory

However, the long term has been more favorable to Cournot. The turning point analyzed in several contributions to this book was the 1870s (see [26, Chap. 4, 27, 35]). Notwithstanding divergences in interpretations and/or in assessing the significance of Cournot's contributions, his name and *Researches* became an inescapable part of the development of the so-called "marginalist revolution." From Walras to Edgeworth, Marshall and Fisher, not to forget Bertrand, acknowledging and/or discussing Cournot's *Researches* became obligatory steps towards further contributions.[35] At the end of the 19th century, translations of *Researches* were available in Italian (1878), English (1897), and German (1924).

It is impossible to discuss and even identify within the limited scope of this chapter the many different channels through which Cournot's influence permeated different domains of modern economics.[36] This influence goes far beyond the introduction of rigorous mathematical reasoning, which is already a major contribution. There is also his foundational role in the development of the Walrasian as well as the Marshallian concepts and models of market equilibria[37]; his law of demand was a major source of inspiration for early econometricians (e.g., Henry Moore), etc. And there is of course his approach to duopoly and, more generally, to the different market structures that permeated numerous studies, particularly from the early 1920s to 1951, when Nash published his results ((Nash 1951); for a discussion of that period, see Menard, 1980a). Although it could be misleading to make Cournot a precursor of game theory, it is impossible to not acknowledge how his reasoning on conjectures in a duopoly overlaps with what game theory would develop over a century later.[38]

[35] In a very different perspective, Cournot has also had a significant influence on philosophy in France, and particularly on the philosophy of sciences. See Deschamps and Martin (ed.) [15].

[36] For a detailed analysis, see Menard [27].

[37] However, Cournot never considered utility or scarcity as relevant foundations to the analysis of equilibrium (see his letter to Walras, Sept. 3, 1873).

[38] Leonard [24] is right in pointing out that interpreting Cournot's analysis in terms of a Nash equilibrium is a retrospective reading that can be distortive. Reading Cournot in terms of game theory required … the development of game theory and the demonstration by Nash of the equilibrium associated with his name!

6 Conclusion

Most results summarized in Sects. 3 and 4 likely seem trivial to a modern reader with a basic training in microeconomics. There is a good and simple explanation to this state of affair: *Researches into the mathematical principles of the theory of wealth* provided the building blocks of the marginalist revolution. *Researches* did so through the development of an analogy between economics and physics that supported the introduction of mathematics, and more precisely of calculus, into economic reasoning. However, in his book Cournot did more than that. Using the tools thus made available and thanks to his high competence in mathematics, he developed a rigorous formal analysis of themes that were discussed in often vague and confusing terms at that beginning of the 19th century. And the situation remained so for decades, relegating Cournot's contribution to almost total obscurity.

Indeed, notwithstanding the richness of the contributions of what became known as "classical economics," it is important to measure the enormous gap from a theoretical as well as a methodological point of view between *Researches* and that literature. Reading Smith, Ricardo, or Say, who were leading figures in the economics of that period, one can measure how far they were from the productive usage of mathematics in economic analysis. And the gap is also huge between Cournot's contribution to the modeling of economics and the attempts to introduce mathematical reasoning made by authors (e.g., Canard, Whewell) long forgotten and for good reasons.

Although it took decades for *Researches* to permeate economic theory, the "spirit" of this book already inspired what became known as the French school of economic engineers of the mid-19th century while the rigor of its method and concepts had to wait the breakthrough made by Jevons, Walras, Marshall, Edgeworth, Fisher and others to be fully or let's rather say **almost** fully acknowledged. However, thanks to the developments that followed the marginalist revolution and later on the introduction of game theory, it is now possible to even better assess the major shift in economic thinking and the rich results that economic theory owes to Cournot.

Acknowledgements Special thanks to Professor Toru Maruyama for his invitation to contribute to this volume and for his highly inspiring insights.

References

1. Appleby, J.O.: Economic Thought and Ideology in Seventeenth-Century England. Princeton University Press, Princeton (1978)
2. Bertrand, J.: Compte rendu de « La théorie mathématiques de la richesse sociale » de M. Walras et des « Recherches sur les principes mathématiques de la théorie des richesses » de M. Cournot. Journal des savants **97**, 499–508 (1883)
3. Breton, Y.: Les économistes libéraux français et l'emploi des mathématiques en économie politique 1800–1914. Oeconomia **5**, 25–63 (1986)
4. Bru, B.: Théorie des fonctions et théorie des richesses chez Cournot (Chap. 2) (Deschamps and Martin (eds.)), pp. 37–58 (2020)

5. Bulletin des sciences mathematiques, physiques et chimiques (also known as Bulletin de Ferusac) (1826–1829). Several short notes by Cournot.
6. Canard, N.F.: Principes d'Economie Politique. Buisson, Paris (1801)
7. Chevalier, L.: Classes laborieuses et classes dangereuses à Paris pendant la première moitié du XIXe siècle. Plon, Paris (1958)
8. Cournot, A.A.: Recherche sur les principes mathématiques de la théorie des richesses. Hachette. Reed, Paris (1838). Vrin, Paris (1980); English translation by N. Bacon, Researches into the Mathematical Principles of the Theory of Wealth. Macmillan, New York (1897)
9. Cournot, A.A.: Traité élémentaire de la théorie des fonctions et du calcul infinitésimal. Hachette. Reed, Paris (1841). Vrin, Paris (1984)
10. Cournot, A.A.: Exposition de la théorie des chances et des probabilités. Hachette. Reed, Paris (1843). Vrin, Paris (1984); English translation by O. Sheynin, Exposition of the Theory of Chances and Probabilities. Scholar Press, Berlin (2019)
11. Cournot, A.A.: Principes de la théorie des richesses. Hachette. Reed, Paris (1863). Vrin, Paris (1981)
12. Cournot, A.A.: Revue Sommaire des Doctrines économiques. Hachette. Reed, Paris (1877a). Vrin, Paris (1982)
13. Cournot, A.A.: Souvenirs (E. Botticelli (ed.) in 1913). Hachette. Reed, Paris (1877b). Vrin, Paris (2010)
14. de Condorcet, N.M. *Essai sur l'application de l'analyse à la probabilité des decisions rendues à la pluralité des voix*. Paris: Imprimerie Royale.
15. Deschamps, M., Martin, T. (eds.): Cournot, économie et philosophie. Matériologiques, Paris (2020)
16. Dos Santos Fereira, R.: La modernité des Recherches (Chap. 4) (Deschamps and Martin (eds.)), pp. 79–82 (2020)
17. de Fontenay, R.: Principes de la Théorie des Richesses par M. Cournot, Journal des Economistes **43**, 231–251 (1864)
18. Herschel, J.F.W.: Traité d'astronomie. French translation by A.A. Cournot (with an addition: Sur la distribution des orbites cométaires dans l'espace). Librairie Paulin, Paris (1834)
19. Jaffé, W.: Walras' Theory of *Tâtonnement*: a critique of recent interpretations. The Journal of Political Economy **75**(1), 1–19 (1967)
20. Jaffé, W. (ed.): Correspondence of Léon Walras and Related Papers, vol. 3 Vols. North Holland Pub., Amsterdam (1965)
21. Kater, H., Ladner, D.: Eléments de mécanique; French translation by A.A. Cournot, with an addendum (Chap. 23, De la mesure des forces et du travail des machines). Librairie L. Matthias, Paris (1834)
22. Keynes, J.M.: A Treatise on Probability. Macmillan & Co, London (1921)
23. de Laplace, P.-S.M.: Théorie analytique des probabilités. Courcier Imprimeur-Libraire, Paris (1812)
24. Leonard, R.J.: Reading cournot, reading nash: the creation and stabilisation of the nash equilibrium. Econ. J. **104**(424), 492–511 (1994)
25. Le Van Lemesle, L.: Guillaumin. Editeur d'économie politique. Revue d'Economie Politique **95**(2), 134–149 (1985)
26. Menard, C.: La Formation d'une rationalité économique. Flammarion, Paris (1978)
27. Menard, C.: Equilibre, Asymétrie, Conflit: Un siècle de théorie des prix en équilibre partiel (1838–1940), 356 pp. Ph.D. in Economics. Sorbonne University, Paris (1980a) (unpublished)
28. Menard, C.: Three forms of resistance to statistics: Say, Cournot. Walras. Hist. Polit. Econ. **12**(4), 524–541 (1980b)
29. Menard, C.: La physique des prix: Le modèle de Cournot. Cahiers d'Economie Politique, **7**, 51–66 (1982)
30. Menard, C.: The machine and the heart. In: Mirowski, P. (ed.) Social Concept. Special issue, pp. 81–95. Yale University Press, New Haven (1989a)
31. Menard, C. (with Philippe De Ville): An insolent founding father. Eur. Econ. Rev. **33**, 494–502 (1989b)

32. Nash, J.: Non-cooperative games. Ann. Math. **54**(2), 289–295 (1951)
33. Poisson, S.D.: Recherches sur la probabilité des jugements en matière criminelle et matière civile; précédées des règles générales du calcul des probabilités. Bachelier, Imprimeur-Libraire, Paris (1837)
34. Quételet, A.: Sur L'homme et le développement de ses facultés. Essai de physique sociale. Bachelier, Paris (1835)
35. Sigot, N.: La réception de l'œuvre économique de Cournot. In: Martin, T. (ed.) Actualité de Cournot, pp. 125–149. Vrin, Paris (2005)
36. Walras, L.: Éléments d'Économie Politique Pure, ou Théorie de la richesse sociale. Librairie Corbaz, Lausanne (1874); English Translation Jaffé, W.: Elements of Pure Economics; or, The Theory of Social Wealth. A. M. Kelly, New York (1954)
37. Whewell, W.: Mathematical Exposition of Some Doctrines of Political Economy. Smith (Printer to the University), Cambridge (1829)

Carl Menger's Alternative to the Classical Theory of Value and Distribution, and Marginalism

Heinz D. Kurz

Abstract In his *Grundsätze* (1871), Carl Menger specified as the main task of his book to surpass and render obsolete the classical economists' and especially Ricardo's approach to the theory of value and distribution. He sought to accomplish this in terms of an approach that starts from the needs of the isolated individual (Robinson Crusoe). This was the causal-genetic starting point of his approach. He then analysed the ways by means of which the individual tried to satisfy the needs as best as possible. Menger went on via a number of analytical steps to investigate what happens in markets and how prices get established and income distribution decided. This approach was taken to be diametrically opposed to that of the classical economists, who are said to have started from production and have tried to explain prices in terms of the quantities of labour needed directly and indirectly in the production of the various commodities, viz. the "labour theory of value". The chapter asks essentially three questions: First, did Menger succeed in this or did he misunderstand in important respects the classical approach to the theory of value and distribution? Second, can he, together with William Stanley Jevons and Léon Walras, be bunched together in a so-called "marginalist school", or are there fundamental differences between the contributions of the three authors? Third, can Menger be considered one of the founders of the so-called "marginal revolution"? It will be argued that the answer to all three questions is no. First, Menger's view of the classical authors and especially Ricardo is flawed in a variety of respects and his interpretation of their advocacy of the labour theory of value mistaken. Second, Menger's theory differs in important respects from those of Jevons and Walras and its causal-genetic perspective is difficult to sustain, which becomes clear in view of his failure to solve the infamous "imputation problem". Third, both marginal utility and marginal productivity theory had been elaborated before Menger by representatives of the "German

I should like to express my sincere thanks to Professor Toru Maruyama for his valuable comments and suggestions on an earlier version of this chapter. Any remaining errors and misconceptions are, of course, my responsibility. In this chapter I draw freely on earlier works of mine on Carl Menger and the "Austrian" economists more generally; see, in particular, Kurz [16], [18], [19].

H. D. Kurz (✉)
Graz Schumpeter Centre, University of Graz, RESOWI EF, 8010 Graz, Austria
e-mail: heinz.kurz@uni-graz.at

T. Maruyama (ed.), *Marginal Revolution in Economics*, Monographs in Mathematical Economics 6, https://doi.org/10.1007/978-981-99-4342-5_3

Use Value school" and by Johann Heinrich von Thünen. The chapter concludes by emphasizing that Menger's theory was neither "marginalist" nor "mathematical".

1 Introduction

Carl Menger's magnum opus, the *Grundsätze der Volkswirthschaftslehre*, was published in 1871 [32]) and was soon considered, together with William Stanley Jevons' *Theory of Political Economy*, also published in 1871 (Jevons 1871), and Léon Walras' *Éléments d'économie politique pure*, published in 1874 (Walras 1874), one of the foundational works of what was called the "marginal revolution".[1] The essence of this "revolution" was seen to consist in refuting the teachings of the British classical political economists, especially David Ricardo, and putting in its place the marginalist doctrine, based on the twin concepts of diminishing marginal productivity in production and diminishing marginal utility in consumption. The main target of criticism was the classical authors' surplus-based explanation of profits, which the "innovators" sought to replace by a marginal productivity of capital explanation. Careful examination shows that the latter amounted to a generalization of the classical theory of intensive diminishing returns and intensive rent from homogeneous land indiscriminately to all factors of production, including capital. Accordingly, the marginalist theory wishes to explain, in conditions of free competition, all distributive variables—the general (and uniform) rate of profits, the wage rates of different kinds of labour and the rent rates of different kinds of land—in terms of a *single* principle only: the *principle of the relative scarcity* of the respective factor of production under consideration. The relative scarcity of a factor is reflected by its marginal productivity, therefore the name of the theory: *marginal theory of value and distribution.* The mathematical tool that found its way into economics via this theory was *infinitesimal calculus.* This tool is said to be applicable to the employment of all productive factors, including capital, which, while consisting of heterogeneous produced (and reproducible) means of production, has to be treated as a single magnitude. This is so, because in competitive conditions each and every owner of capital goods is seen to participate in the sharing out of the sum total of profits in the economy as a whole in a long-period equilibrium, on which the theory focuses attention, like the holder of shares of a joint stock company: profits are distributed in proportion to the value of the capital invested. Knut Wicksell [54], who by the end of the nineteenth century had elaborated the most advanced version of the long-period marginalist theory, pointed out: the economy's endowment of the "quantity of capital" whose relative scarcity was supposed to contain the key to an explanation of the general rate of profits (or rather interest, as he preferred to call it, following the Austrian

[1] In the following, all isolated page numbers refer to Menger's *Grundsätze* (1871), while page numbers in square brackets refer to the English translation of it [34]. Unless otherwise stated, emphasized parts in quotations are Menger's. Translations from German texts referred to in this essay are mine.

economists) has to be given as a *sum of value* expressed in terms of some numéraire. This set the endowment of capital apart from the endowments of the other factors of production, that is, the different kinds of labour and land, which are all given in their own *technical* units.

Not surprisingly, this and other implications of the suggested generalization of the principle of scarcity to explaining all distributive variables were not immediately clear to its early advocates. An understanding of the various difficulties that had to be met filtered in only slowly and subjected the marginalist approach to repeated change and adaptations not only in special details, but also in fundamental design.[2] However, major representatives of the marginalist revolution, i.e. Jevons and Walras, but also Böhm-Bawerk, insisted that if the new theory was to be an alternative to the classical theory of value and distribution, it also had to explain crucial economic variables such as the competitive rate of return on capital, wages, the rents of land and normal prices—only better than Smith or Ricardo, where "better" meant first and foremost: in a coherent way. The third so-called marginalist revolutionary, Menger, cannot be mentioned in this context, because in his book we do not encounter the concept of a *general* rate of profits (or interest) or an unambiguous explanation of profits in terms of the marginal productivity of capital. If the latter is taken to be the characteristic feature of the marginal revolution, as is done in this chapter, then Menger's contribution does not pass the litmus test. However, there are additional reasons why it is difficult to count Menger as belonging to the triumvirate, which will become clear in the following. While Menger worked out a version of marginal utility theory on the basis of contributions of the "German Use Value School", apart from scattered remarks that may be interpreted as pointing in the direction of the concept of marginal productivity, he did not put forward a similarly elaborate version of the second pillar upon which marginal theory began to rest after to the "marginal revolution". In particular, the *Grundsätze* do not contain a theoretical edifice, firmly erected on the two pillars, with the theory of capital and profits as the work's crowning achievement. Therefore, Menger can only partially be considered a marginalist.[3]

Before we continue with our main theme, a few preliminary remarks are in place, which exemplify the kind of problems we are facing when asking whether marginalist

---

[2] The most important change in fundamental design was, of course, starting in the 1920s, the development of temporary and intertemporal equilibrium models, which were supposed to overcome the difficulties besetting the long-period versions of marginalism; see Garegnani [5, 6], Kurz [15: Chap 11] and Petri [35]. In this essay temporary and intertemporal general equilibrium theory will not be dealt with. The focus will be on long-period marginalist theory, championed by Jevons and Walras and also major Austrian economists, such as Eugen von Böhm-Bawerk.

[3] Several ideas Menger entertained have been anticipated by others (see below, especially Sect. 6; see also Hayek's account of Menger's works and the sources he tapped [8]). Essentially the same can be said about Jevons and Walras. I agree with Professor Toru Maruyama that what really matters is how the three authors used the respective ideas and elaborated on them, the value they added, so to speak. While Jevons and Walras were keen to use mathematics in order to establish new analytical results—they were, for example, concerned with deriving optimality conditions in utility and profit maximization—Menger did not use mathematics as a productive tool in economic theory, but referred to mathematics only incidentally, viz. his reference to financial mathematics.

theory succeeded in surpassing classical theory, or failed to do so. When the "innovators" originally proposed their alternative to the classical doctrine, they appear to have been convinced that marginal productivity theory was generally valid in explaining income distribution and not tied to special technical conditions. However, as Wicksell and Philip H. Wicksteed were to point out towards the end of the nineteenth century with reference to Euler's Theorem, this was not so: if marginal theory was to explain the distribution of the social product without leaving any surplus or deficiency, then this presupposed linear homogeneity of the production technology. This was a mathematical requirement, which had, of course, an economic meaning. Another significant precondition in order for the marginalist argument to be applicable is that all factors of production must be fully employed.

When we now turn to the classical economists, we see that none of these assumptions plays a role in their reasoning. Neither Smith nor Ricardo started from the premise that the economy realizes continually the full employment of its productive resources; it suffices to consult the chapter "On Machinery" in Ricardo's *Principles* (*Works* I: Chap. 31). And as is well known, Adam Smith's argument in *The Wealth of Nations* [41] revolves around a continually deepening *social division of labour*, which cannot be captured in terms of constant returns to scale. Marked increases in output cannot sensibly be assumed to be brought about by simply staging up given processes of production, using the same kinds of inputs in the same proportions, only in larger quantities. A deeper social division of labour will rather be accompanied by a change in the proportions and kinds of inputs employed. The division of labour, Adam Smith emphasized, depends on the "extent of the market" (WN I.iii.1). According to Allyn Young [57: 529] this is Smith's central "theorem" and "one of the most illuminating and fruitful generalisations" in economics: it is a cornerstone of the doctrine of *dynamically increasing returns*. Smith stressed that the division of labour and the technological and organizational change it engenders propagate themselves in a circular and cumulative way so that change becomes progressive and persistent. We may interpret Smith as implying that there are constant returns to scale neither with regard to single industries nor with regard to the economic system as a whole. Such an assumption may only be acceptable as a simplifying hypothesis with regard to small changes in output.

The case under consideration shows how certain economic preconceptions, such as the one underlying the marginal theory of value and distribution, enforce or at least suggest the use of certain substantive assumptions like that of full employment or the use of certain mathematical tools. These imply in turn restrictions on the basic economic reasoning entertained. Mathematics is therefore not only a tool that allows one to elaborate an argument in a consistent manner (which it certainly also is), it also forces the argument to comply with the scope offered by the mathematical tool. Since the subject matter of economics is extremely complex (too complex for a highly gifted physicist like Max Planck!), economists will typically (if not notoriously) experience a mismatch between the mathematical tools available to them and the complexity of the problems they try to tackle.

The point I wish to make is that the adoption of certain tools, mathematical and other, in order to deal with given problems is not "innocent" in the sense that the tools do not affect what we can see and achieve. There is indeed the danger, as one commentator observed astutely, that as soon as we have mastered some tool, we are in danger of being mastered by it. This danger is discernible, for example, in the unfortunate inclination of some authors to see the contributions of earlier authors through the lens of the theories they hold and typically consider to be superior to earlier ones. This can be illustrated, for example, with regard to Léon Walras's reception of the theories of the classical economists and especially Ricardo's theory of rent (see [22]). Walras simply could not imagine that there was another way of approaching economic problems than the demand-and-supply approach he was familiar with. In this chapter we shall see that also Menger's view of the classical authors was distorted because he assessed them exclusively in terms of the principle of scarcity.

A final remark seems to be appropriate. In order to locate the contribution of an author in the overall genealogy of a theory, one ought to first specify the theory's origins, followed by a discussion of the direction the work on it took and then, if possible, of the presumable terminal point of its development. Such a procedure allows one to identify the state of the theoretical world when the author under consideration entered it and the state when he or she left it. In the simplest possible case, the difference between the two states is exclusively to be imputed to the author. However, this is rarely to be found in reality, because parallel to the author others have worked on the theory independently of, or in close cooperation with, him or her. We then find ourselves quickly entangled in difficult problems of attributing achievements or failures to the people involved. As will be seen below, the "Austrian" economists were well aware of the intricacies that "imputation" problems might pose. It is not claimed that this chapter contains the suggested genealogy or that it provides a compelling imputation of ideas to Menger and other economists. However, an attempt will be made to throw some light on the backward and forward intellectual linkages of his contribution and thus illuminate somewhat a fascinating period in the history of our subject.

The composition of the chapter is the following. Section 2 provides a brief account of Menger's view of the process of civilization, which is characterized by a growing knowledge of humans and their improving control over nature. The "law of chance" is progressively replaced by "economic laws", which can be studied and put to good use in economic policy. Menger, it is argued, was concerned with elaborating an "economic theory in time" [9]. Section 3 then turns to some major marginalist authors' largely diverging assessments of the classical British economists' theories of value and distribution. While Jevons, Menger and Walras opined that these theories were flawed beyond remedy and therefore ought be abandoned, Alfred Marshall and Knut Wicksell stressed the value and usefulness of at least parts of them and opted for a judicious amalgamation of classical and marginalist elements. Section 4 argues that both views are difficult to sustain because they miss the genuine significance of the classical approach to the theory of value and distribution. This is then explained with reference to what the classical and the marginalist authors took as data or independent variables in their respective analyses and what as unknowns or dependent variables. The perhaps most important difference between the two is taken to consist in the attempt of the marginalists to generalize the principle of intensive diminishing returns, which Ricardo had reserved for an explanation of the rent of homogeneous land, to all factors of production, including heterogeneous capital goods. This generalization turned out to meet with serious difficulties, first noted by Wicksell. Section 5 then discriminates between different versions of marginalism and specifies Menger's position against this background. It is pointed out that in Menger we do not encounter the concept of a competitive general rate of profits (interest), which with Adam Smith's *Wealth of Nations* at the latest had become a central concept around which much of the argument of both classical and marginalist authors revolved. Section 6 deals in some detail with anticipations of the two most important marginalist concepts—marginal productivity and marginal utility—and shows that both concepts have already been known a long time before the so-called "marginal revolution". With regard to the former concept, the reference is to Johann Heinrich von Thünen, with regard to the latter, to contributions of the so-called "German Use Value School" and especially of Karl Heinrich Rau. While Menger was familiar with these contributions, he did not make full use of them. Section 7 summarizes briefly Menger's time-phased conception of production. Different from François Quesnay and the classical economists in most of their analyses, he does not consider production as a circular, but as a linear or unidirectional process of final duration, a sort of one-way avenue of limited length, leading from "goods of higher order" to "goods of first order". Section 8 deals with Menger's causal-genetic and successivist approach to value and distribution. This consists in taking the "prospective prices" of goods of first order, or consumer goods, as known, decided by the needy individual prior to, and entirely independently of, the prices of goods of higher order. This then leads him to the problem of the "imputation" of the given consumer prices to the inputs or factor services that cooperate in their production. Friedrich von Wieser criticized Menger's respective reasoning and insisted that in a circular flow framework, which is the appropriate one vis-à-vis the modern industrial system, the strict separation of consumer goods from all other goods makes no sense, and that a solution of the imputation problem requests at any rate putting up a system of simultaneous equations and solving it. Section 9 turns to Menger's theory of income distribution and especially to his theory of capital and profits, which Eugen von Böhm-Bawerk considered to be fundamentally mistaken: to consider the utilization of a capital good as an independent carrier of value in addition to the capital good itself, whose present value is just the sum of the discounted stream of payments of its utilization, involves double counting. Section 10 scrutinizes critically Menger's criticism of the labour theory of value, on which he thinks the classical theory stands or falls, and any other conceptualization of prices as reflecting inter alia the "substances" transferred to, or embodied in, commodities. It is argued that Menger's respective views are difficult to sustain. Section 11 concludes that Menger's own theory of value and distribution is too opaque and possessed of too many loose ends and inconsistencies to serve as a measure in terms of which the correctness or otherwise of competing theories could be judged.

2 In Search of an "Economic Theory in Time"

Menger sees the history of humankind as a process of civilization driven by a continuous accumulation of knowledge about humans themselves and the world in which they live. In its course, labour productivity increases, humans extend their control over nature, their time horizon expands, economic affairs get more and more subjected to competitive conditions, uncertainty is reduced and the predictability of future events improved. Hicks [9] qualified Menger's theory as an "economic theory in time " and contrasted it with an "economic theory out of time", exemplified by that of Léon Walras. As civilization progresses, economic laws manifest themselves more and more firmly, slowly bringing them to full fruition. In much of his analysis, Menger focuses attention on this advanced stage of civilization. While he is typically credited with emphasizing the time-consuming and error-prone nature of all economic activity and the fundamental uncertainty that clouds the future, interpreters often appear to have missed the fact that according to Menger things look much less bleak in this advanced stage.[4]

Menger divides the process of civilization broadly in three stages. The first stage is one in which humans can barely satisfy their basic needs of food, clothing and shelter. The attention focuses on how to survive, but poor knowledge and a lack of tools to cope with the opportunities and dangers nature supplies imply deep uncertainty. The existence of the human race is constantly endangered because of an inimical environment and deadly conflicts over scarce resources. Transaction costs are high or even forbidding, and exchange is correspondingly rare and sporadic.

The second stage is one in which knowledge about needs and the means to satisfy them is much improved, and so is productivity. In addition to basic needs, now also more sophisticated ones—amenities—can be served. The division of labour is a great deal more developed and still increasing, and while interdependent markets begin to play a growing role in the economy, their institutionalization is feeble due to still considerable uncertainty and significant transaction costs. There is substantial price dispersion at any moment of time and across time, and non-competitive forms of economic relationships still prevail. Menger throws some light on the state of affairs at the first two stages in terms of his studies of isolated exchange and monopoly. With regard to bilateral monopoly he anticipates Böhm-Bawerks's concept of limit pairs (*Grenzpaare*) in horse-trading, which define the interval in which the dispute over the price of the horse might eventually be settled. At these stages, the gains from trade, in Menger's view a major source of rising incomes per capita, cannot be fully realized.

It is only at the third stage that humankind manages to control nature more effectively and also improve significantly its understanding of its needs and the means and ways to satisfy them. Uncertainty gradually gets reduced without ever vanishing, and the time horizon of humans expands. The need to care for provisions in a more distant

---

[4] Menger's unswerving faith in human progress was indeed remarkable and perhaps only topped by socialists who, like Friedrich Engels [4], saw the paradise on earth in reach. For a genuinely pessimistic view, see the concluding passages of Max Weber's *Protestant Ethic* [53].

future is a challenging task and leads to several improvements in the organization of the production and distribution of commodities. As human knowledge grows, "true" needs gain in importance compared to "imaginary" ones despite the fact that luxuries can now be afforded by more and more people. The number and diversity of needs and of goods capable of satisfying them increases, the security of people improves, the institution of the market gets more and more firmly established, and competitive forms of economic transactions begin to gradually gain the upper hand. While at the previous stages there was notorious conflict and severe rivalry amongst individuals and groups of people, often giving rise to violent confrontations, now competition enforces cost minimizing, but typically peaceful behaviour of agents. In advanced civilizations with a well-established system of property rights, with public authorities enforcing the rule of law and with a firm entrenchment of interdependent markets etc., free competition will become the prevalent market form. Competition is therefore a result of socioeconomic development, not its starting point. Menger was highly critical of the inclination of marginalist economists to start from the assumption of even "perfect" competition, without ever asking whether its numerous preconditions were actually met.[5] His reasoning has therefore not without justification been called "non-Walrasian".

In civilized economies, echoing Adam Smith, Menger insisted that the activities of speculators and arbitrageurs will stabilize the movement of prices and reduce price dispersion. Due to these developments, economic laws are gradually manifesting themselves and can be studied. While in times of rapid and disruptive economic change and fundamental uncertainty, rational calculation of costs and benefits is difficult, if not impossible, since the constraints in respect of which agents try to optimize are swiftly changing, in stable competitive conditions the principle of cost minimization becomes dominant.[6] In the third stage, Menger opines, "effective prices" will obtain. These prices resemble strongly what the classical authors called "natural prices", conceived of as centres towards which market prices tend to gravitate or around which they oscillate. Menger also speaks of "economic prices", a concept which corresponds to the classical economists' view of natural prices being cost-minimizing. However, in contrast to the classical, but also major marginalist economists, whether "Austrian" or not, in Menger we do not encounter the corresponding tendency towards a uniform *general rate of profits* as part and parcel of

[5] As Adam Smith famously put it, the "wretched spirit of monopoly" (WN IV.ii.21) never sleeps. It deserves to be mentioned that when with regard to the third stage Menger takes competition to prevail, he appears to have in mind the classical concept of "free competition", which presupposes the absence of significant barriers to entry into and exit from markets, and not the later marginalist concept of "perfect competition", which presupposes very large numbers of agents on both sides of the market. The concept of the contestability of a market power can be said to capture an important aspect of the classical point of view.

[6] There is, however, the following tension affecting Menger's perception of the different stages of society. Uncertainty is arguably high in the "early and rude state of society", to use Adam Smith's concept, for the reasons given. However, the third stage is typically characterized by a speeding up of technological and organizational progress, which has the potential of significantly disrupting economy and society and increasing uncertainty, as Schumpeter was to emphasize. Yet, surprisingly, in Menger's story this fact plays no role.

the competitive process.[7] The idea that in conditions of free competition the rate of profits tends to be equalized across different industries is in fact advocated in all major schools of economic theory: it is the linchpin of the analyses of the classical and also major marginalist authors, including Jevons and Walras. Since contrary to the latter, Menger did not entertain the concept of a general rate of profits, the so-called marginalist "innovators" also for this reason cannot be said to form a homogeneous group.

Differences of opinion amongst marginalist authors and Menger become visible also when taking a closer look at their assessments of the contributions of the classical economists and especially Ricardo. Here we see that there are marked differences between the judgments passed on Ricardo by Jevons, Menger and Walras on the one hand and later marginalists, such as Alfred Marshall and Knut Wicksell, on the other.

3 Marginalist Authors on the Classical Theory of Value and Distribution

Amongst the early marginalist economists, or economists who paved the way towards marginalism, we encounter both authors who maintain that the classical theory of value and distribution is flawed and beyond remedy and authors who consider it to be basically correct, but incomplete and in need of improvement. To the former belong Jevons, Walras and also Menger.[8] According to Jevons, the doctrine of the classical authors and especially Ricardo, based on "mazy and preposterous" assumptions, "is radically fallacious; it involves the attempt to determine two unknown quantities from one equation" [10: 258]: profits and the level of output, given the real wage rate. Ricardo and his followers are said to have shunted the car of economics onto the "wrong track". A fresh start was needed, which required a fundamental break with their doctrine. Léon Walras reiterated the criticism that the latter is underdetermined: "In the language of mathematics one equation cannot be used to determine two unknowns" [52: §368]. Hence a major criticism levelled at the classical authors by the first two critics is that their theory cannot explain anything, because the number of unknowns exceeds the number of equations put forward to ascertain them. While using less drastic language than Jevons and avoiding any comment on the question of the determinateness of their (literary) reasoning, Menger, too, expresses the need to supersede the doctrines of Adam Smith and David Ricardo. He is especially critical of, first, the labour theory of value, with which, he maintains, the classical doctrine stands or falls, and, secondly, Ricardo's "mistaken" theory of rent.

To the second group belong inter alia Alfred Marshall and Knut Wicksell. Marshall calls Ricardo the head of the "brilliant school of deductive reasoning", and explains:

---

7 Whilst the classical economists spoke of rate of profits, the Austrians typically spoke of rate of interest, meaning, however, essentially the same thing in the context under consideration. In the following, I use the two terms interchangeably.

8 On Walras' criticism of the classical economists, see Kurz and Salvadori [20, 22–26], [22].

"His strong constructive originality is the mark of the highest genius in all nations. ... Ricardo's power of threading his way without slip through intricate paths to new and unexpected results has never been surpassed" [30: 629 and n. 1]. About Ricardo's approach to the theory of relative prices Marshall writes: "Ricardo's theory of cost of production in relation to value occupies so important a place in the history of economics that any misunderstanding as to its real character must necessarily be very mischievous". Contrary to a widespread opinion, Marshall insists: "the foundations of the theory as they were left by Ricardo remain intact" [38: 416–17].[9] And Knut Wicksell emphasizes: "the way in which Ricardo develops his argument ... is a model of strictly logical reasoning about a subject which seems, at first glance, to admit of so little precision". And: "Ricardo's theory of value is, one finds, developed with a high degree of consistency and strictness" [54: 34 and 40]. He adds:

> Since, according to Ricardo, wages represent a magnitude fixed from the beginning, and since—as he later shows—the level of rent is also determined by independent causes, the cause of capital profit is already settled. *It is neither possible nor necessary to explain capital profit in other ways, if the other assumptions are sound.* (Ibid.: 36–37; emphasis added)

Hence, while the former authors conceive their works as strict antitheses to classical economics, the latter conceive their own works as broadly compatible with it.

In the following I will largely side with Marshall and Wicksell and argue that much of the criticism Menger et al. levelled at Ricardo cannot be sustained. I will, however, not endorse the view, propagated especially by Marshall, that marginalist theory is simply a continuation and in fact a completion of classical theory, which, while already possessed of a fairly well developed production or supply side, is said to have failed to elaborate an equally sophisticated consumption or demand side. I will argue instead that the two types of theory are fundamentally different.

This can be seen at a glance by turning the attention to the different analytical structures of the two kinds of theory and the different sets of data or independent variables from which classical and marginalist authors start their reasoning in the theory of value and distribution.

4 Preparing the Ground: Independent Versus Dependent Variables in Classical and in Marginalist Theory

Close scrutiny shows that despite numerous differences between, and tensions amongst, the doctrines of the classical authors, they share a common element. This element defines analytically what is meant by "classical" economics (see [20], [47], [14], [25], [27], [42], [43]). The authors grouped under this heading all seek to explain the unknowns or dependent variables—the general rate of profits in the economy,

[9] Since these lines were written a long time after the publication of the statements of Jevons, Menger and Walras cited in the above, Marshall's defence of Ricardo can be read as being directed also at these critics.

the rents paid to the proprietors of the different qualities of land and the ordinary or "natural" (relative) prices at a given time and place—in terms of the following givens or independent variables:

1. The gross outputs of the various commodities produced during a year.
2. The set of alternative methods of production, reflecting the current state of technological knowledge in the economy, from which cost-minimizing producers can choose.
3. The amounts of the various qualities of land (and other natural resources) available in the economy to be used in production.
4. Real wages (or the share of wages in the national product).

As Wicksell saw clearly, and Sraffa [43] showed conclusively, from this set of independent variables the unknowns or dependent variables can indeed be determined in a coherent manner (see also [20: Chaps. 1 and 13]). The criticism put forward by some of the marginalists (see below) that the classical theory of value and distribution is underdetermined is based on a severe misunderstanding of its analytical structure.

As regards marginalist authors, they seek to explain the dependent variables—all distributive variables, including the real wage rate(s) and the competitive (uniform) rate of profits, relative prices, the quantities produced of the various commodities and the quantities employed of the various factors of production—in terms of the following givens or independent variables:

a. The needs and wants (or preferences) of agents.
b. The endowment of the economy in terms of given quantities of all factors of production: labour(s), capital, and land(s).
c. The distribution of property rights amongst agents.
d. The set of alternative methods of production, reflecting the current state of technological knowledge in the economy, from which cost-minimizing producers can choose.

Item 4 in the first list of givens clarifies that the classical authors took one of the distributive variables, typically real wages (or the share of wages), as given and thus treated them as a known magnitude at a given time and place. Property incomes, that is, the rents of land and the general rate of profits were then ascertained *residually* in terms of the *surplus product* that remained after all used up means of production and wage goods in the support of workers had been deducted from gross output levels. This *asymmetric* treatment of wages, on the one hand, and rents and profits, on the other, is in striking contrast to marginal theory, which treats them *symmetrically* in terms of the supply and demand schedules or correspondences established on the basis of the givens a–d. While the classical authors put forward a *surplus-based* explanation of property incomes, the marginalists sought to elaborate a *scarcity-based* explanation. The classical authors employed the principle of scarcity essentially only in explaining the rent of land, a resource that could be taken to be naturally given and essentially non-producible, whereas capital goods (and also humans and therefore labour) could be produced and reproduced and their quantities adjusted to the social requirements for use. The two theories therefore represent different perspectives on the economic world and explain, and determine, the distributive variables and relative prices in markedly different ways. Therefore marginalist theory cannot be considered a simple continuation of, and improvement upon, classical theory.

While shortly after the publication of Ricardo's *Principles* in 1817, the analytical structure of his surplus approach to the theory of value and distribution was still well understood, as is exemplified by the writings of authors such as James Mill or Robert Torrens, shortly afterwards it quickly met with a growing lack of understanding or of misunderstanding. At the time of Karl Marx, who may be said to have recovered parts of it from under thick layers of misinterpretation, it was no longer understood and interpreted more and more through marginalist lenses. It hardly needs to be stressed that the different analytical structures of alternative theories of value and distribution only became clear in recent times essentially thanks to works of Piero Sraffa [42, 43].[10] For a long time the issue under consideration was clouded in uncertainty, if not confusion, generated by frequently heated debates; even nowadays it is still insufficiently understood in the profession. Someone who was brought up with the idea that all economic problems have to be couched in demand and supply terms will try to read the classical economists through the scarcity lens. He or she will therefore be surprised, for example, to find *no given* "quantity of capital", whose relative scarcity is supposed to determine the rate of profits along marginalist lines. And with a *given* real wage rate, there would be no room for the supply and demand approach to determine it. The accusations, reported in Sect. 3, that Ricardo's theory is underdetermined illustrates the lack of understanding or plain misunderstanding of classical theory and the attempt to interpret it with a marginalist perspective in mind. Thus, the classical economists' treatment of the real wage rate in given circumstances as a known magnitude has led to the ridiculous view that they endorsed an "iron law of wages". This interpretation makes as little sense as would one that takes a given endowment of capital in marginalist theory as involving an "iron law of capital", according to which the endowment of capital can never change. Of course, in some other parts of their analyses the classical authors discussed the factors affecting the real wage rate (most notably, capital accumulation, technical progress and the scarcity of natural resources), and the marginalist authors those affecting the size of the capital stock (most notably, saving and investment). The only question that matters is, whether the classical or the marginalist authors, following their very different lines of argument, succeeded in proposing a promising path towards a substantially convincing and logically coherent solution of the problem of value and distribution they were struggling with. We shall come back to this below.

We turn now briefly to variants of marginalist theory. The most prominent of them are arguably the following.[11]

[10] Several authors, some with a Russian background, anticipated partly Sraffa's findings. As to the contributions of Vladimir K. Dmitriev and Ladislaus von Bortkiewicz, see the respective chapters by Gehrke and Kurz in Avtonomov and Hagemann [1].

[11] Since, as has already been stated in the above, the temporary and intertemporal equilibrium variants of marginalism were elaborated only some fifty years after Menger's *Grundsätze*, they will

5 Variants of the Marginal Theory of Value and Distribution

Temporal "Austrian" theory. While Menger [32] shares several of the typical elements of the "Austrian" variant of marginalist theory, he can hardly be considered a proper marginalist author. Things are much clearer with regard to Eugen von Böhm-Bawerk, notwithstanding the heated controversy he had with John Bates Clark and others, and at least the early Friedrich von Wieser. All Austrian economists were convinced that *time* plays a crucial role both in production and consumption. Menger stresses that all economic activity takes time, with the consequences of human decisions and actions today emerging only in a more or less distant future. Uncertainty therefore figures prominently in the writings of Austrian authors. As regards production, Austrian authors, beginning with Menger, developed time-phased schemes to capture the time structure of production and the horizontal and vertical interconnectedness of the various stages of production. They also elaborated, again beginning with Menger, the idea that processes of production that need more time from beginning to completion are more capital-intensive and are typically possessed of a higher overall labour productivity. In their view, time is capital, and capital is productive. Menger's hint was elaborated by Böhm-Bawerk in terms of his "law of the superiority of more roundabout processes of production". As regards consumption, some of the Austrians, including Menger and Böhm-Bawerk, assumed that consumers' decisions express a positive rate of time preference. However, Wieser and Josef A. Schumpeter [39] strongly opposed this view. They objected that a positive rate of interest causes of necessity discounting and therefore a positive time preference instead of being caused by it.

Atemporal Walrasian theory. Léon Walras [52] was variously accused by Austrians of advocating an atemporal theory, but this does not follow, as they argued, from his representation of the system of production in terms of simultaneous production equations within a general equilibrium framework, which lists all actually operated processes of production next to each other. This is perfectly compatible with the view that production takes time, because the inputs entering any one industry have to be produced in the preceding period by some other industries. Critics of Walras, including some of the Austrians, were also strongly opposed to the method of general equilibrium and the simultaneous determination of all dependent magnitudes, including prices and the distributive variables. This contradicted their causal-genetic preconception and thus one of the most sacred elements of their construction. According to this view, consumers, not producers, are the sovereigns of the economic system and shape its properties. Böhm-Bawerk also saw the mathematical method employed by Walras as being incompatible with a causal-genetic explanation. His brother-in-law, Friedrich von Wieser, did not agree with him and elaborated

be left out of consideration here. For an account of their distinguishing features compared with the older marginalist theories and the motives that led to their elaboration, see again Garegnani [6], Kurz and Salvadori [20: Chap. 14] and Petri [35].

a marginalist theory combining the twin principles of marginal utility (we also owe to him the term "*Grenznutzen*") and marginal productivity.

Knut Wicksell's synthesis. The Swedish economist Knut Wicksell [54, 55], was deeply impressed by Böhm-Bawerk's temporal and also by Walras's general equilibrium approach and attempted to cross-breed the two in a single theory that combined each one's strengths and overcame its weaknesses. His construction may be said to represent the highest stage of development of marginalism in the nineteenth century. Wicksell succeeded inter alia in clarifying the properties the theory has to meet, which includes linear homogeneity of the production relations in order for the distribution of the product to just exhaust the product, neither more nor less. He also identified the limits of the theory and pointed out an "insurmountable obstacle" it faced when trying to treat capital on a par with different kinds of labour and land. While the latter could be expressed each in its own technical unit, the "quantity of capital", consisting of heterogeneous capital goods, in competitive conditions had to be conceived of as a single magnitude expressed in terms of a *value* magnitude, each particle of which yielded its proprietor the same rate of return. In Sect. 6 we will deal with this issue in greater detail.

Macroeconomic marginal productivity theory. All authors from Jevons to Walras and from Böhm-Bawerk to Wicksell conceptualized marginalist theory as a genuinely microeconomic theory. However, they were also keen to present the theory in a compact form, making use of aggregate concepts; viz. Böhm-Bawerk's "average period of production" of the economy as a whole, designed to give the capital–output ratio for the economic system. But they took pains (whether successful or not) to render their measurements to meet strict theoretical standards. They typically did not use macroeconomic concepts such as "aggregate" production functions other than for illustrative purposes. (Such functions can in fact be consistently derived from micro units only in exceedingly special conditions.) It was in particular John Bates Clark [3], who established the macro version of marginal productivity theory and successfully promoted its diffusion; Schumpeter [40] called him "the master of American marginalism". Clark's continuing influence on the profession is exemplified in terms of the remarkable prominence of Robert Solow's model of economic growth in macro theory.

Here the focus will be on the temporal Austrian approach, Walras' general equilibrium theory and Wicksell's attempt to combine the two. Menger's contribution has arguably the closest affinity especially to the first one, which does not render him ipso facto a dyed-in-the-wool marginalist. The following will identify more carefully the specific nature and features of his contribution. This requires a brief discussion to what extent Menger benefitted from soundings of marginalist ideas long before the alleged "marginal revolution". We focus exclusively on authors whom Menger cites in the *Grundsätze*: Johann Heinrich von Thünen with regard to the concept of marginal productivity especially of capital, and the "German Use Value School" with regard to the concept of marginal utility.[12]

6 Anticipations of Marginalist Concepts Prior to the "Marginalist Revolution"

When Jevons, Menger and Walras entered the scene, and some of the "innovators", but not Menger, proclaimed that a "revolution" was under way, it apparently had escaped their attention that both concepts—marginal productivity and marginal utility—had already been well known for quite some time. While with regard to the former, the obvious reference is to Thünen's *Der isolirte Staat* [49], with regard to the latter it is the work of Karl Heinrich Rau; see especially volume I of his *Lehrbuch der Politischen Ökonomie*, published first in 1826, followed by several further editions [37]. We shall, however, see that the contexts in which the two concepts had been elaborated, and the meanings that had been attributed to them, were somewhat different from those after their transmigration into a marginalist framework. Concepts invented by some authors are hardly ever simply adopted by others without some adaptation in order to fit them into their new environment.

Thünen's theory of capital and interest. Thünen was an uncompromising objectivist, in whose analysis subjectivist elements play hardly any role whatsoever.[13] He admired the classical economists, especially David Ricardo, and sought to put economic principles on a solid basis by formulating them mathematically and using only assumptions that reflected real facts. He would in all probability have been strictly opposed to Milton Friedman's instrumentalism (and shared perhaps Paul Samuelson's criticism of it), according to which assumptions do not matter; what matters exclusively is the degree of correctness of the predictions of the models based on them. The realism of the results derived, Thünen was convinced, depended irrevocably on the realism of the assumptions postulated.

In the Second Part of the *Isolated State* he puts forward several versions of what is known as David Ricardo's "corn model". In the first version, Thünen refers to production in a tropical country, in which workers produce a means of subsistence (batata) without any produced means of production: he speaks of "capital-less production". (Since wages are advanced at the beginning of the production period, they do of course constitute wage capital.) Since the amount of batatas which workers produce

[12] See Chipman's magesterial study of the German Use Value School [2], which had been inspired by earlier works of Erich Streissler (see, e.g., [44]). For a summary account of the contributions of the Germans and the Austrians, see Kurz [16]. See also the valuable collection edited by Priddat [36].

[13] In a letter of August 1899 to John Neville Keynes, Alfred Marshall, who admired Thünen and was strongly influenced by him, wrote: "you know von Thünen's *métier* was that of agricultural reformer. His abstract economics came in by the way. He was up to his eyes in facts about rye and manure and so on". Cited in Kurz [12: 117 n. 5].

per year are taken to exceed what they need to feed themselves and their families, there is a surplus left after the harvest. If this surplus is appropriated by the owner of the plantation in terms of interest (profits), then the rate of interest, r, equals the total product, Q, minus the sustenance of workers, S, divided by S, that is

$$r = \frac{Q - S}{S}.$$

Thünen then asks what would happen if more labour was to be employed per unit of homogeneous land. The reference is obviously to the case of *intensive diminishing returns*. Without further ado, he states that labour productivity would tend to diminish. With a given real wage rate per unit of labour employed, the surplus per unit of labour produced would decrease and so would the rate of interest.

Thünen then leaves the tropical world and turns to the situation in Europe, where apart from labour also produced means of production or capital goods (horses, ploughs and so on) are employed in the production of rye, and workers now receive a subsistence wage in terms of rye. This brings us to Thünen's next version of the corn model. The problem he faces now is how to deal with (physical) capital and additions to it? Interestingly, Thünen opines that capital goods do not really pose a problem, because they are simply "the accumulated product of labour", that is, "labour done". Heterogeneous capital goods, he insists, can be reduced to (homogeneous) labour or to rye (on which workers subsist) and thus rendered commensurable and added up.

In this way he believes to be able to circumnavigate, first, the problem of whether the Δ capital under consideration consists of a little more of exactly the same capital goods in the same proportions as the original capital. In this case the marginal capital could be *physically* compared with the original capital and there could be no dispute about the size of the addition and therefore the marginal increase of capital. But why should a proportional increase of the quantities of all capital goods be the best way to increase the "quantity of capital" employed? There is no presumption that this is in fact the case. Yet if the amounts of some capital goods are increased by relatively more than those of some other capital goods, or if some capital goods are increased while others are decreased, or if some other capital goods enter the system, while others exit it, what then? How do we know that the aggregate capital stock is now larger than before, and by how much? Obviously, in this case the various capital goods must be rendered commensurate first by multiplying their quantities by their prices. But which prices? Thünen decides swiftly to use labour values, although since Ricardo (at the latest) it is known that, strictly speaking, labour values apply in highly special conditions only:

(i) in the case in which the rate of interest (profits) is nil and the entire product is distributed to workers in terms of wages (setting aside the problem of rent);
(ii) in the case in which the production of all products exhibits the same proportion of direct labour to means of production across all stages of the production process (in this case capital can be treated as if it was a single composite commodity).

Beyond these cases, the labour theory of value applies at best only approximately. Since Thünen did not tie his argument to any one of these two cases, his labour value-based reasoning met with criticism, as we shall see below. The interesting point to be made here is that major later authors, such as Karl Marx and Eugen von Böhm-Bawerk, while belonging to vastly different schools of economic thought, had recourse to exactly the same device to cope with the heterogeneity of capital goods. The need to express the economy's endowment of capital, whose relative scarcity was meant to determine *the* general rate of interest, as a single quantity presupposed rendering capital goods commensurate. These authors felt that this could be to done by using labour values as aggregators without pre-empting the generality of the argument.

Capital (C), consisting of means of sustenance of workers and means of production, both expressed in labour terms, Thünen goes on, helps producers "to increase the efficacy of the actually performed labour" [49: 333]. In order to highlight this property, Thünen asks what would happen if a little more capital were to be employed per unit of land. He answers this question by simply putting in the place of a diminishing marginal productivity of labour in the capital-less economy a diminishing marginal productivity of the "atom of capital employed last"—his expression for the Δcapital—in an economy with capital. He actually states, that "the need [!] has emerged to find a series for the [marginal productivity of capital], whose elements will progressively diminish, and this request is met by the geometric series, whose base number is a fraction, such as 9/10, $(9/10)^2$, $(9/10)^3$, $(9/10)^4$, ..." [49: 346]. Interestingly, while Thünen is keen to establish empirically intensively diminishing returns in agriculture on his estate, with regard to production that involves capital he relies simply on what he calls a "*conclusion by analogy*" ([49: 345]; emphasis added).

In a further step, Thünen then notes that an isolated increase of the employment of physical capital will be accompanied by an increase of the marginal productivity of living labour. This is, however, difficult to reconcile with a given and constant real wage rate: the wage rate is bound to rise. This reflection points in the direction of a fully-fledged marginalist theory of value and distribution, in which all distributive variables—wages, rate of profits and rent(s) of land—are explained indiscriminately in terms of the marginal productivities of the respective factors of production—labour, capital and land(s). At the same time it draws the attention to a problem we have already encountered and which caused Knut Wicksell a lot of headache. Thünen had claimed that there is a *perfect analogy* between all factors of production. Wicksell objected: while labour and land can be given in terms of their own technical units, this is not so with regard to capital: "all these requisites [i.e., produced means of production] have only one quality in common, namely that they represent certain quantities of exchange value, so that collectively they may be regarded as a *single sum of value*" ([55: 145], emphasis added; see also [13]). Yet the fact that the "quantity of capital" available in the economy, which consists of heterogeneous capital goods, can only be expressed as a *sum of value* (in terms of some standard of value) destroys the alleged analogy between the three kinds of income—wages, profits and rents—and the corresponding factors of production—labour, capital and land.

What bothered Wicksell most was the fact that with a change in the "quantity" of capital, the quantities of labour and land remaining the same, the distributive variables—the wage rate, the interest rate and the rent rate(s)—are *all* bound to change and with them the prices of the various products. This will, however, in all probability affect the magnitude of the value of the quantity of capital in the system (as well as the magnitude of the Δcapital in the marginalist thought experiment contemplated). Yet this blurs the picture and questions the solidity of the ground on which the marginalist doctrine rests: if the "quantity of capital" cannot be defined in a purely physical manner that is independent of the way, in which the product is divided between capitalists and workers, we are on shaky ground. If the quantity depends on distribution, how could the former determine the latter?

A summary. Let us briefly take stock of what has been said. The beginnings of Thünen's approach to the theory of value and distribution were deeply rooted in the classical and especially Ricardian surplus-based point of view, which conceived the rate of profits in competitive conditions as given by the ratio of a society's surplus product and the social capital advanced at the beginning of the production period. And like the classical authors from William Petty to Adam Smith, he felt the need to render heterogeneous products constituting the surplus and the capital employed commensurable with one another. The classical authors had sought to accomplish this by reducing the two bundles of commodities either to the direct and indirect amounts of corn (or rye) needed to support workers during production (Petty) or to the corresponding amounts of labour performed (Ricardo). Thünen adopted their procedure. He then boldly postulated an analogy between the labour–land and the capital–land model. Asking what would happen with regard to output (ΔQ) and the general rate of profits (r), if a little more of capital (ΔC) was to be employed on a given plot of land, he "concluded by analogy" that

$$r = \frac{\Delta Q}{\Delta C}.$$

Contrary to Thünen, Ricardo nowhere postulated such an analogy. The reason for this is that he saw clearly that any two capitals are typically "not the same in kind" (Ricardo, *Works* IX: 360) and that their relative values typically depend on the distribution of the product between wages and profits and change with the distribution. While he felt that he could use labour values as a makeshift solution when ascertaining the numerator (surplus) and denominator (capital) in the expression giving the general rate of profits (see above), he was clear that the quantity of capital in the economic system is a magnitude that cannot be given independently of the rate of profits and therefore cannot be treated as an independent variable. While for obvious reasons Ricardo could not anticipate the difficulties besetting the later marginalist "conclusion by analogy", he may be credited with having scented that there were problems with treating capital as a given magnitude. His intuition was indeed remarkable, as several commentators rightly stressed (see in this context [17: 243–44]).

We now turn briefly to Menger and ask whether in the *Grundsätze* we encounter the idea that the marginal product of capital determines the rate of profits? The answer is no. Menger was not possessed of a sufficiently elaborate concept of marginal productivity. While what he calls "loss principle" (*Verlustprinzip*) points in its direction by asking what the reduction in output would be if some input was available only in a smaller quantity, he does not elaborate on it and in particular does not coherently make use of it in his theory of income distribution.[14] Most surprisingly, as has already been pointed out, in Menger we do not even encounter the concept of a general rate of return on capital, which in the case of universal free competition (the workhorse of much of classical and marginalist theory) would tend to be uniform across all sectors of the economy. Since Adam Smith's *Wealth of Nations* at the latest, this concept had gradually become a focal point around which much of economic analysis revolved. Not providing an alternative to the classical explanation, it is entirely unclear how Menger could believe to be able to surpass classical political economy. The fact that he saw no need to justify his outlandish position is also peculiar vis-à-vis the works of his acolytes, especially Böhm-Bawerk and Wieser.

Next we mention briefly anticipations of marginal utility theory, focusing attention on contributions of the "German Use Value School".

Anticipations of marginal utility theory. Menger dedicated his *Grundsätze* "with respectful esteem" to Wilhelm Roscher, a towering figure of what is known as the older German Historical School. Menger in fact insisted that "the most general principles of our science", have "in no small degree"—he added: in fact "almost entirely"!—been established "by the industry of German scholars" (X [49]). Numerous references to contributions of representatives of the "German Use Value School", most notably Gottlieb Hufeland, Karl-Heinrich Rau, Bruno Hildebrand, Karl Knies and Roscher substantiate this remarkable claim. Since Menger himself left no doubt about his deep indebtedness to the German authors, one can only wonder how the myth could emerge that he was one of the originators of marginal utility theory. As Chipman [2: 42] stressed vis-à-vis overwhelming evidence, the credit does not go to Menger, Jevons and before them Hermann Heinrich Gossen, but to the aforementioned Germans, who had developed "the essential ideas of the marginal revolution" in the first half of the nineteenth century.[15]

The majority of German economists, to whom Menger refers, were Smithians of sorts, who did not wish to fundamentally break away from the doctrines of the classical economists, but rather correct and further elaborate on them. They were

[14] Menger, as has been mentioned, cites Johann Heinrich von Thünen, but makes insufficient use of the latter's findings. It was only Menger's student Wieser who combined the marginal productivity principle and the marginal utility principle in a fully-fledged marginalist analysis; see Wieser [50] and the assessment of Wieser's contribution in Kurz and Sturn [26].

[15] The concept of marginal utility had been anticipated by the Swiss Daniel Bernoulli in the context of a discussion of the "St. Petersburg paradox" (see Kurz [15: 165–66]); alas, his work was ignored for a long time. In this context it deserves to be mentioned that according to Gossen [7] the really scarce thing to people is the time at their disposal. He therefore saw the problem of utility maximization as subject to a time constraint in combination with the insight that consumption takes time, a fact that has been overlooked by many interpreters.

also no advocates of a strictly "individual" point of view, but, brought up in the cameralist tradition of German *Staatswissenschaften* (literally: sciences of the state), stressed the importance of collective, public and cultural needs above and beyond private ones. Their reasoning concerning the "*Nutzwerth*" (use value or utility) of goods typically started from Smith's famous diamond and water "paradox", which they re-interpreted, thereby changing unconsciously its original meaning.[16] Rau [37] et seq.) eventually replaced Smith's concept of "use value" of a commodity, that is, its generic capability to satisfy certain needs or wants, by the concepts of utility and marginal utility, which are taken to depend on the quantity consumed of the commodity. German authors typically assumed that needs and wants were hierarchically ordered and therefore came close to advocating what today are called "lexical" or "lexicographic preferences".[17]

It is important to emphasize that according to Rau, commodities are typically not simple things, but possessed of various characteristics, or of the same characteristics in different proportions. His argument thus foreshadows Kelvin Lancaster's [28] view that what consumers wish to acquire are not so much the commodities themselves as the characteristics they possess. These characteristics are obviously tied to the materials or substances productively consumed in the course of the production of commodities and therefore to what these contain or "embody". This observation led Rau directly to the concept of substitution in consumption. To the extent to which commodities share some common characteristics, or, in the language of Rau, represent the same "species", they can replace each other in consumption (and production). Think, for example of the caloric contents of various kinds of food.

As we shall see, Menger denied that such substances had any role to play in explaining values and prices (we shall come back to this questionable claim in Sect. 10

[16] In this context a misunderstanding of Smith's respective argument, which is also to be found in Menger, ought to be cleared up. Contrary to his critics, Smith was convinced that not all cases of consumption can be described in terms of a relationship between a single agent and the realm of things or goods: the display of a diamond, for example, is a social act that cannot be captured in terms of a simple subject–object relationship, as in the case of water. A diamond is first and foremost a device by means of which a person tries to signal his or her wealth and social position to other people. The diamond ought to be seen by them, not so much by its proprietor. Smith spoke aptly of a "parade of riches" (WN I.xi.c.31). Acts of conspicuous consumption involve third parties and therefore subject–object–subject relationships.

[17] Lexicographic preferences defy, however, a representation in terms of utility functions. A hierarchy of needs recurs in Menger's classification of the process of civilization in essentially three stages (see Sect. 2 above). In the first stage, humans can barely satisfy their basic needs of food, clothing and shelter. Their attention focuses on how to survive, but poor knowledge and a lack of control over nature imply fundamental uncertainty. The second stage is one in which knowledge about needs and means to satisfy them is larger and so is productivity. In addition to basic needs, now also more sophisticated ones can be served; we could speak of amenities. It is only in the third stage that humankind manages to control nature more effectively and also improve significantly its understanding of its own needs and how to satisfy them. With the growth of human knowledge, "true" needs and goods gain in importance compared to "imaginary" ones, the diversity of needs and the goods available to satisfy them increases, the security of people improves, the institution of the market gets more and more firmly established, and types of competition begin to permeate the economic system. We could say that "luxuries" gain in importance.

below). This comes as a surprise, not least given his insistence that economics belongs to the natural sciences and his emphasis on "imaginary" needs and goods: if goods were simple things, possessed of a singly quality only, it would be (more) difficult to explain why wrong views about this quality could persist for long periods of time. The existence of imaginary wants and goods raises also the question of whether, in Menger's view, the pursuit of self-interest is always beneficial to the individual or even to society as a whole. According to Menger, individual agents cannot generally be assumed to be fully informed about the properties or characteristics of commodities and therefore, assessed in terms of the ideal of full information, are bound to make mistakes, that is, buy and consume commodities that are much less beneficial to them than expected and may even be harmful. In this case some sort of paternalism cannot be rejected out of hand. Contrary to a widespread interpretation, Menger was no unqualified libertarian.

Did Menger entertain a cardinal or an ordinal concept of utility? While I am not aware of any passage that expresses ordinality in a clear-cut manner, there are passages that can only be read in cardinal terms. It suffices to refer to a comment of Menger on one of his numerical illustrations. He stresses: "Thus when I designate the importance of two satisfactions with 40 and 20 for example, I am merely saying that the first of the two satisfactions has *twice the importance* of the second to the economic individual concerned" (163n* [183n], emphasis added; see also 158n* [179]). There is also the fact that with regard to the estimation of the aggregate wealth of a nation, Menger insists, against Rau, that the right measure is in terms of aggregate utility. This presupposes, however, a cardinal concept of utility. He agrees with Rau, though, that marginal utility diminishes with rising levels of income. This is also the reason why in his view simply adding up incomes does not give an appropriate measure of the wealth of a nation.

While the Austrian economists, including Menger, are rightly famous for emphasizing that all economic activity takes time, this fact plays only a relatively minor role in their treatment of consumption compared with the role it played in Gossen [7]. As has been mentioned in the above, in Menger (128 [153]) we encounter the idea that humans are possessed of a positive rate of time preference. Here it ought to be mentioned that this idea was originally developed in the context of a critical re-assessment of the canonical prohibition of interest taken by representatives of the School of Salamanca in the sixteenth century and by the Flemish Jesuit Leonnard Lessius. Böhm-Bawerk then forged the idea into an alleged "law of the higher estimation of present needs compared with future ones"—the so-called second "Ground" in his theory of capital and interest [46]. We have already heard that Wieser questioned the implied causality: once there is a positive rate of interest, rational agents are bound to "discount" [56: 134].

Next we have a closer look at Menger's time-phased conceptualization of production.

7 Production in a Time-Phased Perspective

Very much like Smith and the classical economists more generally, Menger conceived of production as a time-phased process, in which inputs are transformed into outputs following a well-defined time profile that reflects cost-minimizing behaviour, given the level of technical and organizational knowledge. Menger and the classical authors would have received with astonishment and disbelief the assumption of "instantaneous production" entertained in modern marginalist macroeconomics—Menger actually poked fun at the idea that production could be carried out by a "mere wave of hand" (22 [68]). He conceived production as a *linear* or *unidirectional* process of *finite duration*—a sort of one-way avenue of limited length—leading from inputs, or "goods of higher order", via some intermediate goods, or "goods of lower order", to final goods or means of consumption, that is, "goods of first order". The services of labour, land, machines and so on are considered goods of higher order. The time-phased system reflects a set of forward and backward linkages.

A comparison with the classical perspective on production is appropriate. According to the classical authors, one form of representing production is in terms of a series of inputs, typically amounts of labour, that eventually materialize in an output. However, since their basic understanding was shaped by François Quesnay's conceptualization of production in the *Tableau Économique* as a circular flow, the series of dated inputs is infinite, not finite, as in the unidirectional case of Menger (or Böhm-Bawerk).[18] However, the classical economists were convinced that the series are convergent. This applies to the amount of labour needed directly or indirectly in the production of a commodity no less than to the amount of a particular commodity, for example "corn" or iron, used in its production. In the case of labour, the sum total of all the amounts of labour expended during the commodity's production sum up to what Ricardo called the labour "embodied in" or "bestowed on" the commodity. In modern verbiage, this amount of labour (or of any other input), or labour value, represents the vertically integrated labour content of a commodity. In this perspective commodities reflect something real—the *amounts of labour* (or the *physical real costs* in terms of some *particular* product) by means of which the "difficulty of production" (Ricardo) is overcome in given circumstances.

These considerations led the classical economists directly to the problem of the values or prices of commodities. Normal or natural prices, they argued, are bound to reflect the amounts of labour (or physical real costs) embodied in commodities. Menger was strictly opposed to this view (see Sect. 10 below). He did not dispute, of course, that production involves costs, but was of the opinion that the prices of the goods of first order are to be given priority over those of goods of higher

[18] In agriculture, animal husbandry and so on, production is inevitably circular. In the *Grundsätze* Menger provides several examples of circular cases, but in his more abstract representation he clings to unidirectional production. Also in manufacturing circularity abounds, with iron entering directly or indirectly in the production of all commodities, including itself. As we shall see in the following section, Wieser was to criticize Menger for ignoring a characteristic feature of modern industrial systems.

order, with the latter derived from the former. The subjective utility estimations of consumers are taken to fix directly the values and prices of consumption goods—totally independently of incomes and costs and thus production relations. These prices are then treated as known magnitudes when it comes to the determination of all other prices, including the distributive variables and therefore incomes. Menger's approach implies a causal-genetic, hierarchical and *successivist* analytical structure, in which (some) prices are explained in terms of (some other) prices, and in which the chronological order of things is turned upside down: *consumption rules over production*. The upshot of his argument is the problem of the "imputation" of the prospective prices of consumer goods to the inputs cooperating in their production.

Before we have a closer look at how Menger deals with this problem, it deserves to be stressed that different from (other) marginalist authors he does not explain each and every price symmetrically in terms of the forces of demand and supply conceived of as schedules, relating the price of the good or service and the quantity supplied or demanded in the market. He therefore also does not construct the theory of production and supply in complete analogy to, and parallel with, the theory of consumption and demand, with the concepts of substitution amongst factors of production and of marginal productivity on a par with the concepts of substitution amongst goods and marginal utility. He thus bypasses an objection that is close at hand, namely, that there is no presumption why such different things as human nature and technology should be possessed of similar properties. While Menger does not postulate a fundamental symmetry between the forces of demand and supply (as for example Marshall was going to do), whose existence can rightly be questioned, his successivist approach is beset by other difficulties to which we now turn.

8 Menger on the Problem of "Imputation" and Wieser's Criticism

Menger devotes hardly any energy to explaining how the prospective prices of goods of first order are supposed to come about, nor to how they can be imputed to inputs. He was apparently convinced that expressing the basic idea underlying his argument was good enough to insinuate that a solution existed and made economic sense. However, things are quite otherwise, as especially the attempts of his former students Böhm-Bawerk and Wieser to crack the nut Menger had left show. While Böhm-Bawerk can be said to have struggled in vain with the problem inherited from his master, Wieser [50, 51] deserves credit for three achievements: first, he pointed out serious deficiencies of Menger's suggested solution; secondly, by elaborating on his criticism he (re-)discovered the classical surplus-based concept of the rate of profits; and, third, he came close to "solving" the imputation problem within the framework of a developed marginalist theory.

Deficiencies of Menger's solution. In his book *Der Natürliche Werth* (1889), Wieser at first retained Menger's assumption of unidirectional production, although he was convinced that it did not adequately portray the situation of modern industrializing economies. Assume that there are n goods of first order, whose values are already known, just as Menger had stipulated. Assume then that there are m goods of higher order needed in the production of the goods of first order whose input quantities in the different lines of production are also known. Then a system of linear equations can be put up, in which the values of the goods of higher order are the only unknowns (see Wieser [51: 85–88]). Wieser concluded that a solution of the imputation problem, even within a Mengerian unidirectional framework, requests formulating, and solving, a system of *simultaneous* equations. This presupposes knowledge of the appropriate mathematics. Apparently Menger had not been aware of this: a simultaneous instead of a successivist approach was unavoidable.[19]

Wieser then pointed out that in order for a solution to exist, the number of goods of first order, n, must be equal to the number of goods of higher orders, m, that is $m = n$. For if $m < n$, the system would be overdetermined, whereas with $m > n$ it would be underdetermined. But why should the number of goods of first order be exactly equal to the number of goods of higher orders? There is no convincing reason in support of this. Yet even if this condition happens to be met, this does not mean that all difficulties have gone, for there is no presumption that the solution makes economic sense: a price could, for example, be negative, which makes no sense in a system with essentially only single production (i.e., no joint production), as contemplated by Menger.[20]

Menger's treatment of the imputation problem therefore cannot be sustained and fails to provide the much-needed solid basis for the rest of his argument to stand on.

The problem in a circular flow framework. In *Über den Ursprung und die Hauptgesetze des wirthschaftlichen Werthes*, Wieser [50] had already insisted that the imputation problem ought to be dealt with in the context of a circular system of production.

---

[19] The education of early Austrian economists typically took place in Faculties of Law. They were therefore not particularly versatile in mathematics and yet had to cope with the growing importance of mathematics in economics (cf. the contributions of Gossen, Jevons or Walras). Some Austrian economists responded to this challenge by trying to make a virtue of necessity and opposed the use of mathematics in economics in terms of some "deep" philosophical reasoning. Menger does not belong to them: he knew some mathematics, especially financial mathematics, and was not principally opposed to the use of mathematics. In several recent works, Linsbichler has dealt with the kind of philosophy Austrian economists endorsed and the occasionally highly problematic attitude of some of them towards the use of formal methods in economics; see, in particular, Linsbichler [29]. A general antipathy towards the use of mathematics in economics, not uncommon amongst some "Austrian" economists, is clearly not helpful, given the problems these economists were concerned with.

[20] A joint-products framework would allow for "bads" or "discommodities", to use the terms Jevons had introduced. In such a framework negative prices make perfect economic sense, if bads or waste have to be disposed of in a costly manner: to the negative price of a bad corresponds the positive price of the disposal activity. In this context it is worth mentioning that Menger in one place (51–52 [95]) deplores that even "the most insignificant kinds of scrap (*Abfälle*) are … not available to us in such great quantities that we could not employ still greater quantities of them". The contention that there is an insufficient supply of waste comes as a surprise, unless one assumes implicitly that waste products are *not* bads, but actually goods at least for some people or in some employments.

In such a system "Not even the crudest tool is made without the help of another tool" [50: 115]. To this case he then turns and insists that in it Menger's grouping of prices in two strictly separate categories, one containing goods of first order and the other goods of higher orders, cannot be sustained. There are in particular "cost goods" (*Kostengüter*) [51: 108], which are both outputs and inputs. But how can one start from *given* values of certain products as outputs without involving also the values of the same kind of products as inputs? Do not *all* values have to be considered *unknowns* and determined *simultaneously*?

Secondly, and most interestingly, Wieser for a few sections [51: §§37–40] leaves the Austrian world of *ubiquitous scarcity* of all goods and enters the world of their *general reproducibility*. A simple version of the latter we have already encountered in von Thünen's work (see Sect. 6 above); a much more complex version is contained in John von Neumann's famous model of economic growth, which incorporates joint production, fixed capital and the problem of the choice of technique [48].

In a circular flow framework, Wieser maintains, there will be a positive rate of interest, if and only if more can be produced of various (possibly all) commodities than what is being used up of them as cost alias capital goods. That is, a positive rate of interest presupposes a *physical social surplus*. In this case, "a net return out of the total result has to be imputed to capital goods, *just as if they reproduced themselves directly together with a surplus*" [51: 130]; emphasis added). The surplus is the material substratum of interest alias profits. In physical terms, the general rate of interest (profits) is the ratio of the vector of commodities that constitute the surplus product in the numerator and the vector of commodities that constitute the capital advanced in the denominator [51: 139]. However, since surplus and capital do not in general exhibit the same commodity composition, it seems to follow that they cannot be directly compared, but rather have to be rendered commensurate with the help of the prices of commodities. Yet since these depend on the rate of interest, and the rate of interest depends on them, the prices of both commodities *and* the rate of interest can only be determined simultaneously. (A similar kind of argument we already encountered in the above with regard to Wicksell's reception of Thünen's theory of capital and interest.)

Wieser's reasoning may be said to echo the surplus approach of the classical authors, especially Ricardo and Robert Torrens, elaborated in the 1820s. However, contrary to what has just been stated, closer examination of the issue at hand shows that the rate of profits (interest) can be ascertained independently of prices. In order to do so, the system of production has to be hypothetically re-proportioned without affecting its mathematical properties in such a way that the vectors representing the surplus product and the capital advanced in its production are multiples of each other, that is, they are linearly dependent.

Let us now normalize a single-products system by setting gross outputs of all commodities equal to unity, then the system of production equations can then be written in the following way:

$$(1 + r)\mathbf{Ap} = \mathbf{p}.$$

Here $\mathbf{p}$ is the k-dimensional vector of prices in terms of some standard of value, $\mathbf{A}$ is the $k \times k$ input matrix giving the material inputs (means of production) augmented by the means of sustenance in the support of workers, and r is the general rate of profits. This is an eigenvalue equation with λ as the largest real eigenvalue of matrix $\mathbf{A}$ and $\mathbf{p}$ as the associated eigenvector. Obviously,

$$r = \frac{1 - \lambda}{\lambda}.$$

The general rate of profits is positive, if and only if $\lambda < 1$.

In this perspective, the general rate of profits depends on the technical conditions of production and real wages, reflected by matrix $\mathbf{A}$. The higher are real wages, the larger is λ and the smaller is the rate of profits. This is Ricardo's "fundamental law of distribution" with which he confirmed Smith's conjecture that for a given system of production there is an inverse relationship between the general rate of profits and the share of wages: "The greater the portion of the result of labour that is given to the labourer, the smaller must be the rate of profits, and vice versa" (Ricardo, *Works* VIII: 194). Ricardo's intuition was shown to be correct by numerous authors, including Georg von Charasoff at the beginning of the twentieth century and then John von Neumann [48] (see [20: 387–390 and 403–407]). It found its definitive expression in Piero Sraffa's construction of the Standard system, in which the rate of profits "appears as a ratio between quantities of commodities irrespective of their prices" [43: 22].

Implications of Wieser's findings. We may conclude by drawing the implications of Wieser's above reasoning for economic theory. First, a purely subjectivist perspective in the theory of value and distribution cannot be sustained. As Wieser's argument shows, in the conditions postulated relative prices and the rate of profits can be determined in purely objectivist terms. Secondly, a successivist approach has to give way to a simultaneous one. Menger's idea that the prices of consumption goods can be taken as essentially given from the outside by needy individuals, and the prices of inputs (capital goods, labour and so on) then determined via some imputation algorithm, leads nowhere. Third, the "quantity of capital" available in an economy, which consists of different kinds of capital goods, cannot be taken as given independently of, and prior to, the determination of the prices of commodities and the rate of profits: prices and the rate of profits (and the value of capital invested) have rather to be determined simultaneously. Therefore the rate of interest cannot be conceived as reflecting the relative "scarcity" of the endowment of capital, because it has already to be known in order to ascertain the magnitude of the latter. Fourth, Böhm-Bawerk's concept of the "average period of production", designed to aggregate heterogeneous capital goods, cannot generally be sustained, because it also depends on this rate. Abstracting from this fact by ruling out compound interest, as Böhm-Bawerk does, contradicts the assumption of competitive conditions, which is supposed to be analysed. Fifth, a positive rate of time preference, which Menger (128 [153]) postulates and Böhm-Bawerk then takes up, cannot be considered the "reason (*Grund*)" of a positive rate of profits (interest). As Wieser shows, the rate of profits can be determined without any reference to the rate of time preference.

Imputation and Euler's law. Finally, as regards a "solution" of the imputation problem in strictly marginalist terms, in a section entitled "The value of cooperating factors of production", Wieser [50: 177–179] assumed that the inputs of the various factors of production—x types of labour (l), y types of land (b) and z types of capital goods (c)—each input multiplied by its marginal product, would just add up to the amount of the i-th product produced, q_i. In obvious notation:

$$q_i = \frac{\partial q_i}{\partial l_{1i}} l_{1i} + \ldots + \frac{\partial q_i}{\partial l_{xi}} l_{xi} + \frac{\partial q_i}{\partial b_{1i}} b_{1i} + \ldots + \frac{\partial q_i}{\partial b_{yi}} b_{yi} + \frac{\partial q_i}{\partial c_{1i}} c_{1i} + \ldots + \frac{\partial q_i}{\partial c_{zi}} c_{zi}.$$

However, Wieser was not aware of the fact that this holds true only for linear homogeneous production functions, as the Swiss mathematician Leonhard Euler had shown, whose finding Knut Wicksell and Philip Wicksteed introduced into economics.

We now have a closer look at Menger's treatment of the problem of income distribution. Since according to him a solution of the problem of imputation was *uno actu* a solution of the problem of income distribution, and since his treatment of the former did not fare all too well with his Austrian acolytes, there is perhaps no reason to expect things to be different with regard to his theory of income distribution and especially his explanation of profits.

9 Menger's Theory of Income Distribution and Böhm-Bawerk's Criticism

Menger assumes throughout, at least implicitly, that all productive resources in the economy are continuously fully employed. Since consumers wish to satisfy their needs as best as they can, and since according to Menger the "*capacity* of human needs to *grow*" is "infinite" (38 [82–83]), there is no individual and a fortiori no collective point of satiation. The distributive variables are thus assumed to be explicable in terms of a single principle only: that of the relative scarcity of the corresponding factors of production. However, Menger does not present his argument in marginal productivity terms, either because he was not sufficiently familiar with the concept or because he saw no way how to integrate it into his overall construction.

Wages. Menger devotes barely three pages to the determination of wages. Wages are said to reflect the scarcity of labour, but no details are given what this means in well specified circumstances. While Antoine-Augustin Cournot and then Rau in the 1830s and early 1840s had already developed the concept of "demand curve", the determination of the wage rate in terms of demand and supply, which was soon to become the familiar workhorse of marginalist theory, is absent in Menger's book. The reader is only told that the services of labour are not always "goods or even economic goods simply because they are labor services; they do not have value as a matter of necessity" (150 [170]). This remark appears to be directed at the classical authors, but Menger fails to provide any evidence that these actually held the incriminated view. They did not. Ricardo had made this very clear:

> If a commodity were in no way useful,—in other words, if it could in no way contribute to our gratification,—it would be destitute of exchangeable value, however scarce it might be, or whatever quantity of labour might be necessary to procure it. (Ricardo, *Works* I: 11)

Menger adds to his above remark: "many labor services cannot be exchanged by the laborer even for the most necessary means of subsistence" (150–151 [170]). He concludes: "Neither the means of subsistence nor the minimum of subsistence of a laborer, therefore, can be the direct cause or determining principle of the price of labor services" (151 [171]). Again, to the best of my knowledge no classical author has ever disputed this. Menger's implicit criticism is tilting at windmills.

Rents. While Ricardo has variously been credited (together with Thomas Robert Malthus and Edward West) with having firmly established the principles of extensive and intensive differential rent, Menger attacks Ricardo frontally.[21] The value of land and of land services, he insists, is subject to the same "general laws" as all other goods and services. Menger even "protests" against Ricardo, who is said to have "mistakenly" considered differential fertility (or location) to be the true principle of rent, whereas the true principle is, as always, scarcity. He chastises Ricardo for allegedly claiming that in the case of land of uniform fertility and in that of marginal land there can be no rent: "It is evident rather that even the … least fertile pieces of land in a country where land is scarce would yield a rent, *a rent that could find no explanation in the Ricardian theory*" (147 [169]; emphasis added).

Also this criticism cannot be sustained. In the chapter "On Rent" Ricardo writes: "On the first settling of a country, in which there is an *abundance of rich and fertile land, a very small portion of which is required to be cultivated* …, there will be no rent" (Ricardo, *Works* I: 69; emphasis added). This defines the context within which intensive diminishing returns have to be discussed, but in the case referred to, due to a lack of sufficient effectual demand no intensive cultivation is yet needed. Ricardo states explicitly that the land under consideration is *not* yet scarce. In the sequel he then emphasizes that only land that actually *is* scarce will fetch a rent, because "no one would pay for the use of land, when there was an abundant quantity not yet appropriated, and, therefore, at the disposal of whosoever might choose to cultivate it" (ibid.).[22] One can only wonder how Menger arrived at this misconception.[23]

---

[21] The various aspects of Ricardo's contributions are dealt with in great detail in Kurz and Salvadori [24].

[22] Ricardo apparently assumed implicitly that all land was in private (or public) property. This could safely be assumed subsequent to the Enclosure Movement in Britain.

[23] Surprisingly, Schumpeter [40: 934] supports Menger's criticism of Ricardo in lukewarm terms, but then, even more surprisingly, turns around and insists that in his rent theory Ricardo "did not assert anything that was positively wrong".

Intensive diminishing returns and marginalism. Some further observations on the principle of intensive diminishing returns are useful, because, as has already been emphasized, marginalism consists essentially in an attempt to generalize the *principle of intensive rent* from agriculture and homogeneous land indiscriminately to all industries in the economy and all factors of production alike. While it treats different kinds of labour and different kinds of land differently (see [20: Chap. 11]), each yielding its own specific rate of remuneration, it sees the need to treat heterogeneous and reproducible capital goods in a homogeneous way, because in competitive conditions all capital goods tend to yield their owners a uniform rate of return on the supply prices of the capital goods. The real question therefore is this: Is the generalization possible or does it meet with insurmountable difficulties?[24]

Ricardo, in the *Principles*, focused attention on extensive diminishing returns and extensive rent, because they are empirically ubiquitous and analytically much less challenging than intensive diminishing returns. But he also dealt with the latter. While his analysis marked a considerable step forward, it was far from perfect; see the critical account in Kurz and Salvadori [20: Chap. 10, Sect. 6].

Menger, in his criticism of Ricardo, assumes that all available land is actually employed and is in fact scarce. However, there are good reasons to presume that this is hardly ever the case: just think of plots of land whose fertility is very low and cannot produce enough to feed those who cultivate it; we would have a situation similar to the one Menger contemplates in his discussion of wages above. In the hypothetical case of a huge area of homogeneous and fertile land and a relatively small population, it would make no sense to cultivate the entire land by distributing workers and other resources evenly across it. Cost-minimizing behaviour would rather prompt producers to employ only a part of the land and leave the rest idle, with the employed part cultivated in terms of the method that exhibits the highest labour productivity.

Menger mocks Ricardo's definition of rent as the compensation paid to the owner of land for the use of its "original and indestructible powers" (Ricardo, *Works* I: 69). Yet there are good reasons for Ricardo to have developed his argument subject to this premise. First, it allowed him to differentiate strictly between rents and profits, "for it is found, that the laws which regulate the progress of rent, are widely different from those which regulate the progress of profits, and seldom operate in the same direction" (Ricardo, *Works* I: 68). Ricardo in fact showed that when capital accumulates and the population grows, but there is no further technical progress, the rents of land tend to increase, whereas the general rate of profits tends to fall. Secondly, the above premise allowed him to treat land as an inexhaustible resource and to distinguish it sharply from exhaustible ones, such as coal or ores. Land could, of course, also get

[24] The emergence of the short-period variants of marginalist theory, temporary and intertemporal equilibrium theory, can be interpreted as an indication of the difficulties mentioned. However, this does not mean that capital theoretic problems affecting its long-period variant could be avoided in this way; see Petri [35: Chap. 8].

exhausted as a consequence of the way in which it is used, but for the sake of clarity the cases of inexhaustible and exhaustible resources ought to be strictly kept apart.[25]

We come now to the part of Menger's theory of income distribution—his theory of capital and profits—that was rejected even by his most ardent follower, Eugen von Böhm-Bawerk.

Profits. According to Menger, capital, that is, the aggregate of the produced means of production available in the economy, is "productive" and interest (alias profits) is the remuneration of its proprietors for the utilization of capital. Böhm-Bawerk [46] therefore called Menger's theory a variant of the "utilization theories" of capital and interest advocated earlier on by Jean Baptiste Say and the German economist W. B. F. Herrmann. A necessary condition for interest to be positive, Menger stresses, is that capital services are scarce. Menger, like Smith and Ricardo before him, focuses attention on competitive conditions. These enforce cost-minimizing behaviour, which Menger confirms. Interestingly, he does not also confirm that cost-minimizing behaviour involves a tendency towards a uniform rate of profits.

Menger refers to the value of goods of higher order as their value "*in der Gegenwart*", translated as "*present* value" (135 [158]; emphasis in the original). He adds: "The *present* value of these goods of higher order by themselves is therefore equal to the value of the respective product minus the value of the services of the capital employed" (135 [158]). This statement is difficult to understand, since Menger states that the "value of the services of the capital employed" does *not* constitute the value of the capital good itself. Yet according to the then already conventional net present value method of ascertaining the value of a stock of capital goods today, *all* future net payments of the services of the stock have to be discounted to the present. By excluding the services of capital goods from this, is there actually anything that Menger is in fact discounting? It was Böhm-Bawerk in *Geschichte und Kritik der Kapitalzinstheorien* [46], the first part of his *Kapital und Kapitalzins*, who spotted the confusion.

Apparently, he argued, to Menger capital represents two separate goods: "capital goods proper", on the one hand, and their "utilization", or services, on the other. While the prices of the capital goods are said to refer only to the capital goods proper, interest is taken to be a separate price paid as a compensation for their use. However, to consider the utilization of a capital good as an independent carrier of value in addition to the capital good itself involves, Böhm-Bawerk interjected, a serious misconception and actually amounts to double counting. The value of a capital good, he insisted, involves all the (discounted) payments for the future "utilizations" or services of it for the rest of its life. These services are inextricably tied to the capital good itself and do not exist independently of it. While Böhm-Bawerk sought to be as diplomatic as possible in order not to antagonize his peer, his final verdict was merciless: "the entire theoretical orientation that culminates in his [Menger's] work" on capital and interest was a failure [46: 195].

[25] See in this regard the formalization of Ricardo's theory of exhaustible resources and how it compares with the Hotelling Rule in Kurz and Salvadori [21, 23].

Several Austrian economists, including Schumpeter [40: 925], shared Böhm-Bawerk's criticism (but not necessarily also his alternative explanation; Schumpeter clearly did not), as did representatives of other currents of economic thought. The Russian mathematical economist Georg von Charasoff, for example, commented perspicuously on the intellectual conflict between the two Austrians: while Menger at first assumed that the value of the product and that of the inputs employed are equal to one another, he then realized to his great dismay that this left no room for profits. Yet his "belated attempt to explain the fact of profits was so clumsy that it was criticised unsparingly and dismissed by his most talented pupil and apostle" [47: XX]; see also the discussion in Kurz [11: 72–79].

Menger's theory of income distribution is beset by serious difficulties. The criticisms his pupils and then colleagues levelled at it drew a permanent wedge between him and them. As will be argued below, this was probably the main reason why Menger published very little especially on economic theory during half a century until he died in 1921; he did especially not accomplish to compose the remaining three volumes which originally were intended to complete his magnum opus, of which the *Grundsätze* were supposed to be only the "First, General Part"; and he did not even revise this part, as he had notified to friends shortly after its publication.

10 Menger on the Labour Theory of Value and on "Objectivism"

Menger criticizes the classical economists for advocating an objectivist approach to the theory of value and distribution and the labour theory of value, which is said to be misleading in several regards. Most importantly, it is said to get the causality in the theory of value wrong by starting from production and the labour expended, whereas it ought to start from the needs of consumers and their subjective estimations of things that are capable of satisfying them. It implies that the source of value, which the classical economists erroneously assumed to be the labour embodied in a product, resides actually in the estimation of a good by the consumer. The labour theory of value is also taken to wrongly assume that there is only a single factor of production, labour, whereas there are several. Since all classical propositions are said to presume the validity of the labour theory of value, and since in Menger's view the latter is fatally defective, classical economics cannot be sustained and has to be abandoned.

The classical economists on value. Menger's criticism evokes the following observations. First, it is a widespread myth, advocated not only by Menger, but also by followers of the classical authors, that the labour theory of value was the linchpin of the classical theory of value and distribution.[26] While on the surface it looks as if it was, close scrutiny shows that this is not so. Actually, none of the classical

[26] There are reasons to presume that this interpretation of the classical economists gained prominence with the publication of Marx's *Capital* and his claim that *the* source of all value is (abstract) labour, a claim not to be found in works of classical authors.

economists explained (natural) prices in *strictly* labour value terms other than in special historical or analytical circumstances. Adam Smith entertained such a theory only in the case of the "early and rude state of society" prior to the appropriation of land and the accumulation of capital, that is, for a little more than a single page in *The Wealth of Nations*. Outside that state relative prices are said to deviate from relative labour values, but Smith had relatively little to say about the magnitude of these deviations. Ricardo studied the problem of value in much greater depth and improved our knowledge about it considerably, but he, too, was unable to provide a fully comprehensive theory. He identified, however, the special cases in which the labour theory of value was able to explain relative prices in a correct way (see Sect. 6). Not possessed of a fully elaborate theory, he decided to rely on the labour theory of labour as a simple makeshift solution, which, he speculated, provided a reasonably good approximation to a correct theory. It was perfectly clear to him that relative prices and the general rate of profits do not only depend on the methods of production and therefore on the labour actually expended in production, but also on the distribution of the product between workers and capitalists. He saw that a good deal of the difficulties of the subject had to do with compound interest effects, which reflect the fact that production takes time.[27]

Secondly, while Menger rejects the labour value-based explanation of relative prices, he fails to show conclusively what is wrong with it. Since his own explanation does not get much beyond the proposition that prices equal all costs incurred and incomes paid in the course of the production of commodities, we are given no more than an adding-up scheme that is not much different from the one Adam Smith had proposed. Since Menger does not explain in sufficient depth how (relative) prices are determined in given circumstances, or what the corresponding (relative) labour values would be, he is unable to say whether the two are different, and by how much. Menger is of course right in criticizing vulgar applications of the labour theory of value in contemporary literature.

Ironically, other Austrian and marginalist authors, while also criticizing the classical economists, arrived at the result that in long-period equilibrium relative prices are *exactly* equal to the relative amounts of labour embodied in the various commodities. They thus arrived at a result that is orthogonal to Menger's contention. A brief look at their reasoning must suffice.

Austrian and marginalist economists on labour value-based equilibrium prices. In order to be able to maintain that in (long-period) equilibrium relative prices equal relative labour values, one or the other of the two special conditions that Ricardo had identified must apply: either (i) a zero rate of profits (interest) or (ii) equal input proportions across all operated processes of production (i.e., what in Marx is the case of "equal organic compositions of capital"). Böhm-Bawerk, for example, constructed his "average period of production"—the device by means of which he

[27] It therefore comes as a formidable surprise that many Austrian economists and especially Böhm-Bawerk, who rightly stressed the important role of time in all economic activities, supressed this role in the theory of value and distribution by assuming a zero rate of interest. We come back to this below.

sought to aggregate heterogeneous capital goods—by assuming a zero rate of interest, which ruled out compound interest effects. He did so even though his ultimate aim was to determine the actual rate of interest in the system, which was supposed to be larger than, or at most equal to, the given rate of time preference. His device amounted to using labour values as aggregators of capital goods, just as Thünen had done several decades earlier (see Kurz and Salvadori [20: 435–438]). Böhm-Bawerk's fierce criticism of Marx notwithstanding, his approach in the regard under consideration is essentially the same as Marx's, who, in determining the general rate of profits, ascertained the "organic composition of capital" of the economy as a whole in labour value terms (instead of in terms of prices of production).

A few examples should suffice to illustrate labour value-based reasoning in marginalist and Austrian economists. William Stanley Jevons [10: 182–189] concluded his respective disquisition by insisting that "we have proved that commodities will exchange in any market at the ratio of the quantities produced by the same quantity of labour" (182). Similar statements are to be found in Wieser [50: 159–160], Wicksteed [56: 717–718] Böhm-Bawerk [45: 329–330] and Clark [3: 390]. What these authors rejected was the special causal structure they attributed to the classical economists, not the putative results of their studies—the equality between relative prices and relative labour values. Ironically enough, several of the early marginalist authors were not aware of the fact that with a positive rate of profits this result does not hold unless Ricardo's condition (ii) is met.

Progress in economics is slow and occasionally what is called progress actually involves regress.

Menger and Marx. It is perhaps interesting to notice that both Menger and Marx advocated *essentialist* positions in the theory of value and distribution. The former traces value back to abstract utility, the latter to abstract labour. Both rejected explicitly the idea that commodities represent chemical or physical properties or physical real costs. Menger, in a passage in the *Grundsätze*, compared the quantities of goods of higher order that are combined to produce a good of first order with "chemical reactions, where only a certain weight of one substance combines with an equally fixed weight of another substance to yield a given chemical compound" (139 [162]).[28] With reference to the substitutability amongst certain inputs, Menger rejected the idea of fixed proportions of substances contained in a commodity. While this objection is obviously valid, it does not follow, as Menger appears to think, that substances do not matter at all when it comes to ascertaining relative prices. As soon as the problem of the choice of technique is decided and methods of production get operated to produce particular commodities, the substances involved can be known and the costs incurred in production make themselves felt in the prices realized in competitive markets.

Interestingly, just four years prior to Menger, Marx in volume I of *Das Kapital*, published in 1867, also opposed the idea that the use values productively consumed in the production of a commodity could determine exchange value in analogy to a

[28] In 1808 John Dalton published his atomic theory, which became the basis of chemistry. Several economists, including Maffeo Pantaleoni, investigated whether, and to what extent, this theory could bear fruit in the theory of production and value.

chemical compound whose weight is determined by the weights of the elements that constitute it. He wrote: "So far no chemist has ever discovered exchange value either in a pearl or a diamond" [31: 87]. While readers may join Marx in finding this jibe funny, both Marx and Menger failed to take properly into account the properties of the *system of production* as a whole, which revolves around the full socioeconomic metabolism and therefore the production of commodities by means of commodities. Taking these properties into account, commodities are seen to represent something real—some physical real costs—which cannot but have an impact on their prices.

Substances "embodied" in commodities. In his book, Menger stresses time and again that the values and prices of commodities have nothing to do with the "internal properties (*innere Eigenschaften*)" (60–61 [101]) of commodities. Value, Menger insists, is a purely subjective thing, it is "nothing inherent in goods, no property of them, nor an independent thing existing by itself" (86 [120–121]; see also 119 [146]). From this, does it follow that as regards the values of goods, the subjective estimations of consumers matter exclusively, whereas the objective characteristics of the goods that are effectually demanded and actually produced do not? Menger in fact infers this and insists with implicit reference to the classical authors: "Objectification of the value of goods, which is entirely *subjective* in nature, has nevertheless contributed very greatly to confusion about the basic principles of our science" (86 [121]).

Interestingly, Menger is unable to retain this radical position throughout his book. Consumers, he admits, care about what they get when buying a commodity—its "inner texture" (*innere Beschaffenheit*) (115 [141]), as he puts it. And if in the course of the process of civilization, as Menger is convinced, imaginary needs and goods are gradually losing importance, this can safely be assumed to mean that humans improve their understanding of the inner texture of goods. Menger in fact says precisely this:

> As a people attains higher levels of civilization, and as men penetrate more deeply into the true constitution of things [*das wahre Wesen der Dinge*] and their own nature [*sic*], the number of *true goods* becomes constantly larger, and ... the number of imaginary goods becomes progressively smaller. It is not unimportant evidence of the connection between accurate knowledge [*wahre Erkenntnis*] and human welfare that the number of so-called imaginary goods is shown by experience to be usually greatest among peoples who are poorest in true goods. (4 [53–54]; emphasis added)[29]

However, much more importantly, producers can be expected to know the amounts of particular services and substances that are needed, and paid for, per unit of output, in the production processes they operate. They are bound to care for the "*causal connections* between goods" (*Causal-Zusammenhang der Güter*) and seek "to discover the economic laws to which they are subject" (7 [56]). Whatever the

[29] The following warning seems, however, to be necessary: better information about the (real) characteristics of commodities does not *eo ipso* lead to better decisions. For example, a given (and incomplete) level of information about the real characteristics of two commodities that are substitutes may nevertheless result in the choice of the (objectively) superior good. Partially improving the level of information of the consumer with regard to the characteristics of the two commodities may distort the consumer's ranking and lead to a choice of the inferior good. Partially more and better information therefore need not lead to (objectively) better decisions.

subjective estimations of consumers, whatever the reasons that prompt them to buy certain goods rather than others, what matters with regard to value, and what matters exclusively, is which goods they *actually do buy*. In normal conditions, to which Menger refers, the factual decision to purchase certain commodities rather than others activates the system of production and decides about which processes will be operated and which inputs will be productively consumed. These inputs the classical economists interpreted as "transmigrating" (wholly or partly) into the commodities produced and as constituting their physical contents. The trajectory the economy is assumed to follow according to Menger is towards a "*general* economic situation" (240 [248]), in which people are fairly well informed, information differentials are modest, markets are well "organized" and institutionalized, and competition is close to "complete" on both sides of the market (ibid.). In such conditions, Menger opines, "effective prices" will obtain. These are "economic prices" that cover all costs of production. "Speculators" (*Arbitrageurs*) (243 [251]) are said to quicken the movement to such a state of affairs. The set of relative prices brought about by competition and cost-minimizing behaviour therefore cannot but reflect the quantities of matter and energy (substances and services) that in the course of production get transformed from one configuration to another. Menger in fact emphasizes that "it is necessary in the manner of *all other empirical sciences*, to attempt to classify the various goods according to their *inherent characteristics* [*innere Gründe*]" (7 [56]; emphases added). Some objectification of goods is therefore both needed and possible, and actually taking place in everyday life.

11 Concluding Observations

The main propositions put forward in this chapter are the following. First, the chapter recalls that the concepts of marginal productivity and marginal utility had already been elaborated in the first half of the nineteenth century. Therefore one can hardly speak of a "marginal revolution" *strictu* sensu championed by Jevons, Menger and Walras. Secondly, Menger cannot really be counted as belonging to the "triumvirate" mentioned—there is no *unité de doctrine* discernible. The views of the three authors differ from one another in important respects—with Menger as the odd man out. While he endorses a variant of marginal utility theory as it had been developed by the German Use Value School, most notably Rau, he provides neither a clear concept of the other building block of marginal theory, marginal productivity nor, of course, how the two could be integrated into a single, coherent whole. While other marginalist authors (and especially Jevons) prided themselves with having replaced Ricardo's surplus-based explanation of the general rate of profits by a marginal productivity of capital approach, Menger advocated a version of the old utilization theory of profits, which, however, is marred by inconsistencies and contradictions. In view of the fact that the marginal productivity explanation of profits was widely considered to be the centrepiece and gem of the marginal revolution, it is particularly difficult to count Menger as a marginalist "innovator". Third, and closely related, one can only wonder what made Menger think to have been able to surpass the classical theory of value and distribution without being able to show that its theory of profits was wrong and elaborating a convincing alternative to it. The need to do precisely this was well understood by authors such as Jevons, Walras, Böhm-Bawerk and Wicksell. They sought to accomplish this in terms of a generalization of the principle of intensive diminishing returns and intensive rent from homogeneous land to all other factors of production, including capital, which, in conditions of free competition, was supposed to yield its proprietors a uniform rate of return on the capital invested. Capital owners were therefore seen to participate in the sharing out of profits like shareholders of a joint stock company, each in proportion to the size of the share he or she possesses. The overall "quantity of capital", consisting of numerous heterogeneous capital goods had therefore to be given as a single magnitude, expressed as a sum of value in terms of some numéraire. This destroyed, however, the targeted perfect analogy between the three types of factors—labour, land and capital—as Wicksell bemoaned. And it led to a situation, in which the quantity of capital could not be given independently of the general rate of profits (interest), and whose value changed with a change in this rate, although physically it remained unaltered.

Menger's attempt to attribute predominance to needy individuals in terms of his concept of "prospective prices", which consumers are supposed to ascertain entirely independently of the rest of the system, confronts him with insurmountable difficulties. His case shows that it is futile seeking to elaborate a theory of competitive value by starting from the sphere of consumption and telescoping the principles taken to rule in it to the sphere of production. What matters in the last instance as regards relative prices and income distribution is the metabolism of society as a whole and the "laws" governing the system of consumption and production actually in use. The needs and wants of people play an important role in it, but are to a significant extent shaped by society and the groups and classes to which people belong. Asymmetric power relationships between different fractions of society exert their influence on the distribution of the product and together with the available technological knowledge determine relative prices of commodities.

The criticisms which especially Böhm-Bawerk and Wieser levelled at Menger's treatment of the "imputation problem" and his theory of capital and interest can safely be assumed to have deeply affected Menger. Böhm-Bawerk's attempt at appeasement by claiming that his own theory of capital and interest was fully in the tradition of Menger's, must have looked to the latter like mockery. As Schumpeter famously recalled, Menger "severely condemned it from the first. In his somewhat grandiloquent style he told me once: 'The time will come when people will realize that Böhm-Bawerk's theory is one of the greatest errors ever committed'" [40: 847, n. 8]. Even in his obituary of Böhm-Bawerk, who passed away in 1915, Menger's anger and bitterness are still palpable.

To conclude, we might ask: Was Menger's inability to revise the first edition of his book and compose the other three volumes he had in mind the result of what must have been to him a painful intellectual defeat, inflicted upon him by his closest followers? It is safe to assume, I think, that he eventually realized that Böhm-Bawerk and Wieser's objections could not be refuted. Was there any possibility to remedy the blunders he had committed and salvage at least his overall construction? And if the answer was yes, could he elaborate a theory that stood up to what Böhm-Bawerk and Wieser, both benefitting from some of his ideas, had accomplished in the meantime? Or was he the victim of a second mover advantage seized by his former students? There is reason to think that he was. His article "Zur Theorie des Kapitals" of 1888 (reprinted in [33], vol. III) may be seen as a half-hearted attempt to strike back. Menger appears to have understood that the game was over.

References

1. Avtonomov, V., Hagemann, H. (eds.): Russian and Western Economic Thought. Mutual Influences and Transfer of Ideas. Springer, Berlin (2022)
2. Chipman, J.: German Utility Theory: Analysis and Translations. Routledge, Abingdon and New York (2013)
3. Clark, J.B.: The Distribution of Wealth. A Theory of Wages, Interest and Profits. Macmillan, New York (1899)
4. Engels, F.: Umrisse zu einer Kritik der Nationalökonomie. In: Deutsch-Französische Jahrbücher. 1./2. delivery, Paris. Reprinted in MEGA (Karl Marx/Friedrich Gesamt-ausgabe), vol I/2, pp. 467–494. DeGruyter, Berlin (1844)
5. Garegnani, P.: Il capitale nelle teorie della distribuzione. Giuffrè, Milan (1960)
6. Garegnani, P.: On a change in the notion of equilibrium in recent work on value and distribution. In: Brown, M., Sato, K., Zarembka, P. (eds.) Essays in Modern Capital Theory, pp. 25–45. North-Holland, Amsterdam (1976)
7. Gossen, H.H.: Entwickelung der Gesetze des menschlichen Verkehrs, und der daraus fließenden Regeln für menschliches Handeln. Friedrich Vieweg & Sohn, Braunschweig (1854)
8. Hayek, F.A.: Einleitung (C. Menger (ed.)) (1968), pp. VII–XXXVI
9. Hicks. J.: Some questions of time in economics. In: Tang, A.M., et al. (eds.) Evolution, Welfare and Time in Economics. Essays in Honor of Nicholas Georgescu-Roegen. Lexington Books, Lexington (1976)
10. Jevons, W.S.: The Theory of Political Economy. Macmillan, London ([1871] 1965). Cited after the reprint of the 5th edn. Kelley, New York (1965)
11. Kurz, H.D.: Auf der Suche nach dem „erlösenden Wort": Eugen von Böhm-Bawerk und der Kapitalzins. In: Schefold, B. (ed.) Eugen von Böhm-Bawerk, Geschichte und Kritik der Kapitalzins-Theorieen. Verlag Wirtschaft und Finanzen, Düsseldorf, pp. 45–110 (1994)
12. Kurz, H.D.: Über die Knappheit und eine missglückte Analogie zwischen Arbeit, Boden und Kapital: Thünens Theorie der Produktion und Verteilung. In: Rieter, H. (ed.) Johann Heinrich von Thünen als Wirtschaftstheoretiker. Studien zur Entwicklung der ökonomischen Theorie, vol. XIV, pp. 115–151. Duncker & Humblot, Berlin (1995)
13. Kurz, H.D.: Wicksell and the problem of the "missing" equation. Hist. Polit. Econ. **32**(4), 765–788 (2000)
14. Kurz, H.D.: Adam Smith on markets, competition and violations of natural liberty. Camb. J. Econ. **40**(2), 615–638 (2015)
15. Kurz, H.D.: Economic Thought: A Brief History. Columbia University Press, New York (2016)
16. Kurz, H.D.: German and Austrian schools. In: Faccarello, G., Kurz, H.D. (eds.) Handbook on the History of Economic Analysis. Schools of Economic Thought, vol. 2, pp. 251–273. Edward Elgar, Cheltenham and Northampton (2016)
17. Kurz, H.D.: The theory of value and distribution and the problem of capital. Eur. J. Econ. Econ. Policy Interv. **17**(2), 241–264 (2020)
18. Kurz, H.D.: Re-reading Carl Menger's *Grundsätze*. Another book that "cries out to be surpassed". Eur. J. Hist. Econ. Thought **29**(5), 877–919 (2022)

19. Kurz, H.D.: Menger's construction under scrutiny. In: On the modernity of Carl Menger: criss-cross views. Rountable conversation. Eur. J. Hist. Econ. Thought **29**(5), 967–992 (2022)
20. Kurz, H.D., Salvadori, N.: Theory of Production. A Long-Period Analysis. Cambridge University Press, Cambridge (1995)
21. Kurz, H.D., Salvadori, N.: Classical economics and the problem of exhaustible resources. Metroeconomica **52**(3), 282–296 (2001)
22. Kurz, H.D., Salvadori, N.: One theory or two? Walras's critique of Ricardo. Hist. Polit. Econ. **34**(2), 365–398 (2002)
23. Kurz, H.D., Salvadori, N.: The 'classical' approach to exhaustible resources: Parrinello and the others. In: Kurz, H.D., Salvadori, N. (eds.) Revisiting Classical Economics. Studies in Long Period Analysis. Routledge, London, pp. 304–316 (2015)
24. Kurz, H.D., Salvadori, N. (eds.): The Elgar Companion to David Ricardo. Edward Elgar, Cheltenham and Northampton (2015)
25. Kurz, H.D., Salvadori, N.: Competition, Value and Distribution in Classical Economics. Studies in Long-Period Analysis. Routledge, London (2022)
26. Kurz, H.D., Sturn, R.: Wiesers "Ursprung" und die Entwicklung der Mikroökonomik. In: Schefold, B. (ed.) Vademecum zu einem Klassiker der Österreichischen Schule, pp. 59–103. Verlag Wirtschaft und Finanzen, Düsseldorf (1999)
27. Kurz, H.D., Sturn, R.: Die größten Ökonomen: Adam Smith. UVK Lucius, Konstanz and Munich (2013)
28. Lancaster, K.: A new approach to consumer theory. J. Polit. Econ. **74**(2), 132–157 (1966)
29. Linsbichler, A.: Philosophy of Austrian economics. In: Reiss, J., Heilmann, C. (eds.) The Routledge Handbook of the Philosophy of Economics, pp. 169–185. Routledge, Abingdon (2021)
30. Marshall, A.: Principles of Economics, 1st edn. (1890), 8th edn. (1920). Reprinted, reset (1977) ([1890] 1977). Macmillan, London
31. Marx, K.: Capital. A Critique of Political Economy, vol. I. Lawrence & Wishart, London ([1867] 1954)
32. Menger, C.: Grundsätze der Volkswirthschaftslehre. Wilhelm Braumüller, Vienna (1871). Reprinted in Menger, vol. I (1968)
33. Menger, C.: Gesammelte Werke (Hayek, F.A. (ed.)), Four vols. J. C. B. Mohr, Tübingen (1968)
34. Menger, C.: Principles of Economics (Transl. Menger (1871) by Dingwall, J., Hoselitz, B.F., Glencoe, I. (eds.)). Free Press, New York (1981)
35. Petri, F.: Microeconomics for the Critical Mind. Mainstream and Heterodox Analyses, two vols.. Springer, Cham (2021)
36. Priddat, B. (ed.): Wert, Meinung, Bedeutung. Die Tradition der subjektiven Wertlehre in der deutschen Nationalökonomie vor Menger. Metropolis, Marburg (1997)
37. Rau, K.H.: Lehrbuch der politischen Ökonomie, vol. 1. Grundsätze der Volkswirthschaftslehre, 2nd edn. (1833), 3rd edn. (1837), 4th edn. (1841). C. F. Winter, Heidelberg 1st edn. (1826)
38. Ricardo, D.: The Works and Correspondence of David Ricardo (edited by Sraffa, P., with the collaboration of Dobb, M.H.), 11 vols. Cambridge University Press, Cambridge. In the text cited as Works, followed by the volume number (1951–1973)
39. Schumpeter, J.A.: Theorie der wirtschaftlichen Entwicklung. Duncker and Humblot, Berlin (1912). English translation 1934
40. Schumpeter, J.A.: History of Economic Analysis. Oxford University Press, New York (1954)
41. Smith, A.: An Inquiry into the Nature and Causes of the Wealth of Nations, two vols. In: Campbell, R.H., Skinner, A.S. (eds.) The Glasgow Edition of the Works and Correspondence of Adam Smith, vol. II.1 and II.2. Oxford University Press, Oxford. In the text referred to as WN ([1776] 1976)
42. Sraffa, P.: Introduction. In: Ricardo (ed.) Works, vol. I, pp. xiii–lxii (1951)
43. Sraffa, P.: Production of Commodities by Means of Commodities. Cambridge University Press, Cambridge (1960)
44. Streissler, E.: Carl Menger, der deutsche Nationalökonom (in Priddat, B.P. (ed.)), pp. 33–88 (1997)

45. von Böhm-Bawerk, E.: Wert, Kosten und Grenznutzen. In: Jahrbücher für Nationalökonomie und Statistik, Third Series **3**(58), 321–367 (1892)
46. von Böhm-Bawerk, E.: Kapital und Kapitalzins. Erste Abtheilung: Geschichte und Kritik der Kapitalzins-Theorieen. Wagner, Innsbruck (1884); Capital and Interest. Macmillan, London. English translation (1890)
47. von Charasoff, G.: Das System des Marxismus: Darstellung und Kritik. H. Bondy, Berlin (1910)
48. von Neumann, J.: A model of general economic equilibrium. Rev. Econ. Stud. **13**, 1–9 ([1937] 1945). Originally published in German as "Über ein ökonomisches Gleichungssystem und eine Verallgemeinerung des Brouwerschen Fixpunktsatzes". Ergebnisse eines mathematischen Kolloquiums **8**, 73–83 (1937)
49. von Thünen, J.H. ([1826] 1990). Der isolierte Staat in Beziehung auf Landwirtschaft und Nationalökonomie (Lehmann, H., in collaboration with Werner, L. (eds.)). Akademie-Verlag, Berlin
50. von Wieser, F.: Über den Ursprung und die Hauptgesetze des wirthschaftlichen Werthes. Alfred Hölder, Vienna (1884)
51. von Wieser, F.: Der Natürliche Werth. Alfred Hölder, Vienna (1889)
52. Walras, L.: Elements of Pure Economics. Irwin, Homewood, Ill ([1874] 1954)
53. Weber, M.: The Protestant Ethic and the Spirit of Capitalism (Transl. by T. Parsons). Routledge, London ([1930] 2001)
54. Wicksell, K.: Value, Capital, and Rent (Transl. by S.H. Frowein). Allen & Unwin, London ([1893] 1954)
55. Wicksell, K.: Lectures on Political Economy (Transl. by E. Classen.), vol. 1. George Routledge & Sons, London ([1901] 1934)
56. Wicksteed, P. H.: Das Kapital: A Criticism. To-day, October 1884, pp. 388–411 ([1884] 1999). Reprinted 1999
57. Young, A.: Increasing returns and economic progress. Econ. J. **38**(152), 527–542 (1928)

Viewing William Stanley Jevons as a Thinker on Social Issues in the Late Victorian Period in Britain

Aiichiro Mogi

Abstract This chapter sheds light on how Jevons approached practical social issues, rather than reviewing Jevons in the evolution of modern economics. Jevons was originally a scientific scholar, and later became a theorist and thinker of political economy. The chapter touches upon Jevons's background, Victorian social issues, Jevons's social programmes, and his viewpoint. The chapter also discusses Jevons as a harbinger of management science.

1 Introduction

William Stanley Jevons was a Victorian "polymath"[1] whose interests covered from chemistry and meteorology to mathematics and formal logic. He has usually been recognised as one of the three contributors to the Marginalist Revolution of economics. Jevons' contribution to this economic subject has been understood from the way that he analysed decisions of economic agents in terms of marginal increments of utility. Economic agents whether in their role as consumers, producers, workers or others, came to be seen as subjects of maximising utility or fulfilling objective functions, away from the labour theory of value of the classical economists.

This chapter rather sheds light on how Jevons approached social problems, particularly social and labour issues on which most of thinkers of the time inevitably took serious concern. For Jevons this topic was a persistent theme since student days in London. Once becoming an academic, he gave lectures at labour union meetings and finally wrote *The State in Relation to Labour* [19], and *Methods of Social Reform* [20] which was posthumously published.

Jevons was a scientific scholar, who first trained as a chemist, and then studied calculus in mathematics and formal logic. However, whenever studying topics in real life, Jevons started from facts found in the field rather than just thought or ideas

[1] Maas [26, p. 26].

A. Mogi (✉)
Former Lecturer, Ritsumeikan Asia Pacific University, Beppu, Japan
e-mail: mogi96@mars.dti.ne.jp

T. Maruyama (ed.), *Marginal Revolution in Economics*, Monographs in Mathematical Economics 6, https://doi.org/10.1007/978-981-99-4342-5_4

of the past. This sort of observation-learning began during his time in Australia. Tackling the topic of *The Coal Question* [16], Jevons first explored data on mining and consumption of resources.

However, as a theorist, on the one hand, Jevons persistently supported the market determination view on wage-setting and opposing industrial action for wage bargaining. On the other hand, Jevons had been sympathetic to working class people who suffered from public policies backed by authoritative political economists. He made efforts to find better policies and rendered them in his opportunities for public speaking. My research interest rests on whether there is consistency between the theorist position and practical judgements. From that point of view, this chapter is a result of the pursuit of what links Jevons had that connected these lines of thought. My conclusion is that he was a forerunner of today's economics with many incomplete ideas;[2] he then gradually changed his approach through his apprehension of the real problems, though keeping theoretical underpinnings. It is also noteworthy that he was a harbinger of management science to which he contributed by way of adding a managerial point of view.

Part of this chapter summarises Victorian reform-minded thinkers on the social issues: From Carlyle and John Stuart Mill to John Ruskin, and Jevons' contemporaries like William Morris and Thomas Hill Green. The chapter touches upon the difference and commonality between such thinkers and Jevons.

The latter part of this chapter discusses Jevons's arguments on social issues, his programmes of social reform in the areas of provision of public goods, finance and tax, and trade union issues. Lastly, this chapter takes the view that he was far-sighted in his ideas regarding the future of working society, particularly when he supported cooperatives based on worker–manager–owner integration. In 2014, a newly created body of economic entity named the *Community Benefit Society* based on cooperatives for community building, a counterpart of the *Community Interest Company* within the concept of the private companies act, was created in the UK. This seems to be a realized version of Jevons' idea.

The chapter aims not only to convey another look at Jevons, who had belief for societal development in common with Victorian reformers but also to identify his inner-connectivity.

[2] Partly because of Jevons's comparatively short life, we could not see potenttial accomplishments in several pioneering areas.

2 Sketch of Jevons's Accomplishments, His Personal Background,[3] and the Historical Context

2.1 Jevons's Accomplishments

William Stanley Jevons is

> generally known as one of the 'fathers' of the so-called marginal revolution in economics of the last decades of the 19th century. In his *Theory of Political Economy* [17], with its 'mechanics of utility and self-interest', he analysed decisions of economic agents by means of the calculus, in terms of deliberations over marginal increments of utility. Economic agents—whether in their role as consumers, producers, or others—came to be seen as maximizing utility functions. Jevons is thus commonly considered to have broken with the labour theory of value of the classical economists. Value came to be identified with exchange value, and Jevons identified this with what we now call marginal utilities, not with costs of production. [. . .]
>
> Jevons is also remembered for his innovative contributions to the empirical (statistical) study of the economy. He much favoured the use of graphs to picture and analyse of statistical data.
>
> [27, p. 616]

In that respect Jevons was "heavily indebted to Victorian ways of practicing the natural sciences" and he tried to apply "mechanical analogies to understand the world. Just as William Thomson, the later Lord Kelvin, argued that the making of a mechanical model was the ultimate 'test' of intelligibility of a natural object, so Jevons relied on mechanical models to understand the social world" (ibid., p. 616). Jevons introduced "index numbers to make causal inferences about economic phenomena such as changes in the value of gold following the gold discoveries in California and Australia" (ibid., p. 616). Retrospectively, Jevons did make initial contributions to the field of positivistic economics where even in the 21st century his works were considered to have potential impact.

Lionel Robbins[4] clearly recognised that "Jevons's analysis was not just rooted in Bentham's hedonism, but in psycho-physiological explanations of human conduct. Jevons's new theory of value was not a theory of rational choice. The common perception of Jevons's utility theory as a theory about individual *decision*-making is therefore not quite accurate" (ibid., p. 616). In other words, Robbins found differences in Jevons's tenacity on the human physiological grounding for building the utility theory compared with counterparts of the Marginalist Revolution, who mainly built their theses on thought experiment.

This chapter touches briefly upon Jevons's personal character traits connecting his belief in human selfishness and valuing individualism. Before this a short sketch of Jevons's early life is presented.

---

[3] Explanations of Jevons's works and personal life are indebted to Könekamp [24], Peart [34], and Maas [25]

[4] Lionel Robbins [38, pp. 1–17]

2.2 *Jevons's Personal Background*

Jevons was born in 1835 as the ninth child in a wealthy Unitarian family. In 1882 he drowned near Hastings in England at the age of only 47. He was survived by his wife Harriet and three children. A good biographical essay is Black and Könekamp's introduction in volume 1 of the *Papers and Correspondence of Willam Stanley Jevons* [22]*)*. Jevons's father, Thomas Jevons, was an iron merchant; his mother, Mary-Ann Roscoe, was the daughter of William Roscoe, a Liverpool banker and important collector of art. His family "formed part of the self-confident Unitarian circles in the heartland of the industrial revolution that shared a belief in rational argument and advancement of science to promote the common good" [27, p. 617]. After finishing a good scientific education, Jevons enrolled at University College London to study mathematics and chemistry. Jevons's youth was not without difficulty. His mother died in 1845. His father's iron business went bankrupt in 1848. His family's financial difficulties affected Jevons's early intellectual career. Although Jevons performed very well at educational subjects, winning a gold medal at chemistry, he had to interrupt college courses but got a good job offer from the Australian Mint as an assayer. He sailed to the new continent in 1854 and stayed there for five years. Jevons's Australian period was crucial in developing his research attention[5] mentioned in the latter part of this chapter.

Samuel Bostaph wrote a very stimulating article[6] on Jevons's intellectual development. By the spring of 1856 in Australia,

> Jevons had identified a particular form of self-interest as a key human motivational principle and that he had developed both a serious interest in economic problems and a particular methodological approach to their solution—an approach combining partial equilibrium analysis with factual application. His conclusions concerning self-interest would later provide mainspring of his presentations of 'Economy as a Calculus of Pleasure and Pain' [17, p. vi and 24], while his interest in economic problems and methodological presuppositions resulted in a decision at the close of 1856 to 'write [a] work on 'Formal Economics'[7] and led to a more serious study of political economy in 1857.
>
> [3, pp. 603–604]

It might fairly be concluded that Jevons's own introspection, stimulated by his providential reading of Bulwer-Lytton's *My Novel*,[8] was part of an important influence on his formation of a self-interest concept. The following subsection refers to Bostaph's same article.

Jevons had become increasingly preoccupied with moral questions during the extended period of self-reflection and self-examination he experienced in Australia

---

[5] Maas [27, p. 617].

[6] Bostaph [3, pp. 603–609].

[7] Jevons [22, Volume 1, p. 28; Volume 7, p. 116].

[8] Edward Bulwer-Lytton was a popular Victorian writer and politician. His works reflect the Victorian societal atmosphere of valuing worldly success through industriousness, on which he cast a satirical touch. His works include *Pelham: or The Adventures of a Gentleman (1828)* and plays such as *Money* (1840).

from 1855 to the middle of 1856. Particularly of interest to him was his own selfish nature, a nature that he abhorred[9] and then accepted, given "the inalterability of a man's disposition and character, and of my own in particular".[10] It was at this point that he received two stimuli that focused his conception of "selfishness". You can see that Jevons himself confessed to the occasion of the reading of Bulwer-Lytton's novel.

In the book Bulwer-Lytton seeks to exemplify the aphorism, "Conduct is Fate" (meaning that the morality of one's actions determines one's ultimate rewards in life) in the lives of two characters, Leonard Fairfield and Randal Leslie. A highly moralistic tone pervades everything associated with those characters and all their actions. Both believe in the saying, "knowledge is power" (purportedly a shorthand version of a Baconian aphorism) and seek their fortunes by acquiring and applying knowledge. The key to their separate development as characters contributing to the thematic structure of the novel lies in what sort of knowledge each seeks and how he applies it. Randal studies men to learn their weakness in order to acquire power over them, "for books may give fame after death, but men give us power in life". Leonard seeks to train his mind and to learn moral virtue in order to rise above his peasant beginnings and become an author. He goes on to fame, fortune, and happiness, while Randal ends disgraced, destitute and an alcoholic teacher of classics in a boys' preparatory school.

Jevons's summary and analysis of that book in July 1856[11] is of great interest in any account of his intellectual development. His focus on the "mystery" of selfishness in the character of Leonard and the consequent purpose of intellectual ambition is both accurate and insightful, and the character in the novel clearly spoke to Jevons's concern with his own character, motives, and goals, and runs consistent with his past thinking on these topics, as reflected in his journal and letters. Further, there are considerable similarities between both the situation and character of Leonard and those of Jevons.[12]

2.3 *Jevons's Life and the Historical Context*

The Victorian era effectively covers Jevons's lifetime. His adulthood coincided with "the age of equipoise"[13] as some historians depict, between 1850 and 1870, during

---

[9] Jevons [22, Volume 1, p. 112].

[10] Ibid., p. 110.

[11] Ibid., pp. 130–132.

[12] Also see Michael V. White's rejoinder to the Bostaph's proposition [3].

[13] Cannadine [6, p. 298] and Matthew [28, p. 8]. This period is symbolised by the word "balance". Therefore such state was achieved in the way that because the fears of the earlier part of this century were passed, the modest modernisation of the political system made by the Whig governments through the Reform Act of 1832 and other legislation in the 1830s had been adequate for the time being. Challenges of the society remained, but so far several forces from the right and the left were balanced.

the long reign of Queen Victoria. In the 1820s and 30s there were responses to the strong protest from anti-protectionists of trade, miserable working conditions under the Industrial Revolution, and hardening worker–capitalist conflicts; important political and social reforms were carried out, such as the Repeal of Corn Laws, First Reform Act, Poor Law Amendment, and several Factory Acts. Most of them were introduced in the early years of Victoria's reign, and gradually took effect and changed society. The 1850s and 60s were the finest period of her reign. The Great Exhibition of 1851 was held in London; it symbolised the most progressive stage of of the UK's economic, industrial, and scientific global ambition. Jevons visited the Exhibition[14] several times, perhaps because of his interest in new things. During this period successive reforms, such as the Second Reform Act (household suffrage replacing the £10 income requirement with a simple occupancy), Public Health Act, Education Act, Civil Service Reforms, etc., were also introduced. In the remaining period of Queen Victoria's reign Britain's economic strength became equalled or partly surpassed by other competing countries, such as the United States and Germany.

Over the Victorian period the centre of society gradually moved from the old establishment of landed interests and patronage to the industrialists and workers for good reasons. The principles of conduct of the industrialists were based on liberalism, fair competition, and the markets. It seems that industrialists were opposed to workers in terms of distribution of value added, but in essence both needed each other. In the mid-19th century the growth of the national economy improved the standard of living of the working classes. Also owing to the election reform, through trade unions and several types of associations, the working classes came to have a say in political processes, not directly, but via approving liberal politicians. Thinkers of a reformist mind, such as to be quoted in the next section, supported such tendencies by way of proposing opinions. Reformers' concerns covered a very wide range of topics, including sustenance of political liberalism,[15] poor people's living conditions, public health, infrastructural development and other important public policy areas. Extending education to working class people is another important agenda which Jevons had been particularly eager to tackle. The general ethos behind industrialists, it is fair to say, was the belief in "self-help", as propagated by Samuel Smiles's 1859 work of same title.[16] Jevons's early beliefs were in the same vein.

So far Jevons's personality had been built by the Victorian human-autonomous, self-help view that more or less filled the minds of middle-class people. For him that view was introspectively strengthened by his persistent will. That is quite akin to his approach to the subject of study upon Benthamite hedonism. But the Victorian

[14] When Jevons visited it at the age of 16, he called the Exhibition the "place of all places", with the sense of wonder and pride in the greatness of Britain which all Englishmen then felt (Jevons [22, Volume 1, p. 51].

[15] Among referred thinkers in this section, William Morris was exceptional who formally supported socialism, however which was different from continental and revolutionary Socialism. J. S. Mill was well known to be recognised as a sympathiser to socialism.

[16] Perkin [35, p. 225].

period had different streams of thought which had entered into the intellectual arena through deep understanding of the reality of the society of that time.

3 Perceptions of Social Problems by Victorian Thinkers

Although the Victorian period in Britain witnessed the height of industrial prosperity to which capitalism contributed, at the same time the frame of capitalism as a system caused several social issues including poverty and disparity among classes, and involuntary unemployment along severe business cycles. This section introduces five eminent Victorian thinkers who stated their criticism of capitalism and rendered thoughts on the future of society.

By the early 1830s the concept of Political Economy became indistinguishable from the idea identified with those of advocates of materialism, wealth accumulation, free trade, and unrestrained economic competition. Such exponents of a "dismal science" in Carlyle's phrase became popular understanding like the determinate iron laws refusing involvement by human agency. In response to such a system two major lines of criticism emerged. According to Elisabeth and Richard Jay,

> First, Political Economy had been developed on the presupposition that the moral sciences, including economics, could be studied according to the Newtonian paradigm for the physical sciences of universal mechanical laws. Whether such an abstract deductive system was applicable to the works of human nature is open to question. Did the concept of "economic man" as an egoistical, rational calculator of material interests ignore motives of duty and service which lay behind even business transactions?
>
> [15, p. 7]

The second line of criticism was aimed at "Political Economy's moral bias towards secular hedonism. … [T]he close connection between most leading economists and the Bentham-Mill school of utilitarianism appeared to confirm the subject as atheistic and materialist. It could be charged with dissolving the distinction between morality and self-interest, and, in Carlyle's famous phrase, reducing all human life to the 'nexus of cash relation'" (ibid., p. 8). Although "hedonism as defined by the early utilitarians like Bentham was a distinctive ethical system concerned with enhancing human welfare and rationality" (ibid., p. 8), J. S. Mill was "at pains to distance the subject from identification with any specific class interest, and developed a radical attack upon the restricted quality of life under capitalism" (ibid., p. 8). We discuss some representative Victorian people who resisted the on-going one-dimensional trait of society: Thomas Carlyle and J. S. Mill, then John Ruskin, Thomas Hill Green, and William Morris.

3.1 *Thomas Carlyle (1795–1881)*

Carlyle was a Scottish literary and political writer. He drew thoughts from an intense admiration of Goethe, Schiller in the German-speaking regions and Samuel Taylor Coleridge in England, then developed into a leading Victorian critic of mechanistic materialism. This contempt for materialism came out as an irritable disdain for science and scientific procedures. It also produced an antipathy to "Benthamism" and accompanying utilitarianism. In *Chartism* [7] and *Past and Present* [8], Carlyle applied himself to what he called "the Condition-of-England question," attacking both laissez-faire and the dangers of revolution that the phrase encouraged, and manifesting with more passion than consistency and sympathy with the industrial poor which heralded the new novels, such as Dickens's *Hard Times*, of social consciousness of the society in those days.[17] The following passage depicts his features.

> The condition of England, on which many pamphlets are now in the course of publication, and many thoughts unpublished are going on in every reflective head, is justly regarded as one of the most ominous, and withal one of the strangest, ever seen in this world. England is full of wealth, of multifarious produce, supply for human want in every kind; yet England is dying of inanition. [...] Of these successful skilful workers some two millions, it is now counted, sit in Workhouses, Poor-Law Prisons; or have 'out-door relief' flung over the wall to them,—the workhouse Bastille being stronger. ...
>
> (Carlyle, *Past and Present* [8], Midas (book I, ch. I))

3.2 *John Stuart Mill (1806–1873)*

Mill made two significant modifications to the socio-economic thought of his mentors in his *Principles of Political Economy*. First, preceding scholars in political economy, living during the early development of industrialism, had produced an economics based on the presumption of scarcity, aimed at maximising growth to cater for what were seen as man's limitless material desires. However, under the influence of romanticism's critique of capitalism, Mill shed light on cases of economic abundance.[18]

In relation to this chapter, Mill's attitude towards labour relations is important. Mill generally held the enfranchisement of labour to be both desirable and inevitable. Fear of the ignorant multitude, or class warfare and evil strife, captivated Mill's mind as much as that of any respectable Victorians; and he expected the evolution of existing experiments in industrial partnerships, profit-sharing schemes, and workers' cooperatives or associations for the future of the world in order to mitigate economic conflict and draw the working class from sullen resentment at the system towards responsible participation and reform. This discussion on cooperatives is relevant to Jevons's concern to that type of institution.

[17] Drabble and Stringer [9, p. 92].

[18] Jay and Jay [15, p. 111].

> In the present stage of human progress, when ideas of equality are daily spreading more widely among the poorer classes and can no longer be checked by anything short of the entire suppression of printed discussion and even of freedom of speech, it is not to be expected that division of the human race into two hereditary classes, employers and employed, can be permanently maintained. [...]
>
> (J. S. Mill [29], *Of the Probable Futurity of the Labouring Classes* (bk IV, ch. 7, part 4))

> Under the most favourable supposition, it will be desirable, and perhaps for a considerable length of time, that individual capitalists, associating their work people in profits, should coexist with even those co-operative societies which are faithful to the co-operative principle. Unity of authority makes many things possible, which could not or would not be undertaken subject to the chance of divided councils or changes in the management. ... and in perhaps a less remote future than may be supposed, we may, through the co-operative principle, see our way to a change in society, which would combine the freedom and independence of the individual, with the moral, intellectual, and economical advantages of aggregate production; [...]
>
> (J. S. Mill [29], *Of the Probable Futurity of the Labouring Classes* (bk IV, ch. 7, part 6))

3.3 *John Ruskin (1819–1900)*

Ruskin changed what people thought about industrialisation and ultimately influenced the Arts and Crafts Movement in Britain. Rebelling against classical styles, Ruskin reawakened interest in heavy, elaborate Gothic architecture during the Victorian era. By criticizing the social ills resulting from the Industrial Revolution and disdaining anything machine-made, Ruskin's writings paved the way for a return to craftsmanship and all things natural.

Ruskin was born at a prosperous family in London, spending part of his childhood in the natural beauty of the Lake District in northwest Britain. The contrast of urban and rural lifestyles and values informed his beliefs about art, especially in painting and craftsmanship. Ruskin favoured the natural, the hand-crafted, and the traditional. He was educated at Oxford University. Ruskin travelled to France and Italy, where he sketched the romantic beauty of medieval architecture and sculpture. He also examined the cottage and villa architecture in England, France, Italy, and Switzerland.

Ruskin was a writer, critic, scientist, poet, artist, environmentalist, and philosopher. As a social critic, later like William Morris who studied the writings of Ruskin, he started a movement to oppose industrialisation and rejected the use of machine-made materials—in other words, the results of the Industrial Revolution.

Ruskin regarded factories as degrading places and he tried to improve the conditions in which the working class lived. Many of his social ideas later became commonly accepted by the end of his life.

> The inapplicability has been curiously manifested during the embarrassment caused by the late strikes of our workmen. Here occurs one of the simplest cases, in a pertinent and positive form, of the first vital problem which political economy has deal with (the relation between

employer and employed); and at a severe crisis, when lives in multitudes and wealth in masses are at stake, the political economists are helpless—practically mute: no demonstrable solution of the difficulty can be given by them, such as may convince or calm the opposing parties. Obstinately the masters take one view of the matter; obstinately the operatives another; an no political science can set them at one.

(Ruskin [40], "The roots of Honour," Section 4, *Unto This Last*: Essay I)

... I desire, in closing the series of introductory papers, to leave this one great fact clearly stated. **There is no wealth but life**. Life, including all its powers of love, of joy, and of admiration. That country is the richest which nourishes the greatest number of noble and happy human beings; that man is richest who, having perfected the functions of his own life to the utmost, has also the widest helpful influence, both personal, and by means of his possession, over the lives of others. ...

(Ruskin [29], "Ad valorem," Section 77, *Unto This Last*: Essay IV)

Comparing Jevons with Ruskin in terms of far-sightedness in thought on human history and future, Jevons looks commonplace. Jevons had attended Ruskin's lecture and admired him for his imagination and nature of topics Ruskin delivered.[19] Notwithstanding, Jevons was a genius but of different kind and perhaps confined to the Victorian context.

3.4 Thomas Hill Green (1836–1882)

Green was a scholar of moral philosophy, Fellow of Balliol College, Oxford, and Whyte's Professor at Oxford University (1878–92). He was influenced by Kantian and post-Kantian German philosophy, as well as Aristotle's conception of the *polis* as a partnership for pursuit of the common good. Green's political philosophy exerted a strong influence on the development of liberalism in Britain in the late nineteenth century. He argued that freedom should be understood as the power to develop our best selves in common with others. He provided theoretical foundations for construction of the British welfare state. Green proposed the true meaning of freedom in an idealistic way:

We shall probably all agree that freedom, rightly understood, is the greatest of blessings—that its attainment is the true end of all our effort as citizens. But when we thus speak of freedom, we should consider carefully what we mean it. We do not mean merely freedom from restraint or compulsion. We do not mean merely freedom to do as we like irrespectively of what it is that we like. We do not mean a freedom that can be enjoyed by one man or one set of men at the cost of a loss of freedom to others. When we speak of freedom as something to be so highly prized, we mean a positive power or capacity of doing or enjoying something worth doing or enjoying, and that, too, something that we do or enjoy in common with others. ...

(*Liberal Legislation and Freedom of Contract,* Green [10])

Green represented a type of reforming mind in the middle of the Victorian period along with more radical thinkers. Jevons and Green were almost contemporary and

[19] Maas [25, pp. 86–92].

both died in their mid-40s; they could be expected to be seen as reaching much more developed stages of thinking if they had lived longer.

3.5 William Morris (1834–1896)

Morris was an English designer, craftsman, writer, painter, and visionary socialist. He opposed the concept of fine art and created a company based on the ideal of a medieval guild, in which the craftsman both designed and executed the work. As a socialist he wished to produce art for the masses, but there was an inherent flaw in his ambition, for only the rich could afford his expensive hand-made products. His ideal of universal craftsmanship and his glorification of manual skill thus proved unrealistic in failing to come to terms with modern machine production. However, his work bore lasting fruit, in England and abroad, in the emphasis it laid on the social importance of good design and fine workmanship in every walk of life.

Morris defined *art* in the following:

> That thing which I understand by real art is the expression by man of his pleasure in labour. I do not believe he can be happy in his labour without expressing that happiness; and especially is this so when he is at work at anything in which he specially excels. A most kind gift is this of nature, since all men, nay, it seems all things too, must labour.
>
> ("The art of the People", Morris [30])

Morris proposed the condition for work which was worthwhile to be taken:

> This then is the claim: *It is right and necessary that all men should have work to do which shall be worth doing, and be of itself pleasant to do; and which should be done under such conditions as would make it neither over-wearisome nor over-anxious.*
>
> ("Art and Socialism", Morris [31])

Morris wanted "socialism" to be formulated as follows:

> ... I will say what I mean by being a Socialist, since I am told that the word no longer expresses definitely and with certainty what it did ten years ago. Well, what I mean by Socialism is a condition of society in which there should be neither rich nor poor, neither master nor master's man, neither idle nor overworked, neither brain-sick brain workers, nor heart-sick hand workers, in a word, in which all men would be living in equality of condition, and would manage their affairs unwastefully, and with the full consciousness that harm to one would mean harm to all—the realization at last of the meaning of the word commonwealth.
>
> ("How I became a socialist", Morris [32])

It is worth quoting E. P. Thompson's view on Morris in the following:

> What was the source of the greatness of Morris—this growing stature which he assumes in the perspective of history? His poetry alone, or his work in the decorative arts—profound though its influence was—would hardly be sufficient to establish his claim to the universal greatness suggested by George Bernard Shaw. As a political organizer his efforts ended in failure. As a theorist of the arts—despite all his profound insight—he failed to construct a consistent system, and muddled his way around some central problems. Did he make

> any major contribution which is marked by the stamp of unquestionable originality and excellence? The answer must be, "Yes". Morris's claim to greatness must be founded, not on any contribution to English culture, but on the quality which unites and informs every described as "moral realism": it is the practical moral insight of his political and artistic writings which gives them life. [...] Morris never sought to disguise the leading part which moral considerations played [...] He was brought to Socialism by his conscious revolt against that mechanical materialism which reduced the story of mankind to an objectless record of struggle for the survival of the "fittest", and which, in his own time, under whatever high-sounding phrases, put profit and not "free and full life" as the touchstone of value. He declared, by contrast: "I am a sentimentalist in all the affairs of life, and I am proud of the title." "I must tell you that my special leading motive as a Socialist is hatred of civilization; my ideal of the new Society would not be satisfied unless that Society destroyed civilization."[20]

These ideas of several representative thinkers of Victorian period influenced streams of thinking among people even in the leagues of industrialists and new breed of entrepreneurs.[21] J. S. Mill had a great reforming mind and left a wellspring of such thought.[22] Ruskin was outstanding in terms of passionate ideas of what human should be.

As for the arts, Jevons loved music, played the harmonium, frequently visited concerts and opera performances. But his attitude towards the arts was different from those referred people roughly classified as "Romantics". He was not a Romantic but a kind of person who kept the idea of a "solitary, intellectual, and high-class conception of music" (Mosselmans and Mathijs [33, p. 148] in which there was a gulf between music as high art and that of popular amusement played at, for example, music halls. Therefore Jevons preached the need for several kinds of education for lower class people.[23] Jevons took a different course to the goal, however, called reforming the society. For him Romantics seemed a different kind of animal.

4 Arguments About Social Issues

Although the "reform" years of 1830s were over, the Chartism movement, a kind of reflection of the 1848 revolutions in Europe, gave birth to many labour disputes around Britain. Because such unrest could be viewed as a result of what capitalism caused, political economists were forced to take sides in their stance against such issues. The following sections refer to Rhead Bowman's works.[24]

[20] Thompson [42, pp. 717–718].

[21] Perkin [35, pp. 218–270].

[22] Persky [36, pp. 140–151].

[23] Jevons [20, "Amusements of the People", pp. 1–27].

[24] Bowman [4, pp. 1123–1142].

4.1 Jevons's Social View

Jevons was brought up in circles concerned with social improvement,[25] It seems contradictory when you compare his view of self-convinced "selfishness" and belief in self-help and at the same time the view of a broad-minded attitude or more attention to social problem solving. His general position on reform was formulated in early adulthood by his family background, based on the Unitarian beliefs he followed consistently throughout his life. The first step in social philosophy, he explained, was "to throw aside all supposed absolute rights or inflexible principles" [19, p. 9]. "His social philosophy was compound of evolution theory, utilitarianism, and the Baconian experience-centred conviction, a philosophy that shunned the natural rights and noninterventionist doctrines of classical political economy" [4, p. 1124]. The thoughts of Charles Darwin and Herbert Spencer guided him to believe that evolution was producing a growing complication of society in terms of industrial, sanitary, and political aspects. Accordingly, "[T]he forms of human activity ... must multiply also the points at which careful legislation and continuous social effort are required to prevent abuse and secure the best utilization of resources" [20, p. 1].

What Jevons found at the heart of the bad tendencies produced by such evolutions was a social and family environment that hindered character building among the working classes. The development of working-class children was a clear example: Jevons's *The Coal Question* rendered "[A]t present it may almost be said to be profitable to breed little slaves and put them to labour early, so as to get earnings out of them before they have a will of their own. A worse premium upon improvidence and future wretchedness could not be imagined" [16, p. xxiv]. Generally, social evolution had created a population of which a large part was "ignorant, careless, improvident, or vicious" [20, p. 196].

The theory of exchange value and income distribution developed in his *Theory of Political Economy* was built up on the assumption that individuals made careful utility-maximising calculations in the process of supplying labour services and exchanging the fruits of their labour. Furthermore, "Anything which an individual is found to desire and to labour for must be assumed to possess for him utility. In the science of Economics we treat men not as they *ought* to be, but as they are" [17, p. 38]. "In improvidence and in ignorance, the majority of the population are involved in a vicious circle. They are ignorant because their fathers are ignorant, and so will their children be ignorant to the end of time, unless the State interferes with a strong hand" [20, p. 146]. He presumed that people including the working classes were potentially capable of rational and provident behaviour supposed in his abstract theory. In that case he seemed to believe their nature and character parallel to middle- and upper-class behaviour. But policy measures aimed at social improvement and the elevation of human character must be appropriate to the particularly evolutionary context within which they were to be applied:

---

[25] Könekamp [24].

> We now know that laws are not good or bad with respect to any invariable standard, but in reference to the changing character of society and man. The successful reformer is one who sees for what legislative change the people are ripe, and concentrates the popular energy upon it.[26]

According to Jevons, enhanced state activities were requisite for social progress, but what principle was to lead public policy and set the limits of legislation? He explained that he had no difficulty in accepting the utilitarian greatest happiness principle as long as the widest and highest interpretation was put upon the term used:

> No laws, no customs, no rights of property are so sacred that they may not be made away with, if it can be clearly shown that they stand in the way of the greatest happiness. [...] I conceive that the State is justified in passing any law, or even in doing any single act which, without ulterior consequences, adds to the sum total of happiness.[27]

Through the progress of policy discussions Jevons reconsidered his early fundamental individualism and endeavoured to improve the concept applicable to the policy issues.[28]

4.2 *Looking Back at Classical Doctrines*

Adam Smith established a tradition in making the distinction in the way that revolving around the production of material goods that could be sent on to labour in a subsequent period and the provision of services wherein production and consumption occurred simultaneously. The Ricardian analysis raised some limits to growth by showing that the economy worked under conditions of diminishing returns to land. But the policy programmes of Smith and Ricardo were essentially identical, with the emphasis upon freedom of trade and enterprise and a minimal use of resource for "unproductive" activities, particularly those of the state. The disinclined stance of the classical school towards government spending was completely consistent with their theory of growth and economic welfare. But it was all but opposing to Jevons's social philosophy and emphasis upon the need for works of public utility: "The utmost benefits may be, and, in fact, are secured to us by extensions of Government action of a kind quite unsanctioned by the *laissez-faire* principle. I allude to the provision of public institutions of various sorts—libraries, museums, parks, free bridges" [21, p. 205].

John Stuart Mill, as a transitional figure between the noninterventionist doctrine of his predecessors and the good government with interventionist doctrine, faced the difficulty of supporting a programme of social improvement while desiring to maintain the classical doctrines that apparently contradicted it. Mill's efforts at reconciliation were not successful in congruous discussion. Eric Roll commented that his

26 Jevons [20, p. 249].

27 Jevons [19, p. 12].

28 This point is important enough to delve. See Bowman [5] and Inoue [14].

economics "left far too many logical inconsistencies to serve as an adequate complement to the philosophy of compromise and steady reform" [39, p. 366]. That is very stern statement, if you think about his wholesome contributions.

Regarding the public works, Jevons found other key elements of his reform programme in conflict with classical thought. His strong interest for extensions of the Factory Acts, for example, particularly as regards women and children, found no counterpart in the view of his forerunners. As Blaug mentions: "All things considered, the Ten Hours camp[29] was not far wrong in regarding 'political economy' with its slogan of 'free agents' as a major obstacle to factory reform" [2, p. 224]. It is true to evaluate the classical economists realised such legislation as had already been achieved but warned against further measures.[30] Concerning class conflict and the prospect of reforming labour-capital relations, the focus of Ricardian economic theory was much related with the form of distributive shares analysis. Therefore, the conundrum could not be solved just by modifying the set of the natural rate of wages theory, the wage-fund theory, and the theory of the inverse relation between wages and profits. Hence, the existing policy programme integral to the old economics, mixed with the "mazy and preposterous assumptions" [17, p. xliv] hinging on the classical theories of value and distribution, "would seem to have provided ample motivation for an individual of Jevons's intellectual power and interest in social improvement to reconstruct the science" [4, p. 1132].

4.3 Changes in Economic Situation

In comparison with the established thinkers of the Victorian era, Jevons argued from a different angle as a scientist that England had achieved relative abundance but had to face eventually economic stagnation or even decline because of the exhaustion of its coal resources. This argument depended largely on the analysis of Jevons's *The Coal Question*, written in 1865 and his first major work. In the book he stated the evidence that England was well-established on the path of growth as indicated by data showing rapid increases of manufacturing, employment, marriages, population, state revenues, profits, wages, and leisure time.[31] As for Thomas Robert Malthus and David Ricardo, they had been too pessimistic and too much inclined to consider poverty and overabundance of population as the normal condition, Jevons contended:

> The momentous repeal of the Corn Laws throws us from corn upon coal [...] For the present our cheap supplies of coal, and our skill in its employment, and the freedom of our commerce with other wide lands, render us independent of the limited agricultural area of these islands, and apparently take us out of the scope of Malthus' doctrine. We are growing rich and numerous upon a source of wealth [...].[32]

---

[29] "Ten hours" working was the topic of dividing point for this discussion at that time.

[30] Blaug [2, pp. 224–226].

[31] Jevons [16, pp. 1–14].

[32] Jevons [16, pp. 173–177].

Therefore the evidence of unparalleled prosperity gave him grounds for arguing that the nation could afford social programmes that would not have been attainable in an earlier period. Prohibiting childbearing women from working in factories and providing them with support by the state was an example: "The growing wealth of the kingdom, and the ever-advancing powers of machinery, allow that to be done now which might not have been done before" [20, p. 178]. In the same way, at the time of argument for free public libraries he mentioned: "Nobody can doubt that there is plenty of money in this kingdom to spend for worse or better" (ibid., p. 31). But there were dark clouds on the horizon. When the price of British coal had risen enough relative to the price of the more abundant foreign coal, the main divisions of the country's industry would be damaged seriously. The change from corn to coal as the country's staple output had merely prevented the Ricardian stationary state for a time: the extensive margin for corn had been shifted abroad but replaced domestically by coal, which eventually would experience increasing costs relative to foreign competitors. England would find it difficult to sell its manufactured goods abroad to pay for its imports of food and raw materials and would cease to grow.

Jevons insisted that material deterioration would stop England's leading influence among nations and cause potentially to mass emigration, moral and intellectual degradation and civil conflict. It was crucially important that proper policy be established promptly. The agenda "seems to me to have an immediate and practical importance [...] the moment we begin to see a limit to the increase of our wealth and numbers, we must feel a new responsibility [...]" [16, p. xxii]. He disagreed with restrictions on the use of coal or interference with foreign trade because such policies he believed would reduce the world's sum of happiness. On the contrary, policy should concentrate on a proper use of the nation's rapidly growing wealth:

> [O]ur duties wholly consist in the earnest and wise application of it. We may spend it on the one hand in increased luxury and ostentation and corruption, and we shall be blamed. We may spend it on the other hand in raising the social and moral condition of the people, and in reducing the burden of future generations. Even if our successors be less happily placed than ourselves, they will not then blame us.[33]

From Jevons's point of view, wise use and consumption of coal was requisite for reasoning for public policy, including education. In two lectures on coal in 1867[34] and 1868[35] he returned to the theme of providing education, housing, libraries, and others. Also a reduction in the national debt would reduce the burden of future generations.[36]

[33] Jevons [16, p. xxv].

[34] *On Coal: Its Importance in Manufactures and Trade, Science Lectures for the People, Lecture IX*, delivered in the Carpenters' Hall, Manchester, 16 Jan., 1867.

[35] "On the Probable Exhaustion of our Coal Mines", being a lecture to the Royal Institution of Great Britain, on 13 March, 1968.

[36] Jevons [16, pp. 354–369].

5 Jevons's Programme of Social Reform

In this section, Jevons's specific proposals for reform will be reviewed. He observed that there was a high probability that any large government property would be mismanaged. He insisted that state ownership and management kept advantages over private enterprise on condition that it was operated under the public eye and where unnecessary capital expenditure was spent. He suggested that you could only see such conditions that were combined in the highest perfection in the Post Office, which was a model of efficiency,[37] and the case of the state parcel post,[38] but the situation was the opposite in the nationalisation of railways.[39] In this respect it is surprising to observe that Jevons had acquired a view of the importance in industrial management in addition to that of political economy at the theoretical level. You can observe examples of Jevons's managerial point of view.

5.1 Public Goods and Education

With regard to state enterprise, Jevons was a strong advocate of supplying various types of public goods (public construction works or services): "Community of property is most wasteful in some cases ... but there are cases of community of production. Community of consumption, on the contrary, is often most commercial" [21, p. 205]. In this category were "all the more immediate needs of society" associated with rapid industrialisation and urbanisation, including boards of health, medical officers, reformatories, free libraries, museums, art galleries, parks and popular amusements, city beautification, halls, highway boards, main drainage schemes, water supplies, purification of rivers, public education, public housing and town planning, improved police, better poor law medical service and "a score of other costly reforms".[40] Together with the above reforms, public education and housing programmes, incorporated with restrictions on the labour of women and children, were particularly important because of their relation to the family environment. Child labour should be banned,[41] and mothers of small children should also be excluded from the workplace and supported by the state.[42] Education should be compulsory and state-supported: "True progress will tend to render every class self-reliant and independent. [...] Money spent in the education of the young has this beneficent effect" [20, p. 191]. He observed that charity generally had the opposite effect. Good housing was critical: "You must know how much the condition of a family is influenced by the cleanliness and comfort of their dwelling. 'As the home, so the people'" (ibid., p. 112). But,

[37] Jevons [20, pp. 279, 280].

[38] Ibid., p. 328.

[39] Ibid., p. 383.

[40] Jevons [20, pp. 28, 202, 21, pp. 204, 205].

[41] Jevons [16, pp. xxiv, xxv].

[42] Jevons [20, pp. 172–175].

"the great mass of the people will not have healthy houses by the ordinary action of self-interest. The only chance of securing good sanitary arrangements is to pull down the houses which are hopelessly bad … and most carefully to superintend under legislative regulations all new houses that are built" [21, pp. 204, 205].

5.2 Public Finance and Taxation

Jevons argued that maintaining public goods and services must be supported by a continuous increase of local taxation.[43] He was an advocate of proportionality in taxation excepting the very poor. Levies on commodities were the preferred method of taxation.[44] According to Jevons a progressive income tax would face practical difficulties and have opposing effects on capital accumulation. Jevons's advocacy of proportionality was dependent on his view of the economic status and character of the working classes: "As things are, they earn well, but they spend badly" [20, p. 6]; "There are comparatively few signs that the wages of the working-classes even when sufficient, are saved and applied really to advance the condition of the recipients" (ibid., p. 205). Bad habits developed over many generations, including drunkenness, meant that the working classes did not spend their rising real incomes wisely. Consequently, he both favoured an increase in local rates and opposed the further alleviation of general taxes:

> It is only with the increase of education and temperance that the increase of wages will prove a solid advantage. Thus, when the really hurtful taxes are removed, it by no means follows that the further remission of taxes leads to the profitable expenditure of income. The money may be spent in a way far more profitable to the whole nation than it will be spent by those whose taxes are remitted.[45]

5.3 State's Function and Its Limit

Jevons found it little effective under the income redistribution category of reform. It is a kind of corollary to his advocacy of proportionality in taxation. Despite that, he argued a programme of public works would have redistributive effects: "A little observation will convince anyone that there are many social devices which carry the benefits of wealth to those who have no wealth. Public ownership is a most potent means of such vulgarisation of pleasures. A public park is open to everyone" (ibid., p. 33). Factory legislation, consumer protection, and regulation of private industry exemplified other areas of reform. He praised the Factory and Workshop Act of 1878 as "one of the noblest products of legislative skill and patience" and which "affords also the best encouragement to proceed with several other measures of like nature"

---

[43] Jevons [21, pp. 202, 203].

[44] Ibid., pp. 231–237.

[45] Jevons [20, p. 205].

including regulation of the working hours of adult men.[46] State inspection of products was appropriate in the circumstance of inadequate consumer knowledge.[47] As far as the monopoly power of business was concerned, Jevons paid little attention to its regulation with the exception of the railways, whose high rates suggested the need for control.[48]

5.4 *Trade Union Policy and Partnership*

From Jevons's point of view, trade union policy and labour–capital relations were areas in need of substantial reform. Hostility, conflict, and riot were the usual consequence of the current relations of labour and capital.[49] Jevons acknowledged that most economists of the day criticised unions' efforts to raise wages through strikes and the restriction of labour and their opposition at times to the introduction of machinery. On the other hand, Jevons knew that they acted beneficially in creating funds to assist poor members and in attempting to shorten the hours of labour and make the workplace more hygienic and safer. He suspected that such goals could not be attained by the workers themselves. The individual labourer "dependent for his living upon his week's wages, is utterly incompetent to enforce any concession from his wealthy employer. Union is the natural remedy" [20, p. 108]. From Jevons's point of view, "[A]ny attempt to heal the discords of labour and capital by destroying unions would be futile and undesirable, The best policy was to put labour-capital relations on a sound footing by way of worker joint-stock companies and profit-sharing via industrial partnerships" [4, p. 1129].

> Partnership of industry are, no doubt, an innovation, having hitherto existed only in exceptional trades and rare experiments; but I assert confidently that *they are an innovation of which the utility is evident and the necessary urgent.* They are equipped not by the restless desire of change, but as the natural sequel of great revolutions in our social condition. Our great factories and our great army of artisans have sprung up within one hundred years and it is quite to be expected that so vast an innovation should lead to other innovations.[50]

This institutional aspect is dealt with later in the chapter.

5.5 *Short Summary—Industrialisation and Urban Infrastructure*

As a summing up, Jevons's social philosophy and programme of reform were aimed at developing solutions to what he believed were critical problems accompanying

---

46 Jevons [20, p. 178].

47 Jevons [19, pp. 40–46].

48 Jevons [20, pp. 382, 383].

49 Ibid., pp. 129–131.

50 Ibid., p. 148.

social evolution, problems that may be lumped into two broad groupings: those associated with the spread of urban industrialism and a serious shortfall of public goods and services necessary to accommodate it, and labour–capital conflict which, in his view, was entering a new, potentially destructive phase with the growth of trade unionism. Urban deterioration had worsened by the 1850s; the conditions in London were a decisive influence on Jevons in his youth.[51] A response in the form of planning, regulation, and the provision of public services had become a top priority. Notably, Edwin Chadwick began serious research on urban problems in the 1840s, and several laws were passed in the 1840s and 1850s aimed at improvement of public health, housing, and amenities. On labour relations, somewhat half of the revolutionary movements of the 1830s and 1840s, which were replaced with practical trade unionism beginning in the 1850s, had effects of strengthening the position of the working classes for the agenda of wages and working conditions. Social division was becoming more apparent in the cities with industrialisation.

6 Jevons's Viewpoint Reconsidered

In his book on economic theory, Jevons's purpose was to establish new theories in the principal areas treated by the then science, value-setting and distribution, which would be compatible with his social philosophy and give a theoretical basis for a programme of social reform. In that process, classical theory and its policy were considered to be changed. His method was to create an entirely new frame of reference from which to analyse value and distribution. The point is whether he succeeded in that project. One can understand his intention in the following:

> The problem of Economics may, it seems to me, be stated thus: *Given a certain population, with various needs and powers of production, in possession of certain lands and other sources of material: required, the mode of employing their labour which will maximise the utility of the produce.*[52]

The inevitable consequences of this grasp of the subject were to focus on individual and micro-foundation of an economic entity's behaviour. From the point of growth perspectives of classical thought, Jevons therefore drastically narrowed the boundaries within which value and distribution were to be analysed. But from a utilitarian perspective, the boundaries of science, including public sector analysis, were broadened because, for example, services were included in a common theoretical framework with material goods.

[51] Könekamp [24, pp. 254–255].

[52] Jevons [17, p. 267].

6.1 Jevons's Logic of Proposition

Because the maximisation of utility was the focal point of the "Calculus of Economics" [17, p. 37], Jevons started by focusing attention on the objects and actions that were sources of pleasure and pain: commodity was "any object, substance, action, or service which can afford pleasure or ward off pain" (ibid., p. 38); labour was "any painful exertion of mind or body undergone partly or wholly with a view to future good" (ibid., p. 168); utility represented "the abstract quality whereby an object serves our purposes, and becomes entitled to rank as a commodity" (ibid., p. 38). In contrast to the classical school, labour was defined in subjective terms, as opposed objectively to the using-up of human energy, and commodity included services. The classical limitation of wealth to accumulable objects would "cut out of our science a full half of its appropriate objects—namely, services of a passing and perishable kind" [21, p. 16]. Jevons included them on the grounds that economics revolved around the problem of utility maximization, but services just as well as material goods yielded utility. Consequently, the classical distinction between productive and unproductive labour was "quite untenable. All Labour is directed to the production of utility—of useful or agreeable effect—a hedonic balance" (ibid., p. 87). Labour generally generated a subjective surplus as opposed to the physical surplus of productive labour in classical thought.

Jevons particularly noted the following type of goods. Durable "inconsumables" included the large mass of commodities that could be used repeatedly and which were therefore subject to a "multiplication of utility". The methods of maximizing their utility "mostly turn upon arrangements for the joint or successive use of things by several persons." He gave three examples: a circulating hire system; common ownership by private groups; and, the most important method, public ownership or "community of consumption": "The extreme cases of the multiplication of utility are probably afforded by institutions or edifices provided for the enjoyment of the public at large, and paid for out of the public revenue, local or general" [21, p. 40].

6.2 Management of Public Sector and Policy

Throughout the discussion above, the connection between Jevons's reconstruction of the science and his position on social reform becomes easy to understand. He would defend subjective valuation and utility maximisation on explicit theoretical grounds, and an expansion of public works concept. "Jevons's long list of 'the more immediate needs of society,' which must be supported by the state were more immediate for his Calculus of Economics: resources devoted to them would yield a greater balance of utility than would alternative private goods. Labour engaged in public services was therefore very 'productive' and community of consumption was 'most economical'" [4, p. 1136]. This kind of reasoning was completely unrelatable to the classical thought. They had said that only labour employed in the private sector was productive;

namely labour employed in the public sector was unproductive. Jevons allowed an economic analysis of the proper allocation of resources between public and private sectors and perhaps within the public sector as well.

While understanding that his reconstruction of the science established a new logic for public sector analysis, he stated that "[I]nstead of one dictum, laissez-faire, laissez-passer, we must have at least one science, one new branch of the old political economy ... But I am quite satisfied if I have pointed out the need and the probable rise of one new branch, which is only to be found briefly and imperfectly represented in the works of Mill or other economists" [21, p. 206]. That new branch would "show what the Government should do, and what it should leave to individuals to do" (ibid.). Jevons tried to reform "the science of economics from a concern with maximising a physical surplus originating in the private sector of a capital accumulating economy to that of a subjective surplus originating in both private and public sectors" [4, p. 1137]. His concept provided "a new logic for public sector analysis and his concept of multiplication of utility: the public sector as a source of utility must increase relative to the private in order to achieve a social maximum over time" (ibid.).

6.3 Management of Industries—Worker–Capitalist Relations

Here this chapter should like to take a not-ordinary angle from the management of economic institutions or industry in order to position Jevons's attention to labour issues or that areas. For that purpose, Charles Babbage (1791–1871) should be mentioned if it goes to the discussion of how to manage industries of those days. Today Babbage is unquestionably considered as a "Father of the computer". He is credited with inventing the first mechanical calculating machine, the "Difference Engine". However, Babbage should be remembered not only as a scientific genius but also as a careful observer of the factory system. On the latter, he wrote important proposals on improvement of factory processes and management.[53] He may well be thought of as a predecessor of Frederick Winslow Taylor in the United States, who was one of the founders of management science and is known for the "Taylor system or Taylorism,"[54] which revolutionized factory management by scientific approaches.

Babbage's lifetime coincided with the height of the Industrial Revolution. Although many modernised factories and workshops were built and began operating, they were mostly family-owned and exposed to successive risk in terms of potential shortage of sustainable management abilities. Hiring outside experts was

[53] Babbage [1, Part 3].

[54] In Britain, however, Taylorism was almost wholly ignored until the period of the Great War, 1914–18. Part of the reason was that the new methods demanded uniformity and were considered to be inhuman. His schemes reduced the need for any skills and that they were but a camouflage for speed-up and slave driving [37, p. 64].

one option, but in those days Britain was at a fledgeling stage of producing professional managers. Babbage's task was to provide good instructions about practical factory operations and what the company management was.

Another noteworthy point is that Babbage played an important role in the development of ideas of cooperation in terms of profit sharing in the economic institution. Jevons studied Babbage's works and quoted his concepts at several instances in his writing.[55] Babbage wanted to show that workers' interests were not separated from those of their masters. Through the profit-sharing scheme, he considered, a "new system" came up as a suitable incentive for cooperation between these two classes.

Profit sharing under Babbage's scheme suggested two important features. Firstly, it would create a different psychological setting in the working place. Since workers would have a direct share in the profits, there would be an incentive to use their talents and skills towards the improvement of the business and to cooperate to guard business interests. Secondly, in order to seek mutually agreeable business objectives, both parties would have to collaborate with and to monitor each other. Particularly, the latter would be beneficial to productivity. Along this line of thought it is noted that cooperation did not demand major changes to the existing social and economic structure. It involved improvements for the business's prosperity. The following messages were found in Babbage's writing:

> The workmen and the capitalist would so shade into each other- would so evidently have a common interest, and their difficulties and distresses would be mutually so well understood that, instead of combining to oppress one another, the only combination which could exist would be a most powerful union between both parties to overcome their common difficulties.[56]

Digesting Babbage's managerial concept, Jevons claimed: "The result which emerges is that the supposed conflict of labour with capital is a delusion" [19, p. 101], because "capital can hardly be accumulated and employed without vivifying industry, and diffusing comfort ad subsistence through the whole body of society" [20, pp. 107–108]. The socially desirable way to improve wages was by increasing output with the aid of capital: "It is not by fighting against capital and against machinery, but by having them on your side" (ibid., p. 118). This could best be accomplished by way of worker joint-stock companies and profit-sharing industrial partnerships:

> [C]ombination should be in a perpendicular, and not in a horizontal, direction. The master is to combine with his men, to be their true leader, and after all the ordinary costs of wages, interest, and superintendence are provided for, the surplus is to be fairly divided among all who have contributed towards it.[57]

Also, Jevons's concern on a joint-stock company is noteworthy, because in the mid-Victorian period it was the right time to reform corporate laws.[58] Reforming

---

[55] Jevons [20, *On Industrial Partnerships*, Section 7, "Mr. Babbage's Proposal of the Partnership Principle", pp. 134–136].

[56] Babbage [1, Section 313].

[57] Jevons [20, p. 127].

[58] Joint-stock company status has a long history in Britain. But of the cases before that time an owner has unlimited responsibility.

at that time was very meaningful to gauge the future of corporations: the engine of the economic value making, a capitalist–employee–manager collaboration, in other words partnership.

> But there is a great future for joint-stock enterprise in one form or another, and when there are many shareholders among the capitalists, I see no reason whatever why the partnership principle should not at once be adopted as regards the men.[59]

In spite of Jevons's worker-assisting proposition, he also added that the current trade-union practices of strikes and opposition to employers' improvement policies were futile and inconsistent with Jevons's own theoretical conclusions that were linked to output and to factors that elevated the skill and productivity of labour, including capital and education.[60] The absence of disruptive strikes, the higher income-generating effects of capital accumulation and education, and direct worker interest in output produced would ensure the success of industrial partnerships.[61]

According to Bowman:

> The theory had the merit of apparent consistency with important contemporary developments, including rising real wages, a reduction in the working day and a growing diversity among workers regarding skill and wage levels. But Jevons's sanguine hopes for industrial partnerships were to be little realized, perhaps because they were inconsistent with the values, aspirations and perceptions of self-interest of both capitalist and workers, which were conditioned above all by property ownership or its absence.
>
> [4, p. 1141]

6.4 Coda

Britain has a tradition of cooperative activities. Different from socialist state-building and movements[62] in other countries, the concept and practice of cooperatives survive today and are instrumental to social and community business developments.

In the UK, in 2014 a term of economic entity was newly created, named *Community Benefit Society* (CBS),[63] based on cooperatives for community building, a counterpart to *Community Interest Company* (CIC) under the private Companies Act,

---

[59] Jevons [20, p. 138].

[60] Jevons [17, pp. 271–272, 18, pp. 31, 57, 58].

[61] Jevons [20, pp. 129–131, 139, 140].

[62] Britain also had several entities and thinkers whose aims were realising socialist or communist models, but their methods and activities were within the parliamentary system.

[63] A CBS is a form of cooperative whose purpose is designed to benefit the wider community. It is a limited company and must be registered with the Financial Conduct Authority rather than Companies House. Anyone who subscribes to the objects of the society can become a member; they do not need to be either an employee or a user of the CBS. Members are entitled to vote for the Board of Directors and may bring proposals to the annual general meeting in relation to the running of the CBS. Members can also stand for the Board of Directors. However, a CBS must not just serve its own direct community (i.e. their members) but also the wider community. Members are considered to have a "share" of a society, but they do not receive any dividend: CBSs are expected to ensure that any profits are used for the benefit of the community. There is thus no secondary

another scheme for social enterprise basically in private hands. That seems to be a realized version of Jevons' idea. This scheme was initiated under Tony Blair's Labour Government. However, it was enacted into law under the Conservative Government. This means that the CBS scheme was endorsed nationwide and on a nonpartisan basis. Jevons supported the scheme of industrial partnership and/or cooperation between capitalist and workers for the reason of supporting workers with determinate will and no objecting to education on the one hand, also for the reason of the need for capitalist management initiatives on the other hand. A CBS is an entity of several related or willing stakeholders for the purpose of realising activities.[64] In a sense both J. S. Mill's and Jevons's dream has come true. Unlike J. S. Mill, Jevons did not have full confidence in the management of cooperatives, but he might have supported this type of cooperative. The reason is that this entity is open to non-members and provides the opportunity of maintaining good management and governance.

7 Conclusion

Jevons's new political economy can be understood as an endeavour to integrate economic theory and public policy relating economy for the purpose of attempting to solve social problems and conflicts. At the time of rapid growth and increasing social complexity, the classical principles of non-intervention and small government, and a preoccupation that the growth would render material welfare, were no longer reliable from Jevons's point of view. New policy directions were needed that must have a sound theoretical basis. He touched upon two main areas of economic policy, labour issues and urban policy against deterioration. Jevons's adoption of the scarcity concept, subjective valuation, and utility maximisation would provide a theoretical basis for innovative thinking and strong argumentation (through lectures and essays on social reform over his career) on both issues. "Jevons's political economy represented an interesting example of the adaptation of a received doctrine, Benthamite Utilitarianism to the problems at hand. Utilitarianism proved to be expedient because from it could be distilled not only a policy of state activism but also the utility concepts for reconstructing economic theory" [4, p. 1142].

T. W. Hutchison also reviewed Jevons in the later stage: "Jevons, the neoclassical pioneer of what M. H. Dobb called 'the Jevonian Revolution', was fundamentally and philosophically anti-dogmatic. Even if Jevons had supported *laissez-faire*, it would not have been as a dogma. ... Jevons's later view also marked *an abandonment of laissez-faire*" [13, p. 376].

purpose for the society. A CBS can opt to have an "asset lock." This means that if the company ceases to exist, any assets will be distributed for the benefit of the wider community and will not be divided among the members. Although a CBS is a registered company rather than a charity, it is in some ways closer to a charity than a business as it operates on a not-for-profit basis.

[64] Leicester Vaughnan College: What is a Community Benefit Society? (https://vaughan.coop/leicester-vaughan-college-community-benefit-society/what-is-a-community-benefit-society/).

Another point I must put is that Jevons seemed to be one of the early and rare cases of managerial economists in Britain. If you read his reports, you would be surprised to find that the course of his discussions is constructed on the minute observations about factory details, following Babbage's way. You can detect such a managerial view in his comments on public policy and social issues.

Jevons was a genius of diversified subjects but had an unexpectedly short life. Because he left many areas of incompletion, Keynes judged that his genuinely innovative and talented works rested only in his early days.[65] But Keynes valued his positivistic and social statistical works. It is a pity that this chapter cannot extend to focusing on that theme. At the same time I believe Jevons's efforts in the policy area should be worthy of revisiting. This chapter looks at them only briefly. Lastly, my personal opinion is that Jevons was one of the eminent Victorians, not in Lytton Strachey's satirical sense [12, 41] but in the sense that although he was basically conservative, he kept a passionate social reforming mind like other great challengers on social reforms in the Victorian period.

Acknowledgements Firstly, I am greatly indebted to Professor Maruyama, general editor of this volume, who invited me to contribute. My sincere appreciation goes to Rhead S. Bowman, Monica Hernandez [11], Harro Maas [25], Sandra Peart [34], and Takutoshi Inoue for their works and other scholars in the area of this subject.

References

1. Babbage, C.: On the Economy of Machinery and Manufactures, 2nd edn. Charles Knight, Pall Mall East (1832)
2. Blaug, M.: The classical economists and the factory acts – a re-examination. Q. J. Econ. **72**(2), 211–226 (1958)
3. Bostaph, S.: Jevons's "'antipodean interlude': the question of early influences, with a rejoinder by Michael V. White and a reply by Bostaph". Hist. Polit. Econ. **21**(4) (1989)
4. Bowman, R.S.: Jevons' economic theory in relation to social change and public policy. J. Econ. Issues **XXIII** (1989)
5. Bowman, R.S.: Policy implications of W. S. Jevons's Economic theory. J. Hist. Econ. Thought **19**, Fall (1997)
6. Cannadine, D.: Victorious Century: The United Kingdom, 1800–1906. Viking, New York City (2017)
7. Carlyle, T.: Chartism. J. Fraser, London (1840)
8. Carlyle, T.: "Midas", Past and Present (Book 1, Chap. 1). In: Traill, H.D. (ed.) The Works of Thomas Carlyle. Cambridge University Press, Cambridge 1843 (2011)
9. Drabble, M., Stringer, J. (eds.): The Concise Oxford Companion to English Literature. Oxford University Press, Oxford (1987)
10. Green, T.H.: Liberal Legislation and Freedom of Contract: A Lecture. Slatter and Rose, Oxford (1881)
11. Hernandez, M.: Inequality as Lack of Co-operation in Economic Thought. The New School for Social Research Working Paper 18/2017, May 2017. Department of Economics, The New School for Social Research (2017)

[65] [23, p. 146].

12. Holroyd, M.: Lytton Strachey: The New Biography. Chatto and Windus, London 1994 (2011)
13. Hutchison, T.W.: The politics and philosophy in Jevons's political economy. Manch. Sch. **50**(4) (1982)
14. Inoue, T.: W. Stanley Jevons (1835–1882): From a Man of Science to an Economist. Discussion Paper No. 33, School of Economics, Kwansei Gakuin University, Japan (2007)
15. Jay, E., Jay, R. (eds.): Critics of Capitalism: Victorian Reactions to 'Political Economy'. Cambridge University Press, Cambridge (1986)
16. Jevons, W.S.: The Coal Question: An Inquiry Concerning the Progress of the Nation, and the Probable Exhaustion of Our Coal Mines, Second edn. Macmillan and Company, London 1865 (1866)
17. Jevons, W.S.: Theory of Political Economy. Macmillan and Company, London 1871 (1888)
18. Jevons, W.S.: Political Economy, Science Primers. Macmillan and Company, London 1878 (1920)
19. Jevons, W.S.: The State in Relation to Labour. Macmillan and Company, London 1882 (1910)
20. Jevons, W.S.: Methods of Social Reform and Other Papers. Macmillan and Company, London (1883)
21. Jevons, W.S.: The Principles of Economics: A Fragment of a Treatise on the Industrial Mechanism of Society, and Other Papers. Macmillan and Company, London (1905)
22. Jevons, W.S.: Papers and Correspondence of William Stanley Jevons (Black, R. D. C., Könekamp, R. (eds.)), vols. 1–7. Macmillan, London (1972–1981)
23. Keynes, J.M.: Essays in Biography. The Collected Writings of John Maynard Keynes, vol. 10. Macmillan St. Martin's Press, London 1933 (1972)
24. Könekamp, R.: William Stanley Jevons (1835–1882) Some biographical notes. Manchester School, **30** September, pp. 251–273 (1962)
25. Maas, H.: Pacifying the workman: Ruskin and Jevons on labor and popular culture. Hist. Polit. Econ. **31**(Supplement) (1999)
26. Maas, H.: William Stanley Jevons and the Making of Modern Economics. Cambridge University Press, Cambridge (2005)
27. Maas, H.: William Stanley Jevons (1835–1882). In: Vernengo, M., Caldentey, E.P., Rosser, B.J., Jr. (eds.) The New Palgrave Dictionary of Economics. Palgrave Macmillan, London (2018)
28. Matthew, C.: Introduction – The United Kingdom and the Victorian century, 1815–2001. In: The Nineteenth Century – The British Isles: 1815–1901 edited by Colin Matthew, The Short Oxford History of British Isles. Oxford University Press, Oxford (2000)
29. Mill, J.S.: Of the probable futurity of the labouring classes (Book IV, Chap. 7). In: Principles of Political Economy with Some of their Applications to Social Philosophy. John W. Parker, London (1848)
30. Morris, W.: The art of the people. In: The Collected Works of William Morris, vol. 22, pp. 28–50. Cambridge University Press, Cambridge 1879 (2012)
31. Morris, W.: Art and socialism. In: The Collected Works of William Morris, vol. 23, pp. 192–214. Cambridge University Press, Cambridge 1884 (2012)
32. Morris, W.: How I became a socialist. In: The Collected Works of William Morris, vol. 23, pp. 277–281. Cambridge University Press, Cambridge 1894 (2012)
33. Mosselmans, B., Mathijs, E.: Jevons's music manuscript and the political economy of music. Hist. Polit. Econ. **31**(Supplement) (1999)
34. Peart, S.: The Economics of W. S. Jevons. Routledge, London and New York (1996)
35. Perkin, H.: The Origins of Modern English Society: 1780–1880. Routledge & Kegan Paul, London and Henley (1969)
36. Persky, J.: The Political Economy of Progress: John Stuart Mill and Modern Radicalism. Oxford University Press, New York (2016)
37. Pollard, S.: Management and Labour in Britain during the Period of Industiralization" (Chap. 1). In: Labour History and the Labour Movement in Britain. Ashgate Publishing Limited, Aldershot, Hampshire (1999)
38. Robbins, L.: The place of Jevons in the history of economic thought. Manchester School **7**, pp. 1–17 (1936)
39. Roll, E.: A History of Economic Thought. Prentice-Hall, Englewood Cliffs (1963)

40. Ruskin, J.: Unto This Last. (First published in Cornhill Magazine.) Penguin Books, London 1860 (1997)
41. Strachey, L.: Eminent Victorians. Chatto and Windus Ltd., London (1918)
42. Thompson, E.P.: William Morris: Romantic to Revolutionary. The Merlin Press, London (1955)

Another Look at the Lasting Utilitarian Imprint in Economic Theory

Bruna Ingrao

Abstract The chapter deals with the lasting imprint of utilitarianism in economic theory since the utility function was introduced as the basic pillar of price theory and economic equilibrium by the scholars of the marginal revolution in the 1870s. As J. A. Schumpeter observed in his *History of Economic Analysis*, the scholars in the marginalist revolution anchored their conception of utility to philosophical ideas of pleasure and choice. In later years economic scholars claimed that economic theory had got rid of this philosophical background to reduce the utility function to a merely logical frame to represent choice and maximization in mathematical models. The aim of the chapter is to assess this claim looking at the changing interpretations of the utility function and the role it performs. The chapter discusses ideas of utility in economics in the light of philosophical controversies, finally comparing them to contrasting ideas of self-consciousness and deliberation through the lenses offered by meditations about pleasure, identity and choice explored in some literary masterpieces.

> E quì do fine al mio discorso, lontano egualmente dal gregge degli epicurei, come dall'insensibilità della Stoa: se avrò fatte cessare rapidamente e con frequenza le sensazioni dolorose di chi mi ha letto; se l'avrò invitato a pensare, ad analizzare l'inesauribile fondo della propria sensibilità, avrò ottenuto il fine che mi era proposto.
>
> (P. Verri, *Discorso sull'indole del Piacere e del Dolore*, 1781)

1 The Philosophical Roots of Ideas of Utility in Economics

In the history of culture, pleasure and happiness are controversial concepts, on whose definition theoretical and theological debates have been going on for centuries and millennia. Ideas of pleasure and happiness were debated in the Enlightenment culture, and in the late 18th century ideas of utility covered an ample range of meanings in philosophical controversies.

B. Ingrao (✉)
Sapienza Università di Roma, Rome, Italy
e-mail: b.g.ingrao@gmail.com

T. Maruyama (ed.), *Marginal Revolution in Economics*, Monographs in Mathematical Economics 6, https://doi.org/10.1007/978-981-99-4342-5_5

From the late 18th century and throughout the 19th century many scholars dealing with value theory in political economy, among whom notably Galiani, Malthus, J.S. Mill, Jevons and Edgeworth, explored the nexus between ideas of utility in economic theorizing and their philosophical roots. Utilitarianism spread in various currents in 19th century thought, and different scholars articulated their vision of human behavior and ethical principles around the utility concept. In Great Britain, through the influence of Bentham and Sidgwick, philosophical utilitarianism had a strong connection with the evolution of economic thought. Edgeworth built a comprehensive utilitarian philosophy including ethics and economics.

In *A History of Economic Analysis* (published posthumously in 1954) J. A. Schumpeter devoted a note to utility theory, where he recognized the philosophical roots of early utility concepts in economics; but he denied any "hedonist postulates or philosophies" in the contemporary "utility theory of value" [59: 1023]. Like other economists in the first half of the 20th century, Schumpeter rejected the nexus between marginal utility theory and philosophical utilitarianism. In the first quarter of the 20th century the so-called "ordinal" revolution claimed to have set economic theory free from unnecessary nexuses with hedonistic visions of behavior, while maintaining the key role of the utility function and utility maximization.

In the second half of the 20th century the dominant approach to utility in economics pursued the aim to restrain the utility function within an instrumental conception of cognitive rationality, abandoning any pretense to theorize about pleasures and pains, or other psychic or behavioral dimensions of choices based on utility maximization. The prevailing evolution of utility ideas in economics has been on the path opened in game theory in the 1940s and later developed in decision theory. However, since the 1970s, there has been an explicit return to the utilitarian tradition in new currents of research in economics that opened the way to generalizing the utility maximizing approach to account for a wide range of choices and actions, or for the whole complexity of human behavior. A new line of research inquired the measurement of "happiness" in relation to income, income distribution and other economic variables. In the last quarter of the 20th century utility theory came to the forefront in experimental psychology; a number of theorists advocated the neuro-physiological records of sensations of pleasure and pains, reviving the old terminology, without necessarily endorsing philosophical utilitarianism as an overall interpretation of human behavior. The return to the utilitarian roots as practiced in experimental psychology influenced new currents also in economics.

This complex picture shows that the status of utility ideas remains ambiguous in contemporary economics; the link between utility concepts in economic theory and the philosophical hedonist roots that Schumpeter denied, is far from having been broken. In the second half of the 20th century, there were still ambiguities as regards the conceptual content of the utility function in contemporary theory. Blind spots in the semantics of utility functions emerged in the debate about the measurement of utility, of which Ivan Moscati provides an ample survey in his book *Measuring Utility* [46]; beyond the controversial measurement issues, they are not often taken openly to the surface as far as the interpretation of human choices is concerned. The utility function is a central theoretical concept in economics, and utility functions dominate economic theorizing through the dominance of assumptions of utility maximization in models in all areas of economics. The utility maximization axiom is predicated to be the necessary foundation of coherent theorizing in mathematical models in economics, failing which the theory is discredited as irrelevant or contradictory.

After the ordinal turn advocated by Pareto and Hicks, among others, and approved by Schumpeter, the common and dominant interpretation is that the utility function is only relevant to embody the rational agent's preferences; but the concept of preferences is per se a controversial philosophical notion that needs to be defined in relation to choice and behavior. As F. Ranchetti observed, the definition of utility functions as mere indexes on preferences orderings that should be considered in isolation from any philosophical conception of utility or any doctrine about psychic processes or behavior, cannot stand a properly philosophical inquiry [57, 58]. The divide does not stand attentive methodological screening. Thus, the pretense to break up ideas of utility from any philosophical background, notably the utilitarian one, did not finally succeed.

The aim of this chapter is to stimulate the debate on the ambiguities of utility ideas in contemporary economics. After remembering the philosophical roots of the early utility ideas in value theory, the chapter explores the revival of the utilitarian imprint in economic research in the second half of the 20th century, and beyond; in the final section utility ideas in economics are looked at by comparison with the semantic fields of pleasure, happiness, choice, action in some literary texts.

2 Pleasures and Pains in Human Experience: Pietro Verri and Jeremy Bentham

In 1781, in his essay *Discorso sull'Indole del Piacere e del Dolore* (*On the Nature of Pleasure and Pain*) Pietro Verri addressed many issues that were later to enter the debate on the definition of utility in economic theory [65]. His perspective was philosophical, and it was rooted in a perceptive psychological approach to choices and judgment in human life. His essay was oriented to assessing whether in human life the sum total of feelings of sorrow and feelings of pleasure gave a net result in favor of pleasurable sensations. Verri addressed such subtle questions as the asymmetric relationships between pleasures and pains, the role of anticipations and thus of hope or fear on expected pleasures or pains, the duration of pains and the conversion of diminishing pains into pleasure, and the way in which this conversion operates on the human psyche. He compared the complexity of passions in human hearts to a flowing current, where different perceptions mix together so that the mind is incapable of retracing all feelings which concurred in the flow. He argued that in human life the sum total of painful sensations is larger than the sum total of pleasurable ones, including the moral sensations. The pessimistic conclusion was redeemed by the thesis that pains, and the effort to escape them, stimulate the search of those pleasures, which arise from fine arts and sciences. The effort to reduce or avoid pains acts as the midwife helping to give birth to the fruits of civilization.

In Jeremy Bentham's thought pleasure calculus is both the dominant motive to action and the norm of rational legislation. It is not just a flow, but rather a precise pointer both in indicating action and in suggesting normative principles. In the first chapter of *An Introduction to the Principles of Morals and Legislation*, Bentham wrote the famous passage where pain and pleasure are portrayed as the "two sovereign masters" of mankind, with an ambiguity of meaning, because their governance covers both the "standard of right and wrong" and the "the chain of causes and effects" guiding human behavior.

> They govern us in all we do, in all we say, in all we think: every effort we can make to throw off our subjection will serve but to demonstrate and confirm it. In words a man may pretend to abjure their empire; but, in reality, he will remain subject to it all the while. The *principle of utility* recognises this subjection, and assumes it for the foundation of that system, the object of which is to rear the fabric of felicity by the hands of reason and of law. Systems which attempt to question it deal in sounds instead of sense, in caprice instead of reason, in darkness instead of light. [9, chap. I: 1, 2]

In his philosophical construction, Bentham included both the "extra-regarding pleasures" and the "self-regarding pleasures"; he did not mean to confine pleasure to merely egoistic feelings or to the selfish nucleus of the human personality. Bentham attempted a sophisticated analysis of the sensations of pleasures or pains, both as physical perceptions and as mind phenomena, or as feelings arising from anticipations and conceptual imaginations. He differentiated the value of pleasures or pains according to various dimensions (intensity, duration, propinquity or remoteness, fecundity, purity) or, when considering groups of people, their total extent [9, chap. IV]. He underlined the pleasures of imagination arising in the mind wandering across past, present and future, and the pleasures of expectation, arising by the contemplation of pleasures in the future and associated with the sentiment of belief [9: chap. V].

The meaning of "utility" is derivative, since it is defined by its reference to pleasure and pain or to similar words (happiness, good, advantage) that Bentham juxtaposes with no substantial differentiation.

> By 'utility' is meant that property in any object, whereby it tends to produce benefit, advantage, pleasure, good, or happiness, (all this in the present case comes to the same thing) or (what comes again to the same thing) to prevent the happening of mischief, pain, evil, or unhappiness to the party whose interest is considered: if that party be the community in general, then the happiness of the community: if a particular individual, then the happiness of that individual. [9, chap. I: 3, 4]

In chapter IV, Bentham explains the rational calculus procedure to evaluate the effects of an act in terms of total utility, computing by summation the balance that will justify the good or bad tendency of that act taking into account all his consequences, even when these involve various persons.

> Sum up all the values of all the *pleasures* on the one side, and those of all the pains on the other. The balance, if it be on the side of pleasure, will give the *good* tendency of the act

> upon the whole, with respect to the interest of that *individual* person; if on the side of pain, the *bad* tendency of it upon the whole. [...]
>
> Take the *balance*; which, if on the side of *pleasure*, will give the general *good tendency* of the act with respect to the total number or community of individuals concerned; if on the side of pain, the general *evil tendency*, with respect to the same community. [9, chap. IV: 52].

From a number of passages and unpublished fragments, as from the frame of Bentham's proposals for institutional reform (e.g., the almost forced work in the system of workhouses), it seems that Bentham did not think that all human beings share the spontaneous capacity for rationally maximizing their utility, that is for pursuing their forward looking, genuine self-interest.[1] The guidance of the legislator is needed to direct choices to the maximizing results, and thus promote collective welfare according to the principle of the greatest happiness.[2]

The complexity of Bentham's doctrine and the mere fact that a large part of his writings remained unpublished,[3] available only at the stage of notes or unfinished papers, makes the overall reconstruction of his philosophical views puzzling. Many controversies arose on the nature of the utility calculus. These touch notably on the cardinality or ordinality of the pleasure calculus[4]; on the symmetry or asymmetry of the perceptions of pleasure and pains[5]; on the evaluation of Bentham as a liberal reformer or, on the contrary, as promoting oppressive social control (as notably Foucault argued in relation to the Panopticon)[6]; on the tension between paternalistic state intervention in reforming legislation and the moderate defense of laissez-faire policies.[7]

In our perspective, the critical question in Bentham's utilitarianism is the unsolved tension between the ambition to classify the full range of human feelings, combining them in exhaustive, systematic ways to describe motives in choice and action, and the priority aim to achieve the *felicificus calculus*. The pretense is to outline the procedure to arrive at the precise statement of the tendency of the pleasure balance to incline towards the side of pleasure or pain. In Bentham's philosophical discourse

---

[1] "That the uncoerced and unenlightened propensities and powers of individuals are not adequate to the end without the control and guidance of the legislator is a matter of fact of which the evidence of history, the nature of man, and the existence of political society are so many proofs" (quoted in Hutchison [30: 303].

[2] "By the principle of utility is meant that principle which approves or disapproves of every action whatsoever, according to the tendency which it appears to have to augment or diminish the happiness of the party whose interest is in question: or, what is the same thing in other words, to promote or to oppose that happiness. I say of every action whatsoever, and therefore not only of every action of a private individual, but of every measure of government" [9: chap. I: 3].

[3] *An Introduction to the Principles of Morals and Legislation* (1789) [9], *Table of the Springs of Action* [7] and *Deontology* (1813–1814 [8]) are his most well-known writings; but his unpublished papers offer many further sources.

[4] See among others Guidi [28] versus Lapidus and Sigot [40], who argue for ordinality.

[5] See Sigot [62, 63].

[6] See among others [66], Crimmins [20], Stokes [64].

[7] See Viner [66], Hutchison [30], Sigot [61].

the idea of utility covered a variety of feelings, but the effort was to reduce all its semantic shades within a compact quantitative frame, to obtain a single balance of pleasures and pains. On the side of complexity the detailed description of nuances of feelings should be taken into account, but on the side of calculus all nuances should be reducible to computable evaluation so that a net summation results, deciding action and suggesting normative prescriptions.

A further theoretical question is the nature of pleasure and pain as forces dominating human behavior. Are they natural forces, which impose their dominance to the will as instincts or biological urges? Or do they impose their dominance through the enlightened reason that recognizes the dictates of the principle of utility for achieving the common welfare? In later proposals for utilitarian calculus in economics, different interpretations of the principle of maximizing utility were suggested, reviving the ambiguities Bentham's formulas left open.

The disputable semantics surrounding the idea of utility, with the related interpretations of human behavior, have recurred in the debates on utility theory in economics since the last quarter of the 19th century. In the next section, by way of contrast, the semantic richness of ideas of pleasures and pains and their balance, surfaces looking at their occurrence in the language of poetry in the early 19th century. By exploring their meanings from the literary perspective, choosing just a few literary texts, new nuances appear in the feelings of pleasure and pain that call into question both their transparency and their neat separation. A poem by S. Coleridge, with which J. S. Mill (a scholar in utilitarian philosophy) was conversant, evokes the dark shadow of the fading urge to enjoy life.

3 A Comparison of Bentham's Balance of Pleasure with the Language of Poetry and Fiction: Wordsworth, Shelley, Coleridge, Dickens

William Wordsworth's *Preface* to *Lyrical Ballads* is a presentation of Wordsworth's ideas on the art of poetry and the tasks the poet should accomplish [69]. We cannot examine it as a literary manifesto, but it is interesting to underline that it introduces the idea of pleasure as a fundamental conceptual tool in understanding the poet's purposes and tasks.

> Nor let this necessity of producing immediate pleasure be considered as a degradation of the Poet's art. It is far otherwise. [...] further, it is a homage paid to the native and naked dignity of man, to the grand, elementary principle of pleasure, by which he knows, and feels, and lives, and moves. We have no sympathy but what is propagated by pleasure: I would not be misunderstood; but wherever we sympathize with pain it will be found that the sympathy is produced and carried on by subtle combinations with pleasure. [69: 605]

Contrary to Bentham's search for a balance of pleasure and pain, Wordsworth underlines the poet's task to produce an "overbalance of pleasure". He declares: "The end of Poetry is to produce excitement in co-existence with an overbalance of pleasure" [69]: 609). Pleasure is the fundamental principle moving human life, a principle encompassing the sympathy with other human fellows and the core, "naked" dignity of human beings. This vital principle admits of a nourishing overflow that enriches life. It is poetry's task to produce such revitalizing, inspiring overflow.

> [...] the Poet [...] ought especially to take care, that whatever passions he communicates to his Reader, those passions, if his Reader's mind be sound and vigorous, should always be accompanied with an overbalance of pleasure. [69: 611]

Wordsworth does not refer to selfish felicific calculus.[8] The overbalance of pleasure is connected to the wider meaning of happiness in Wordsworth's poetry: the plenitude of a good life by the community of feelings with other human beings and the excellence reached in the activities of true value. In her article devoted to J. S. Mill and J. Bentham, Martha Nussbaum quotes Wordsworth's poem "Character of the Happy Warrior" as expressing a conception of happiness that fundamentally follows the vision of *eudaimonia* in Aristoteles' *Nicomachean Ethics* [48]. *Eudaimonia* is the flourishing, rich life for the person.

In Percy Bysshe Shelley's poem "To a Skylark" (1820) the image of joy in its purity is embodied in the flying skylark, the "blithe Spirit", that is "like an unbodied joy", "a flood of rapture so divine", "clear keen joyance". This almost mystical image of joy is compared to the compounded feelings of the human soul, which are uncertain in duration and always polluted with some pains. Even the joy of love is polluted by "love's sad satiety".

> We look before and after
>
> and pine for what is not.
>
> Our sincerest laughter
>
> with some pain is fraught.
>
> Our sweetest songs are those that tell of saddest thought.
>
> Yet if we could scorn
>
> Hate, and pride, and fear;
>
> If we were things born
>
> Not to shed a tear,
>
> I know not how thy joy we ever should come near.
>
> (Shelley, *To a Skylark*, II. 86–95)

---

[8] On ideas of pleasure in the Enlightenment and notably in Wordsworth's poems, see the detailed study in Boyson [13].

The purity of joy is unattainable to the human mind. The strict separation of feelings of pleasure and pain cannot be experienced by the soul; in human experience passions are always troubled by the search "for what is not"; even joyous feelings are upset by negative affections that spoil the plenitude of happiness. If in Shelley's poem human joy is always with some pain fraught, Samuel Coleridge's poetry evokes a more extreme theme: the psychic state of mind where it is impossible to experience any feelings of pleasure. Feelings of joy disappear, becoming devoid of meaning; the soul is unable to experience or even conceive whichever form of happiness.

In his youth J.S. Mill experienced this extreme absence of joy. His father James Mill had imposed on him a very strict regime of studies and mental effort that had affected his psychic stability in years when exercise, friendship, budding love relationships, should be the nourishing source of a healthy evolution towards adult life. In his twenties, John Stuart experienced an episode of deep depression, whose nature he narrated in his autobiography.

> I was in a dull state of nerves, such as everybody is occasionally liable to; unsusceptible to enjoyment or pleasurable excitement; one of those moods when what is pleasure at other times, becomes insipid or indifferent; [...]. In this frame of mind it occurred to me to put the question distinctly to myself: "Suppose that all your objects in life were realized; that all the changes in institutions and opinions which you are looking forward to, could be completely effected at this very instant: would this be a great joy and happiness to you?" And an irrepressible self-consciousness distinctly answered, "No!". At this my heart sank within me: the whole foundation on which my life was constructed fell down. All my happiness was to have been found in the continual pursuit of this end. The end had ceased to charm, and how could there ever again be any interest in the means? I seemed to have nothing left to live for. [45, chap. V: 133–134].

He recognized that the feelings he had experienced in this dramatic period of his life were perfectly described by some lines of Coleridge's *Dejection*, a poem that Coleridge had written in 1802. Coleridge's verses of which Mill speaks in his *Autobiography* [45, chap. V: 134] are the following:

> A grief without a pang, void, dark and drear
>
> A stifled, drowsy, unimpassioned grief
>
> Which finds no natural outlet, no relief
>
> In word, or sigh, or tear.
>
> (Coleridge *Dejection: An Ode*, II: 21–24)

In his earlier career as a scholar, Mill criticized Bentham in a sharp critical essay [44]. As J. Viner underlines, in later years he nuanced his criticism towards Bentham; but he always maintained his attention to the variety and complexity of human feelings.

> By the time Mill was working on his *Principles of Political Economy*, he had swung back a large part though not all of the way to Bentham's political theory and moral philosophy. What was left of his revolt was confined mostly to a continued insistence on recognition of the complete range of human feelings and a consequent endeavor to avoid exaggerating the role of rationalistic hedonism in human behavior. [66: 376]

Mill himself described his point of arrival, pointing to personal happiness as the result of purposeful life devoted to some ideal end, such as the happiness of others, the improvement of mankind or excellence in some art, or some other high achievement. The true satisfaction in life does not come from the conscious search for pleasure; it comes from actively pursuing these high aims. It is a way of life that brings contentment to fill a person's heart. He wrote: "Ask yourself whether you are happy, and you cease to be so. The only chance is to treat, not happiness, but some end external to it, as the purpose of life" [45, chap. V: 142].

Mill's utilitarianism tones down and corrects Bentham's pleasure calculus under the influx of Aristoteles' idea of *eudaimonia.* According to M. Nussbaum, in the later papers collected in *Utilitarianism*, his book published in 1863, Mill traces a distinction among pleasures of various quality, more elevated or more vulgar, which contribute differently to the dignity of human beings, although he reaffirms the principle of the maximum happiness for the highest number as the criterion for just action [48].

It is well known that Dickens in *Hard Times* (1854) makes fun of the pedagogical principles of utilitarian imprint typified by the character of the school master Thomas Gradgrind. However, the distance of Dickens from Bentham and utilitarian discourse is controversial among scholars dealing with Victorian literature. His political evolution has been studied, emphasizing the growing influx of Carlyle on his ideas after an earlier phase of sympathy for Bentham's ideas [26]. In *Oliver Twist* (1837–1839) Dickens is critical of the Act reforming the Poor Laws in 1834 (*Poor Law Amendment Act*), an act inspired by the system of workhouses that Bentham had proposed[9] [64]; but Dickens's criticism of the inefficient judiciary system in Great Britain recalls Bentham's effort at rational reform of the judicial system. The image of female love in Dickens's novels offers a further source of distance from utilitarianism. Among the female characters, Agnes in *David Copperfield* and Amy, the protagonist in *Little Dorrit*, suggest the ideal image of contentment and spiritual plenitude to be enjoyed in loving dear persons with unselfish love and in being firmly faithful to one's duties in life.

We do not wish to affirm here any insurmountable barrier between the different languages of culture; various studies attest that the dialogue was rich.[10] The purpose is to point out to the variety of meanings which pertain to the semantic area that utilitarian philosophy pretended to summarize under the idea of utility, in the radical proposal to compress them all within the utilitarian calculus. The richness of ideas of pleasure, joy, happiness in the full range of their nuances raises questions also when we come to the role of the utility function in economic thought since the last quarter of the 19th century.

---

[9] See *Pauper Management Improved*, first published in 1798, and reprinted in 1802 and in 1812.

[10] Connell [19] explores the views of Romantic writers in relation to the economic and political culture of their times against oversimplified interpretations of opposition to the dismal science. In her book on Victorian literature, utility and political economy, Blake [11] denies any radical break between utilitarianism and literary culture in the Victorian period. On Blake see also Ingrao [31].

4 Building Economics on Utilitarian Foundations

Historian M. Blaug notes that in Great Britain in the 1850s, there was a renewed interest in Bentham's thought [12: 283]. In 1855, Richard Jennings argued the marginal utility principle in his essay *Natural Elements of Political Economy*, expressing it in terms of a law of variations of sensations [33]. W.S. Jevons declared that the calculus of pleasure and pain is at the foundation of political economy since his early essay "Notice of a General Mathematical Theory of Political Economy", which he presented in 1862 at the meeting of the *British Association for the Advancement of Science* and later published under the title "Brief Account of a General Mathematical Theory of Political Economy" [34]. In the essay he evoked the feelings of pleasures and pains as being the springs of human action, but he underlined that economic theory only deals with the pleasures and pains related to the ordinary wants and desires. It does not deal with higher human motives, such as those arising from compassion, conscience, moral or religious views, which should be dealt with in other fields of knowledge.

In *The Theory of Political Economy*, first published in 1871[1879 2nd ed.]), opening the chapter *Theory of Utility*, Jevons wrote: "Pleasure and pain are undoubtedly the ultimate objects of the Calculus of Economics". And he explained: "*to maximize pleasure*, is the problem of Economics"[35: 40]. In the Preface to the first edition of the book he had theorized the mathematical science of political economy to be a "Calculus of Pleasure and Pains".

> In this work I have attempted to treat Economy as a Calculus of Pleasure and Pain, and have sketched out, almost irrespective of previous opinions, the form which the science, as it seems to me, must ultimately take. [...] The nature of Wealth and Value is explained by the consideration of indefinitely small amounts of pleasure and pain, just as the Theory of Statics is made to rest upon the equality of indefinitely small amounts of energy. But I believe that dynamical branches of the Science of Economy may remain to be developed, on the consideration of which I have not at all entered. [35: vii–viii]

To this purpose, Jevons dealt with feelings of pleasure and pain in terms of intensity and duration; he adopted a perfect homogeneity assumption between units of pleasure and pain to be measured on the same axis with opposed signs; they are quantities of the same kind differing only in sign [35: chap. II]. There is thus no asymmetry between pleasure and pain, a subject with which both Verri and Bentham had dealt with. These ideas were graphically expressed and the law of the diminishing degree of utility was stated.

> Utility must be considered as measured by, or even as actually identical with, the addition made to a person's happiness. It is a convenient name for the aggregate of the favorable balance of feeling produced- the sum of the pleasure created and the pain prevented. We must now carefully discriminate between the total utility arising from any commodity and the utility attaching to any particular portion of it. [35: 49]

Jevons was conscious of the difficulties of measuring and even clearly conceiving the elementary, infinitesimal units of pleasures and pains; to overcome it, he resorted to the idea that these units should be measured only indirectly and comparatively, by observing the motions of market prices.[11]

As far as utility is concerned, two approaches coexist in his theoretical discourse: a biological approach and rationalist one. Jevons had read both Jennings and Bentham, and he was inspired by both. He explicitly mentioned Jennings's views on "the physical groundwork of Economics".[12] However, in the theoretical study of maximum utility conditions, Jevons adopted assumptions of perfect rationality more than any behavioral idea of how people choose. He speaks of the ideal choice of "a perfectly wise being" [35: 65]. The rationalist approach is adopted notably in dealing with the forward-looking computation of maximum utility under uncertainty. In a state of uncertainty about the quantities of some good the agent will be able to have in his or her possession at some future date, Jevons proposes the procedure of maximizing the expected utility by weighting the anticipated utility from each quantity of the good available at each future date with the estimated probability to have it in one's own possession at that future date. The agent who makes these computations is a perfectly rational being: "a being of perfect good sense and foresight" in his own words [35: 78].

He wrote: "[...]now it is the inevitable tendency of human nature to choose that course which appears to offer the greatest advantage at the moment" [35: 64]. But he discussed the possible divergence of the ideal rational mind from the effective capacity for rationality of people, when they evaluate present consumption versus future consumption.[13]

The utilitarian foundations of economics were argued in philosophical discourse by F. Y. Edgeworth, who developed utilitarian doctrines both in his philosophical works and in *Mathematical Psychics* [24].[14] According to Edgeworth, the Calculus of Pleasure has two branches: the Economical Calculus at the core of Economic science and the Utilitarian Calculus on which Utilitarian Ethics should be built. The Economical Calculus studies the equilibrium of the "hedonic forces" which

---

[11] "A unit of pleasure or of pain is difficult even to conceive; but it is the amount of these feelings which is continually prompting us to buying and selling, borrowing and lending, labouring and resting, producing and consuming; and *it is from the quantitative effects of the feelings that we must estimate their comparative amounts*. We can no more know nor measure gravity in its own nature than we can measure a feeling; but, just as we measure gravity by its effects in the motion of a pendulum, so we may estimate the equality or inequality of feelings by the decisions of the human mind. The will is our pendulum, and its oscillations are minutely registered in the price lists of the markets" [35: 12–13].

[12] "The writer, however, who appears to me to have most clearly appreciated the nature and importance of the law of utility, is Richard Jennings, who, in 1855, published a small book called the *Natural Elements of Political Economy*. This work treats of the physical groundwork of Economics, showing its dependence on physiological laws. It displays great insight into the real basis of Economics; yet I am not aware that economists have bestowed the slightest attention on Jennings' views" [35: 59].

[13] The rationality of choice was discussed in relation to the different capacities of various social or ethnic groups for reaching rational evaluations, a debate we cannot deal with here [53].

[14] For an ample discussion of utilitarian ethics in Edgeworth's thought, see Ranchetti [56, chapter 9 in Ingrao], Ranchetti 56].

push each agent to move towards maximum individual utility, while the Utilitarian Calculus studies the equilibrium of a system in which the hedonic forces push all agents towards maximum universal utility [24: 15–16].

Edgeworth assumed that the utility function is the function measuring pleasure as a psycho-physical quantity, whose ultimate unit is to be conceived as the just perceivable variation; the unit is a theoretical notion, but it is rooted in physiology. The principle of pleasure in utilitarian ethics is the cornerstone of ethical doctrine, admitting the interpersonal comparison of utility.[15] To the two, integrated fields of utilitarian doctrine, the Economical Calculus and the Utilitarian Calculus, Edgeworth attributed an almost theological meaning, going well beyond economic science per se.

> 'Mécanique Sociale' may one day take her place along with 'Mécanique Celeste,' throned each upon the double-sided height of one maximum principle, the supreme pinnacle of moral as of physical science. As the movements of each particle, constrained or loose, in a material cosmos are continually subordinated to one maximum sum-total of accumulated energy, so the movements of each soul, whether selfishly isolated or linked sympathetically, may continually be realising the maximum energy of pleasure, the Divine love of the universe. [24: 12]

In his metaphorical language, the urge pushing to maximize utility is compared to cosmic forces; the movements of each human soul are compared to the forces of energy in the material cosmos. If utility is ultimately explained in terms of psycho-physical perceptions and urges, the movements of single agents and of whole communities towards maximizing utility are understood as pertaining to the divine order of nature. In ethics, however, the utilitarian calculus establishes a normative doctrine for the general welfare.

In Appendix III in *Mathematical Psychics* "On Hedonimetry", Edgeworth went as far as to define a theoretical instrument to ideally measure the psycho-physical sensations of pleasure. He named it the hedonimeter. The hedonimeter is just an ideal machine, but as such it is perfectly designed to register units of pleasure.

> To precise the ideas, let there be granted to the science of pleasure what is granted to the science of energy; to imagine an ideally perfect instrument, a psychophysical machine, continually registering the height of pleasure experienced by an individual, exactly according to the verdict of consciousness, or rather diverging therefrom according to a *law of errors*. From moment to moment the hedonimeter varies; the delicate index now flickering with the flutter of the passions, now steadied by intellectual activity, low sunk whole hours in the neighborhood of zero, or momentarily springing up towards infinity. The continually indicated height is registered by photographic or other frictionless apparatus upon a uniformly moving vertical plane. Then the quantity of happiness between two epochs is represented by the area contained between the zero-line, perpendiculars thereto at the points corresponding to the epochs, and the curve traced by the index; or, if the correction suggested in the last paragraph be admitted, another dimension will be required for the representation. The integration must be extended from the present to the infinitely future time to constitute the end of pure egoism. [24, Appendix III: 101]

[15] "For moral calculus a further dimension is required; to compare the happiness of one person with the happiness of another, and generally the happiness of groups of different members and different average happiness. Such comparison can no longer be shirked, if there is to be any systematic morality at all" [24: 7–8].

Edgeworth's hedonimeter took inspiration from the newborn discipline of psychophysics that was emerging in experimental psychology at the time, notably in the research and theorizing of E. Weber and G. T. Fechner, who conducted experiments on human subjects to relate intensity of stimulus to reactions, these being conceived as both physical and emotional sensations.[16]

Schumpeter was right when he observed that the utility function was introduced into economics heavily charged with both philosophical and physio-psychological contents. It was so beyond the difficulties of measurement, which were admitted and taken into account. The utility concept claimed its status either directly from the philosophical utilitarian roots, or from appealing to contemporary psychological and physiological doctrines on the variations of sensations, which gave a prominent place to sensations of pleasure or pains; or it was argued appealing to some combination of both fields of knowledge. Among the marginalist scholars, Léon Walras was not inspired by hedonist theories or philosophical utilitarianism imbued with Bentham's imprinting; he never adopted any easy assumption of cardinal utility, but still in 1909 in his text *Economie et Mécanique* he defined pure political economy a "psychico-matematical" science [68].

This brief excursion recalls how at the turn of the 19th century the manifold interpretations of utility ideas included different perspectives to explain maximizing utility in behavioral terms. Utility maximization appeared as rooted in variations of sensations and ultimately in natural instinct on biological foundations. Utility maximization was claimed on cognitive foundations, as in Jevons's reference to the choice of a being of perfect good sense, or in normative rationality in utilitarian ethics, where it is a dictate of the rational mind imposing the norm of choice and action. But it was also associated with physicalist metaphors, comparing the abstract *homo economicus* to a hedonist particle subject to forces as a material point in mechanics, or evoking the tendency towards maximum energy, or comparing the equilibrium produced by market forces to the cosmic movements of planets.

5 The Semantic Change from Pleasure to Utility. The Ordinal Perspective and Beyond

This mixed heritage of sensism and philosophical utilitarianism, vague appeals to experimental psychology, or suggestive but poorly scientific mechanist metaphors, proved to be heavy luggage from which the discipline of economics should free itself to claim a proper scientific status. In the first half of the 20th century the economists contributing to the so-called ordinal turn promoted the separation between utility in value theory and philosophical utilitarianism or psychological doctrines of sensations. The focus moved to a discourse on preferences about alternative consumption bundles, as being the primitive concept to be assumed in economic theory, without exploring the psychic processes of their formation, their social or individual roots,

[16] On psychophysics and Edgeworth discussion of the measurement of utility, see Colander [18].

their development or change. The pretense was to advance towards the exclusive consideration of preferences as they are expressed in a virtually complete ranking of all alternatives that are open in individual choice. The status of utility theory remained ambiguous, notwithstanding the ambition to cut out of economic theory any reference to discourses on the psychology of choice or behavioral psychology.

The difficulties of the separation appear clearly in Pareto's thought. In the *Manuale di economia politica* [51, 52], Pareto chose the word "ophelimity" to indicate the subjective utility as evaluated and perceived by agents. He maintained that the agent's subjective preferences should be exclusively represented by the indifference curves dismissing any ambiguous ideas of utility rooted on unnecessary and confusing metaphysical assumptions. Indifference curves describe preferences by utility index-functions, whose unique role is to rank the preferred choices between alternatives. The word "tastes" remains in the theoretical language of Pareto, but his effort was to cut out all utilitarian assumptions of vague philosophical content.

Unsolved issues appear clearly in his discourse.[17] Pareto's language is imbued with mechanical metaphors as if some elementary psychic force or physiological urge were acting to forcefully push the agents towards the objects of their desires. The Paretian agent seems to move almost mechanically when making his or her preferred choices. The mechanical metaphors suggest that needs and desires act on the agent as a powerful force of attraction that may be balanced only by other, contrasting forces until some equilibrium is reached [49: 132–133]. In other passages, the agent as *homo oeconomicu*s is a hedonist molecule, a metaphor still imbued with an elementary utilitarian substratum. But the *homo oecomicus* is just a provisional theoretical construct in the radical abstractions of economic science, and Pareto underlines that the complexity of human behavior cannot be interpreted remaining within the language of pure economics. Complementary scientific disciplines should address other aspects of motives and beliefs; the overall synthesis is left to the comprehensive language of sociology [50]. Thus Pareto denies any elementary behavioral interpretation of utility maximization; paradoxically, cleaning pure political economy from the metaphysical residuum in utility functions deprives pure political economy of substantial interpretative contents, which are the subject of other sciences, notably sociology.

By emphasizing that the hedonist molecule is a fictional character in theoretical discourse, in his representation of preferences Pareto rejects the procedure to infer indifference curves from the casual observation of choices. He was aware that preferences depend on the assessment of the satisfaction the consumer experienced by consuming goods. Stable preferences imply the familiarity of the consumer with the set of goods which as rational agent he or she has the opportunity to choose. Preferences (and thus choices) depend on the comparative evaluations, which people acquire through their experience as consumers, which in turn depend on social status and customary bundles of consumption. To ask the peasant woman about her preferences between diamonds and rubies is an idle task: she never had the opportunity to appreciate these goods comparing the satisfaction they provide in the social context

[17] On the controversial interpretations of utility in Pareto see Bruni and Guala [15].

in which they are consumed. If people consumption bundles are largely influenced by their customary incomes and status, as Pareto underlines, the indifference curves remain a virtual, logical construction that might be unattainable by direct empirical observation. To circumvent this difficulty, in pure economics Pareto assumes a fictional consumer, who has confirmed his or her preferences by repeated choices, and who is not subject to being disappointed or making mistakes.

In this perspective, the indifference curves on which the maximizing choice is reached are a highly abstract theoretical construct, not at all the result of immediate reports or surveys. In Pareto's thought, the primitive notion of preferences, far from being disconnected from philosophical premises, requires an elaborate epistemological construction. And, of course, the idea of a stable, complete preference ordering demands a philosophical or behavioral frame of interpretation to be applied in interpreting choice and behavior. The idea of preference orderings per se assumes some vision of the way the human mind works in choosing and acting that ultimately requires philosophical and psychological foundations.

In his book *Value and Capital*, first published in 1939, John Hicks followed Pareto in assuming indifference curves as the foundational datum in theorizing optimal choice, reducing utility functions to indexes of the complete orderings of preferences about alternative bundles of goods. In this semantic passage, there was again the effort to expel any philosophical postulates or psychological doctrine on which economics might appear to be dependent. The aim is to protect economic theory within its own boundaries to isolate the theoretical frame of optimal choice on the coherent rankings on choice sets. The utility functions are indexes, which represent orderings of preferred alternatives within the appropriate set of conditions to guarantee consistency and some basic assumptions. Choice embodies the preferences on the choice sets; preferences are stable or their change is not taken into account or explored.[18]

These assumptions remove the problem of understanding preferences in their dynamic relation with decision taking in behavior. The questions related to the transparency of information, the asymmetric information, the cognitive abilities, the coherence of tastes and their change, the disappointment of expectations in consumption experience, discovery or simple curiosity are removed. Some of these issues would emerge in debates in the last quarter of the 20th century, and again in contemporary debates.[19]

In the ordinal turn the utility function became a numerical index to be used as an operational instrument in maximizing, but it was still anchored to tastes about alternative baskets of goods. It was controversial whether it should be devoid of any behavioral content. In the second half of the 20th century, with the emergence of game theory and decision theory, the cognitive reading prevailed, assuming the search for

[18] "Economics has conventionally assumed that each individual has stable and coherent preferences, and that she rationally maximizes those preferences" [55: 11].

[19] See Scitovski's book *The Joyless Economy* [60]. For an exploration of creative consumption, see Bianchi [10]

utility maximization as the norm of rational choice in the normative reading predicated in game theory or decision theory. In the 1940s the utility function had changed, turning into the expected utility function in game theory [67]. The change marked both the return to a (new) conception of cardinal utility and its radical disjunction from any interpretation linking it to pleasure perceptions or states of the mind.

In game theory, losses and rewards are mere numbers, conventional pay-offs, which symbolically summarize whichever advantages or disadvantages may be gained or suffered in abstract interactive situations represented by zero-sum games or by cooperative games admitting coalitions. The new expected utility function framed in rigorous axiomatization converts the maximization of utility into the search for optimal rewards in symbolic monetary lotteries under conditions of uncertainty. Lotteries under uncertainty have a precise mathematical definition, and the expected utility function is built on a system of axioms that permits the definition of a metrics, going beyond the mere complete orderings defined by the theorists of the ordinal turn.

In game theory, and then in decision theory since the 1950s, pleasure disappears; the expected utility function is dematerialized. Alternative lotteries are the symbolic images of gains and losses that are dealt with as such, under conditions of uncertainty, with no nexus to biological urges, feelings, conscious or unconscious motives or cognitive processes, as they had been debated in the utilitarian language of scholars in the marginal revolution. They have no connection with the hedonic molecule in Pareto's theoretical language. Utility loses its sensorial, affective, psychic value, be it explicitly theorized or covertly assumed, to become just an instrumental variable in rational optimization.

In the 1950s and 1960s a technical discussion took place on the appropriate definition of expected utility (as it is shaped by the system of axioms that von Neumann and Morgenstern introduced to define it) in relation to its ordinal or cardinal nature. Controversies arose on whether in its new axiomatic form the expected utility function permits differential cardinal evaluations, or whether it merely allows the rankings of alternatives lotteries. In their mathematical construction von Neumann and Morgenstern seem to defend cardinality; the expected utility function is to be understood as indicating not just the complete ordering of alternatives, but the intensity of the preference between couples of alternatives.

In 1958, Baumol firmly argued that the notion of expected utility as being an index ranking preferences about monetary lotteries had lost any contact with the philosophical discourse on the pleasure and pain calculus, cutting away any residual link to joy as a psychic, affective phenomenon.

> But note that this is not cardinal utility in the old-fashioned sense. Not a word has been said about successive increments of some item yielding diminishing (or increasing) marginal joy. Indeed, in the old-fashioned sense the N-M index is a sheep in wolf's clothing—it is nothing but an ordinal measure, after all! For it has been used to predict, but to predict only rankings of lottery tickets! [3: 666]

The expected utility function had no connection at all with the idea of utility as intensity of feelings, of pleasure or with whichever idea of marginal variations of joy.

> It is not the purpose of the Neumann-Morgenstern utility index to set up any sort of measure of introspective pleasure intensity. Such a measure of "strength of feelings" is totally unnecessary in the theory of games for which the N-M utility theory was constructed. Rather the utility measure was set up for purposes of calculation, or rather of prediction (in the subtler sense of the word) to permit the theorist to *determine in the absence of the player* which of several risky propositions the player will *prefer*. For the solution of a duopoly game must predict which strategy each player will choose (prefer), and the theory must therefore be able to predict how each player will rank risky strategic decisions.
>
> The theory makes this prediction not by measuring subjective utility but by assuming preference patterns for the individual, patterns which have sometimes been described as " rational." [3: 665]

In various interpretations the so-called "instrumentalist view" prevails: the expected utility function is interpreted as a tool to predict choice that has no psychic correlate in the mind of the agent; in other interpretations, the mentalist view persists, with reference to the desire for money and the psychological attitude towards risk [46: 212 ff., 277]. We shall not deal with the debates about measurement, which have been attentively reviewed by I. Moscati, who argues that progressively the distinction between the cardinal versus the ordinal measurement of utility lost its relevance, and the point of arrival was a new indistinction between the ordinal and the cardinal approach reached through more sophisticated conceptions of measurement [46: 209].

In the course of these controversies, a more weighty meaning of utility resurfaces. Economics, after all, is a discipline that has to account for interpretations of human affairs, and to provide advice on policy matters. In [29], Harsanyi, vindicating the perspective of individualistic ethics, claimed the need to provide interpersonal utility comparisons, to build evaluations about social well-being respecting individual preferences. He vindicated the effort to build social welfare functions that could be firmly rooted in factual knowledge.

> In any case, it should now be sufficiently clear that interpersonal comparisons of utility are not value judgments based on some ethical or political postulates but rather are factual propositions based on certain principles of inductive logic. [...] Thus if we accept individualistic ethics and set public policy the task of satisfying the preferences of the individual members of the society (deciding between conflicting preferences of different individuals according to certain standards of impartial equity), our social welfare function will always tend to take the form of a sum (or mean) of individual utilities; but whether the weights given to these individual utilities have an objective basis or not will depend wholly on the extent of our factual (psychological) information. [29: 319–320, 321]

More than thirty years later, M. Allais complained that cardinal utility had been excluded from the economic discourse; he advanced the proposal for going back to cardinal utility to reach evaluations on collective welfare and distribution policies, which necessarily require interpersonal comparisons of utility [1, 2]. According to Allais, Savage, on the contrary, had dismissed cardinal utility as a mythology [2: 104].

If the utilitarian philosophy that imbued Jevons' or Edgeworth's writings was put aside, an ambiguous semantics emerges, oscillating between persistent mentalist interpretations and the effort to reduce the expected utility function to a merely instrumental role to express rankings. But the ambiguity remains, since the new variable is applied to build models representing human choices in a variety of fields, not just in theoretical bets about monetary lotteries.[20]

6 The Explicit Return to the Lexicon of Utility and Happiness

It is difficult to conceive choice in the comprehensive sphere of human affairs without analyzing the psychology of motives and decision making within the social context in which people act. Preferences, which should be the primitive foundation on which economics is built in full independence from any psychological doctrine of behavior, are a complex construct; they require both philosophical interpretation and some psychological doctrine to acquire meaning. The tension was already evident in Pareto's thought. In contemporary decision theory, a tension is again present between the descriptive side of choice (a positive side that cannot be cancelled from the conceptual space of economics), and the merely virtual representation of choice, pretending to exclude any theorizing about how choice happens and how preferences take shape. The contemporary controversies touch on the contrasting interpretations of preferences as either naïve, primitive desires or as sophisticated judgments, involving the exploration and weighting of the whole spectrum of alternatives [47]. In turn, the assessment of these issues involves a complex evaluation of the epistemological status of decision theory.

As far as the subjects dealt with in economics are concerned, the need persists for building narratives of human actions, as in the long tradition of accounting for motives, emotions and reasonings in human behavior that marks other human sciences and the humanities at large. Because of the very nature of their discipline, economists offer narratives about economic events and behavior; the narrative pressure pushes research away from merely instrumental ideas of utility towards psychologically founded interpretations. Preferences, the independent basement on which to build maximizing choices in economics, require conceptual examination, and the philosophical foundation cannot be cut out from economics. It is not easy to account for the motives of economic behavior without involving a weighty appeal to the utilitarian doctrine, or to its rejection. Since the 1970s, new narratives of utility emerged to explain behavior both in the economic sphere and in wider social contexts. Two major currents of research went back to the old utilitarian imprint as regards both the behavioral interpretations and the policy issues.

The explicit return to the utilitarian tradition characterized the line of research aiming at extending the utility maximizing model to a new range of narratives. The method should be applied not just to choices about symbolic monetary rewards, but

[20] The wide field of applications was the ambition of Morgenstern in his collaboration with von Neumann on game theory, as he argued in a number of his articles about the role of game theory in the social sciences.

to more carnal affairs. Gary Becker, Nobel prize in economics in 1992, was the most notable scholar to promote this return. Becker aimed at revising, and reviving, the utilitarian tradition in economic thought enlarging its range of applications, and reinterpreting the utilitarian approach as a general approach to human behavior [4, 6]. He proposed an economic approach to choice based on a new form of utilitarian calculus "weighing the advantages and disadvantages of alternative actions" [5: 402]. He denied looking only at self-interest as the core motive. As he explained in his Nobel Lecture, his aim was to extend the analysis of individual rational choice "to incorporate into the theory a much richer class of attitudes, preferences, and calculations" [5: 401]. The rational weighing of alternatives in maximizing could be imagined as incorporating all kind of preferences accounting not only for self-regarding, but also for other-regarding feelings. Preferences incorporate, e.g., the prejudices about some social group if they might prove relevant in the analysis of discrimination, or they include the feelings of duty towards members of one's own family or circle.

In his Nobel Lecture, Becker argued the reasons for his research approach, explicitly quoting Bentham.

> I was not sympathetic to the assumption that criminals had radically different motivations from everyone else. I explored instead the theoretical and empirical implications of the assumption that criminal behavior is rational (see the early pioneering work by Bentham [1931] and Beccaria (1797[1986]), but again "rationality" did not necessarily imply narrow materialism. It recognized that many people are constrained by moral and ethical considerations, and did not commit crimes even when they were profitable and there was no danger of detection. However, police and jails would be unnecessary if such attitudes always prevailed. Rationality implied that some individuals become criminals because of the financial and other rewards from crime compared to legal work, taking account of the likelihood of apprehension and conviction, and the severity of punishment. [5: 390]

Benefits and costs are not visualized in terms of pleasures and pains, as in the old lexicon; but their semantic field share much with the original Benthamite frame. In Becker's language the utility calculus came back as the comparison of benefits and costs in all life choices, much as in the original work of Bentham that was explicitly mentioned as a source of inspiration. Becker's rational choice analysis builds on maximizing behavior to deal with investment in human capital, the allocation of time, discrimination against women or other groups, and notably family behavior.[21]

> The point of departure of my work on the family is the assumption that when men and women decide to marry or have children or divorce, they attempt to rise their welfare by comparing benefits and costs. So they marry when they expect to be better off than if they remained single, and they divorce if that is expected to increase their welfare. [5: 395]

The maximization of utility is the paradigm to narrate behavior in the various situations of life, from marriage to education to crime, and even drug addiction. The abstract reference to lotteries in decision theory was put aside in favor of going back to the stylized description of human purposes and reactions, all of them framed to be

[21] For this analysis see e.g. Kahneman, Krueger [37].

told in terms of maximizing utility; but utility was not addressed per se, conceptually, as a psychic state or a rational norm, as in the philosophical constructions of the 18th and 19th centuries. The utility function becomes the concise, operational variable to summarize the whole variety of motives, feelings, emotions, which concur in producing perceived advantages or penalties, these being in turn not explored in their emotional foundations. The whole set of human actions, which involve affections such as love, duty, hate, esteem, contempt, violence, crime, should be narrated within a universal paradigm of rational maximizing choices built around the "utility" variable; whatever the word utility might mean in the new context. The economic semantics of utility became coextensive with and encompassing the whole set of different meanings, which in other disciplines deserve a range of detailed and attentively defined terms to account for behavior. It was a shortcut in both conceptual content and terminology that suggested the return to the dominance of Bentham's two masters, since human motives and actions were all to be read through the synthetic variable named utility.

Another Nobel prize for economics, Robert Lucas, underlined the straightforward simplicity of this approach, compared to explicitly dealing with the variety of the psychic and cognitive processes in different human minds.

> I think the basic view of economics that Hume and Smith and Ricardo introduced, taking people as basically alike, pursuing simple goals in a pretty direct way, given their preferences, where you are trying to explain differences in behavior by differences in the situation people are finding themselves in rather than differences in their culture, their inner wiring, inner workings, their race, whatever, their class, just thinking about people as people and then trying to account for their behavior in terms of how they are responding to their environment, that this is it for economics. [41: 21–22]

Utilitarianism was openly back in economics but still under the ordinality pretense of independence, and with an uncertain epistemological status. Strong statements about behavior and how people react to their environment seemed to require neither explicit philosophical reasoning nor any support from psychological or sociological doctrines, in striking contrast to the research path that Pareto had recommended.

Meanwhile, there was a parallel return to an explicitly utilitarian discourse through the opening of the new field of "happiness research" in economics. The starting point was the examination of the controversial relationship between increasing income and the subjective perception of "happiness". The so-called "Easterlin's paradox" is the statistical observation according to which over time, in the long term, the state of happiness does not increase in a linear relationship to income in comparing different countries with diverse levels of income or in comparing different groups enjoying various levels of income within the same country [22, 23]. The paradox was extensively discussed, and variously explained on the basis of *tapis-roulant* effects for the progressive adaptation to higher levels of welfare (the hedonic treadmill), for increasing aspirations to higher welfare goals (the satisfaction treadmill), or for the perception of relative status as a source of differential welfare in social life (the positional treadmill). Through this channel, "happiness" acquired a new legitimacy in economic literature, and its measure became the subject of academic theorizing and statistical studies. Happiness comparisons, which appeared doubtful to Jevons or

Walras, were revived. The United Nations produces happiness reports, considering comparisons among countries all over the world.

Thus, since the 1970s there has been the return to the lexicon of "happiness" at the core of philosophical utilitarianism; but this lexicon is slippery, instead of being the subject of detailed philosophical enquiry as in the late-18th-century literature. Happiness as captured by statistical indexes fails to be examined with the accuracy that Verri or Bentham devoted to examining pleasures in their multiple dimensions, shades of intensity and multifaced nature. The contemporary literature on happiness speaks of happiness with a variety of words: "individual well-being", "satisfaction", "utility", "subjective well-being", "pleasant and unpleasant affect", "life satisfaction", "happiness".[22] Meanings are multiple and elusive: different words are introduced as "various notions of subjective well-being". The dominant attribution of meaning refers to happiness as the subjective evaluation of the overall satisfaction with one's own life.[23] It is admitted that being in a state of more or less happiness is a subjective evaluation; each person may define this state of grace in his or her own ways.[24] But happiness is measured by numerical indexes and data banks are built to assess the state of subjective well-being among social groups all over the world, failing the inquisitive look that the early philosophers of utilitarianism devoted to the idea of pleasure or the controversial assessment of the balance of pleasures versus pains in human life.

The explicit return of elementary notions of happiness stimulated historical research on the debates in the 18th century. A number of historians examined texts dealing with happiness in the 18th century, with mentions of Aristotelian eudaimonia or civic virtues as sources of relational happiness [14–16]. In the overall production of economics, it is a niche research program, but valuable studies were produced [17].

From the critical perspective, A. Sen and M. Nussbaum developed the approach to the evaluation of personal well-being based on their innovative definition of capabilities. In evaluating welfare, the capabilities approach looks at the opportunities to freely have access to the fundamental spheres of human experience, and thus at the true possibilities of choice open to each individual person in his or her life.[25] An interesting aspect of Sen's and Nussbaum's exploration about the unequal access to the enjoyment of life is that the statements about aims or desires are not necessarily genuine; unequal societies severely limit the poor's aspirations, and downsize the dreams of the discriminated social groups. Social life often silences people's true desires.[26] If this is the case, the subjective evaluation of life's satisfaction is not necessarily the genuine assessment of accomplishment; life experience shapes

---

[22] "Throughout this article, I use the term happiness, subjective well-being, satisfaction, utility, well-being, and welfare interchangeably" [23: 465].

[23] "Happiness" is, for simplicity's sake, in the following interchangeably used with the term "satisfaction with life" and "reported subjective well-being" [25: 918].

[24] " [...] each individual is free to define happiness in his or her own terms" [23: 466–467].

[25] The literature on capabilities is large and we cannot attempt here any review or assessment.

[26] In another perspective, the silencing of desires by repressive mechanisms in social life is dealt with in Foucault's thought.

aspirations under the heavy constraints imposed by social rules and poor resources. The limits people are subject to in their environment distort their view of which joys they may dream of. It is a complex subject, for the scholar runs the risk of compiling a stereotyped list of which factors should contribute to human accomplishment.

Once again, the complexity of preferences surfaces, with the difficulty of disentangling them from ideological horizons.

7 Going Back to Utility in Experimental Psychology

In more recent years the immediate sensory reading of utility has also been explored. Side by side with the emergence of the happiness research program, whose utilitarian roots are plain, a utilitarian trend of research was born through the encounter between experimental psychology and economics [38]. Psychologist D. Kahneman, going back to Bentham's conception of utility, elaborated research tools to capture in experimental tests the fragmentation in time of instantaneous feelings of pleasure or pain, objectively measuring them moment by moment.

Kahneman defined the concept of "experienced utility" in continuity with Bentham's theory, to be distinguished from the notion of "decision utility" adopted in decision theory ((Kahneman et al. 2000)). "Experienced utility" (the observable experience of pleasure or pain as it is immediately perceived or its assessment built up in remembering past experience), should be conceptually distinguished from the expected utility function in decision theory, or from utility as it might be theoretically defined by looking at the orderings of preferences in choice. In [39], Kahneman and his coauthors Wakker and Sarin explicitly claimed that their approach is in continuity with Bentham's theorizing.

> To distinguish the two notions, we shall refer to Bentham's concept as experienced utility and to the modem usage as decision utility. With few exceptions, experienced utility is essentially ignored in modern economic discourse. The rejection of experienced utility is justified by two standard arguments: (i) subjective hedonic experience cannot be observed or measured; (ii) choices provide all necessary information about the utility of outcomes because rational agents who wish to do so will optimize their hedonic experience. [39: 375]

This research line in experimental psychology maintains that the immediate hedonic experience with direct physiological foundations is observable and measurable. In the same essay, Kahneman and his coauthors argued that perceived utility may be measured by objective measures on the real time axis, with the aim to report "instant utility".[27] "Instant utility" as measured moment by moment is to be distinguished from "remembered utility", that is utility as it is subjectively perceived and

[27] "Pleasure and displeasure are attributes of each moment of experience, but the outcomes that people value are normally extended over time. The basic building block of experienced utility in our analysis is instant utility: a measure of hedonic and affective experience, which can be derived from immediate reports of current subjective experience or from physiological indices. Instant utility corresponds to the dimension of 'intensity' in the writings of Bentham, Jevons, and Edgeworth" (Kahnemanetal. [39: 376].

reported about past experiences extending in time, assessing in recollections the intensity of their overall pleasant or unpleasant nature. Since measurement is based on methods that register the subjective perceptions moment by moment (the so-called "moment-based methods"), this methodology leads to a notion of total utility, which is founded on empirical measurement according to scientific methodology, although it records the subjective states of welfare (or suffering) as reported by the persons involved in the experiments. The moment-based approach registers the hedonic experience in objectively real-time measures of the pleasure and pain experienced by the subject, reporting both its pleasurable or painful nature and its intensity. The total utility results exclusively from the record of the moment-utilities during the examined episode [36].

The objective measurement of utility may not arrive at the same evaluation emerging from reports by other surveys on the subjective state of welfare, because it is the accurate record of the physic or affective state of the subject as attested moment by moment according to some signaling, as the subject might signal the perception of a color or a sound, reacting to its varying intensity [36]. It does not take into account the remembered utility about some life episode such as it is perceived and expressed after the immediate experience has already past. In adopting this approach to experimental psychology, Kahneman maintained that it is possible to admit, with reasonable plausibility, the interpersonal comparisons in measuring utility, notwithstanding the well-known difficulties. However, objective happiness as measured by these procedures is not adopted as a notion of global welfare, but just as a partial component of each person's well-being. Kahneman and Krueger [37] advocate a plurality of measures of subjective well-being. They recognize that there is a complex interaction between total utility and remembered utility, because persisting memories sustain in present experience a positive perception of past experiences or, on the contrary, affect present happiness with the shadow of negative recollections from the past. Notwithstanding the appeal to the ultimate fragmented perceptions of instant utility, the processing of past experience in conscious psychic evaluation cannot be severed from the overall state of well-being in human minds.

These conclusions suggest the need for more intricate interpretations of human behavior than the exact measurement of variations in hedonic experience through some updated versions of Edgeworth's hedonimeter. Notwithstanding the references to Bentham, there is no pretense to force the interpretation of human behavior as reducible to the unique dominance of pleasure and pain. Moreover, experienced utility being an objective variable, the reality of utility maximization should be assessed and checked as an empirical proposition, to see if the outcomes of behavior support or fail to support the conjecture of maximizing behavior.

The effort by Kahneman to revive an explicit Benthamite approach to utility theory remained a niche trend in economic research, notwithstanding its success and visibility. The utilitarian imprint, more underground or in more overt presence, was not explicitly vindicated in clear philosophical or psychological language as being at the root of the approach to choice and action in economic theory. For most economists the utility function in economic models is a logical instrument for building up a rigorous, mathematical theory of choice. In the last section we shall look at the covert, ambiguous utilitarian theory of choice to compare it with the narratives of some literary texts, where human choices are focused on nuanced, dramatic psychic processes. By way of contrast, we look at aspects of choice and behavior, which cannot be forced into the straitjacket of rational, forward looking, maximizing utility, and cannot be examined through utilitarian spectacles.

8 Choice as Change and Loss from the Perspective of Literary Narrations

The issues we discuss touch on major literary themes; the specific stories and novels which we quote are but a few examples from a wide set of relevant literary sources. Let us list the points to explore: the fluidity of preferences and the disappointment of expectations; the plural character of the self and the auto-representation of the self; the disjunction of choice from action, and action from choice; the gratuitous acts of benevolence, and the difficulty in tracing a neat line between self-regarding and other-regarding pleasures; the opacity of desire.

Contrary to the assumption that the agent's preferences on alternatives are complete and consistent, the change and instability of preferences in life experience are major subjects in literary masterpieces. The fluidity of preferences is, of course, the main subject in innumerable stories dealing with love and sentimental life; but it is dealt with also in relation to preferences as regards tastes for fashion, lifestyles, friendship, work. Preferences evolve with the evolving path of life, notably when the characters are changing to grow and mature through life experience.

Life experience may bring the radical disappointment of expectations, those expectations which embody the preferences on which the choices had been taken. Let us just remember the title of Balzac's novel *Lost illusions*.[28] Innumerable characters in fiction have to face lost illusions, and to elaborate strategies to face the disappointment of the hopes, on which they had built their search for happiness. It is not just some disappointment imposed by external events, which had not been correctly foreseen in their impact on future life. It is the troubling discovery of the frail, illusory nature of our own desires. The feelings of deep disillusionment about intimate desires, aims, purposes that had been pursued nurturing cherished hopes of happiness is at the center of innumerable stories, from the narratives of simple characters in their common lives to the most moving tragedies. The subject is dealt with in religious doctrines and moral philosophy. The disappointment of expectations evokes the process of discovering true desires from painful life experiences, after wandering in the quest of illusory hopes of pleasure until reaching the true source of contentment within the deeper self. But human persons often fail in their search for happiness, and their life adventure comes tragically towards its end on a feeling of disruption. Happiness, far from being maximized, appears as a dream

[28] *Illusions perdues* in *La Comédie Humaine*, V, Bibliothèque de La Pléiade, Paris: Gallimard, 1977 (1st ed., 1837–1843).

never to be attainable. As Macbeth cries in his despair, life appears "a tale told by an idiot, full of sound and fury, signifying nothing" (Shakespeare, *Macbeth*, Act 5, scene 5, lines 16–27).

In literary characters, the inner self is not a consistent unity of purpose, tastes, choice. The multiplicity of our both affective and cognitive personality is a subject well known and explored in psychology, and again and again dealt with in literary masterpieces. The double or plural self of literary characters suggests the denial of an a priori coherent ordering of preferences dominating choice. In various narratives, manifold psychic identities cohabit within the same character, of which the character may be more or less conscious. On this theme, Virginia Woolf built her novel *Orlando* [32]. The character is torn, and eventually breaks down, because of contrasting aspirations and desires.

In an intriguing literary dialogue, Luigi Pirandello created two characters, *Gran me* (Big me) and *piccolo me* (Little me), who have to cohabit in the same body and mind, accommodating their opposed desires and contrasting choices within their common bunch of daily hours to live and enjoy.[29] *Piccolo me* is prone to enjoying food, parties, love and a relaxed family life with a nice wife; *Gran me* is devoted to scholarly studies and meditation. They dialogue in the inner self of the same man; their preferences clash, but they have to cohabit in his daily life. In many literary works by Pirandello, the self is not a coherent, unique agent; each character is a manifold personality, whose multifarious identity depends on how people in the community to which the character belongs (family, friends, colleagues, fellow citizens) project imaginary constructions of his or her true nature and behavior. Each human person is one, no one and one hundred thousand, as in the title of Pirandello's novel *Uno, nessuno e centomila* dealing with the difficulty to free one's own conscious self from other people's projections. In the novel *Il fu Mattia Pascal*, a man takes advantage of his presumed death to escape his unbearable family life; after experiencing his social death, he frames fluid identities for himself, but always under new constraints from his social environment.

The plural self of literary characters has a crucial time dimension, when it is rooted in the development of each character through life experience, and the troubled search for one's path in life. In Tolstoy's novel *War and Peace*, various characters go through unexpected changes, and at crucial moments of their lives they show new sides of their inner self, till then not yet coming to surface. In a memorable dialogue, princess Marija Bolkònskàja reveals her feelings to Nikolaj Rostov, till then remained unexpressed dreams under the cover of her usual shyness and restraint. Both Natasha Rostov and Pierre Besuchov deeply change their aspirations and attitudes, evolving through hardship and love.

"Very well then I contradict myself, (I am large. I contain multitudes.)" writes Walt Whitman in his poem *Song of myself* (section 51). It is a deeply human,

---

[29] "Dialoghi tra il Gran me e il piccolo me" was written between 1895 and 1906; Pirandello anticipated its subject in a private letter to his future wife in 1894 portraying the conflict as being his own personal conflict. The story was later included in the *Appendix to* Pirandello's collection of short stories *Novelle per un anno* [54].

universal experience, as Borges forcefully signifies in his poem devoted to the Greek mythological god Proteus.

> De Proteo el egipcio no te asombres,
>
> tú, que eres uno y eres muitos hombres.[30]
>
> (Borges, *Proteo*: 13–14)

"You", are of course all of us, as human beings. We, as choosing and acting persons, are all "one" but also "many others". If this multiplicity of the self brings contradictions, or even splits the personality, to be "large" and to contain "multitudes" is also the key to understanding other people by feeling empathy with their feelings and sufferings. Literary work on the fluidity of choice signals the possibility of evolution and development in human experience. Fluidity, flexibility and change are attributes of both emotional and social intelligence. In this perspective, the stability of preferences and choices is a sign of stubbornness and close-mindedness more than of rationality.

A further intriguing aspect of human experience, which is narrated in many literary masterpieces, is also of great relevance. Are choices, even those that are consciously assumed, always resulting in states of mind that may be characterized as maximizing? This is certainly not true of many affective choices, but also of other choices in life. Each choice, by imposing limits, implies constraints, and the loss of what has not been chosen. The path not taken could have been the source of further human development. Choice often leaves scars, cutting out some sides of past experience, hopes, dreams. People choose, but they suffer from their choice, even if they deliberately choose the course they choose. They suffer if the choice cuts away deep affections, or it maims some sides of their personality. It is what happens to Archer Newland, the protagonist of Edith Wharton's novel *The Age of Innocence*, when he chooses to stay with his wife and go on in his established life in the community to which he belongs. There is no going back to unconventional choices.

Some choices are impossibly overbearing. The loss will be forever. Viktor Pavlovic and Maria Ivanovna, two characters in Vasily Grossmann's *Life and Fate*, experience tormenting feelings of tearing apart, because they have to acknowledge that their impossible love cannot go on. Their choice is mixed to desperate feelings of irretrievable loss.

> One thing was plain: he had lost his peace of mind for ever. Whatever happened, he would never know peace. Whether he hid his love for the woman besides him or whether it became his destiny, he would not know peace. Whether he was with her, feeling guilty, whether he was apart from her, aching for her, he would have no peace.
>
> She was still staring at him. Viktor found her look of mingled happiness and despair almost unbearable.
>
> [27: 691]

---

[30] "Proteus the Egyptian should not surprise you, you, who are one, but also many others". The poem is in the collection *La Rosa profunda*.

In the approach to rational, forward-looking maximization, the act is assumed to be coherent with the choice as much as the choice is in conformity with the preferences. Act and choice are not disentangled as two different stages; they are not disjoint in the sequence of mind deliberation and acting. It is assumed that acting reflects the maximizing choice, as much as the maximizing choice embodies the stable preferences on which it is deliberated. The three stages, including the cognitive exploration of the preference set, the consistent optimal choice, the act that follows from it, are assumed to be coherent each with the following one in an instantaneous present. If the mentalist interpretation of choice is refused, as in the interpretations refusing any behavioral reading of rational optimization, choice remains unexplained; but still it is assumed that it predicts the act.

A disturbing, but common experience is the dissociation of act from previous choice. It is a common occurrence in daily life, whenever people are unable to pursue the course that they know they should pursue for the better, be it for their future advantage or to achieve immediate pleasure. People often fail to pursue some chosen course of action, for any of a variety of reasons, indecision, habit, fear, pulsion, social pressure or whatever. The road to hell is paved with good intentions. The time dilates, while the act fails to follow the choice apparently clearly adopted in the mind. In so many stories for weakness of character or other reasons some act is again and again delayed. In Italo Svevo's novel *La coscienza di Zeno*, each of innumerable cigarettes is the last one, according to stated intentions. In fact, the "last cigarette" is continuously delayed.

In many stories the time between choice and act is stretched or suspended. To act in conformity with choice is not just the immediate resolution after some rational procedure of maximizing achieves optimal choice. The act is delayed, or it will never follow. In tragic choices the passage from deliberation to act is a tragic choice per se, because of the irreversible consequences the act will bring. In the sequence from mind deliberation to acting dismay captures the heart, when facing fatal consequences. Macbeth delays killing Duncan, and provokes Lady Macbeth's reproach: "Art thou afeard to be the same in thine own act and valour as thou art in desire?". Some acts change forever a human life. The passage from mind planning to acting opens up a new fate for the man or the woman, who conceived them in desires.

In other stories some crucial act takes the scene, never asking the permission to be acted as a consequence of choice. In human life events simply happen. It is a common experience in daily life: people act as if they were guided by some inner command. In the life of the mariner Billy Budd, the irrevocable event happens without being anticipated in decision. "The next instant, quick as the flame from a discharged cannon at night, his right arm shot out, and Claggart dropped to the deck" [43: 298]. The right arm springs, the fist strikes, the man is killed. The gesture brutally changes the course of Billy's life; he will be hanged. In his cruel fate the gesture that brings the fatal exit is an immediate reaction; no time space divides choice and acting, and no conscious deliberation anticipates it.

The dominance of spontaneous acting over deliberation may impose also acts of benevolent behavior. Acts simply happen of gratuitous pity. Let us read a passage on the inner force of benevolence that affirms its power, beyond clear, conscious choice. The episode is from Vasilyi Grossmann's *Life and Fate*. After German soldiers surrendered to the Red Army, some of them, who are prisoners, are carrying out Russian dead bodies from a cellar, amidst a crowd of Russian people staring at them with hate. One of the German soldiers is a young man, shaking his head convulsively as if he were mad. A woman in the crowd cries at the pitiful sight of the dead body of an adolescent girl. She looks at him, and she seems ready to throw a brick at him.

> The woman could no longer see anything at all except the face of the German with the handkerchief round his mouth. Not understanding what was happening to her, governed by a power she had just now seemed to control, she felt in the pocket of her jacket for a piece of bread that had been given to her the evening before by a soldier. She held it out to the German officer and said: "There, have something to eat."
>
> Afterwards she was unable to understand what had happened to her, why she had done this.
>
> [27: 789–790]

Many writers narrate the mystery of acts of benevolence, kindness, pity, which affirm their right to be acted in instantaneous reaction, or which become constant habits of behavior. Bentham contemplated other-regarding pleasures, which still imply that acting in regard of other people's well-being is oriented to the additional pleasure resulting from anticipating that other people will increase their well-being. It is a selfish motive, though other regarding. Certainly, we all experience the strange mixture of self-regarding and other-regarding pleasures, which we may enjoy in acts of benevolence towards other people. Does true unselfish behavior exist? As we reminded above, some women characters in Dickens's novels are enriched by the gift for disinterested love. In Tolstoy's story *The Death of Ivan Illich*, Ivan Illich's servant offers his care going unnoticed, in utmost simplicity, sheltered in his humble social position and almost materially set back in the secluded room of his dying master. A silent character, he acts to alleviate his master's pains, while the nearest relatives leave the father and husband unattended during his final illness. He attends to him with caring gestures, trying to calm the excruciating aches the sick man is suffering: acts of quiet human compassion, unrewarded, and not demanding any rewards. His behavior rises a deep interrogation on the nature of love going much beyond Bentham's other-regarding pleasures.

The last aspect we wish to focus upon is the opacity of desire, that is, the failure to feel any impulse toward joy, happiness, enjoyment, cheerfulness. The very vital impulse to maximize enjoyment is missing; it fails to support both choice and action. This is also a major literary subject. "I would prefer not to..." says Bartebly, the scrivener in the short story by Hermann Melville [42: 25]. As a literary character, Bartleby has the stature of a hero attempting a metaphysical revolt against his fate; but "I would prefer not to..." is a declaration of retreat from life, joy, experience, curiosity, effort. Bartleby the scrivener denies the right to sovereignty to one of the two masters, which Bentham declares the masters of the human kind, and the ultimate motives accounting for behavior. It denies pleasure the right to dictate the actions he should choose. His attitude contrasts with the impulse of maximizing pleasure, which should dominate human choices. "I would prefer not to..." is his radical rebellion against an oppressed life that appears devoid of meaning, hopes, affections.

Stavroghin, one of the main characters in the novel *Devils* by F. Dostoyevsky, suffers a psychic illness of "indifference" that he describes by saying that his desires are always opaque. He confesses his moral illness in the notes by way of confession, where he narrates the crime he committed by seducing and raping the innocent, ten-year-old girl Matryosha, who was living in the house where he lodged. In this confession he writes:

> I, Nikolai Stavrogin, a retired army officer, lived in Petersburg in 186-, indulging in depravity in which I took no pleasure. [...] Just about then I was thinking of killing myself because of my morbid indifference; well, I am not really sure why. [...] In general I found life very boring at that time; I was sick to death of it.
>
> [21: 460, 463, 470]

He narrates again his malaise in his last letter to Daria Pavlova before killing himself.

> Just as before I'm still capable of wishing to do a good deed and taking pleasure in doing so; but at the same time I wish to do evil and also take pleasure in that. But just as before both feelings are always too trivial and never very powerful. My desires are too weak; they can't be a guide to me. It may be possible to cross a river on a log, but not on a splinter of wood. [...] but the only thing to emanate from me is negation without magnanimity, without power. Not even negation has come from me. Everything is always so trivial and stale.
>
> [21: 754]

The lenses offered by meditations about pleasure, identity and choice in some literary masterpieces suggest caution in adopting any simplified synthesis of motives under the ambiguous cover of utility. The purpose of interdisciplinary dialogue is to restore the value of semantic richness, recovering the complexity of utility ideas in all their nuances, in comparison with the richer semantic fields which explore human feelings. We need the richness of multiple meanings to understand human choice and behavior, and we need to restore the philosophical discourse in economics to discriminate on their uses, or their abuses.

References

1. Allais, M.: Determination of Cardinal Utility According to an Intrinsic Invariant Model (1984). In: Allais, H., Hagen O. (eds.) Cardinalism: A Fundamental Approac, pp. 31–64. Springer, New York (1993)
2. Allais, M.: Cardinal utility. history, empirical findings and applications. Theor. Decis. **31**, 99–140 (1991)
3. Baumol, W.J.: The cardinal utility which is ordinal. Economic Journal **68**(272), 665–672 (1958)
4. Becker, G.S.: The Economic Approach to Human Behavior. Chicago University Press, Chicago (1976)
5. Becker, G.S.: "The economic way of looking at behavior", Nobel lecture. J. Polit. Econ. **101**, 385–409 (1993)
6. Becker, G.S.: Accounting for Tastes. Harvard University Press, Cambridge (1996)

7. Bentham, J.: A table of the springs of action. In: Goldworth, A. (ed.) Deontology, Together with a Table of the Springs of Action and the Article on Utilitarianism. Clarendon Press, Oxford (1815 [1983])
8. Bentham, J.: Deontology. In: Goldworth, A. (ed.) Deontology, Together with a Table of the Springs of Action and the Article on Utilitarianism, pp. 117–281. Clarendon Press, Oxford (1813–1815 [1983])
9. Bentham, J.: An Introduction to the Principles of Morals and Legislation. E. Wilson, London (1823)
10. Bianchi, M. (ed.): The Active Consumer. Novelty and Surprise in Consumer Choice. Routledge, London (1998)
11. Blake, K.: Pleasures of Benthamism. Victorian Literature, Utility, Political Economy. Oxford University Press, Oxford (2009)
12. Blaug, M.: Economic Theory in Retrospect, 5th edn. Cambridge University Press, Cambridge (1997)
13. Boyson, R.: Wordsworth and the Enlightenment Idea of Pleasure. Cambridge University Press, Cambridge (2015)
14. Bruni, L.: L'economia, la felicità e gli altri. Città Nuova editrice, Roma (2004)
15. Bruni, L., Guala, F.: Vilfredo Pareto and the epistemological foundations of rational choice. Hist. Polit. Econ. **33**, 21–49 (2001)
16. Bruni L., Porta P.L.: Introduction. In: Bruni, L., Porta, P.L. (eds.), Economics and Happiness: Framing the Analysis. Oxford University Press, Oxford (2005)
17. Bruni, L., Porta, P.L. (eds.): Economics and Happiness: Framing the Analysis. Oxford University Press, Oxford (2005)
18. Colander, D.: Edgeworth hedonimeter and the quest to measure utility. J. Econ. Perspect. **21**, 215–225 (2007)
19. Connell, P.: Romanticism, Economics and the Question of 'Culture.' Oxford University Press, Oxford (2005)
20. Crimmins, J.E.: Contending interpretations of Bentham's utilitarianism. Can. J. Polit. Sci. **29**, 751–777 (1996)
21. Dostoyevsky, F.: Devils (English translation by M.R. Katz). Oxford University Press, Oxford (1871 [1999])
22. Easterlin, R.: Does economic growth improve human lot? Some empirical evidence. In: Davis, P.A., Reder, M.W. (eds.) Nation and Households in Economic Growth: Essays in honor of Moses Abromowitz, pp. 89–125. Academic Press, New York and London (1974)
23. Easterlin, R.E.: Income and happiness: towards a unified theory. Econ. J. **111**, 465–484 (2001)
24. Edgeworth, F.Y.: Mathematical Psychics. An Essay on the Application of Mathematics to the Moral Sciences. C. Kegan & Paul, London (1881). Reprint New York: Augustus M. Kelly
25. Frey, B.S., Stutzer, A.: What can economists learn from happiness research. J. Econ. Lit. **XL**, 402–435 (2002)
26. Goldberg, M.: From Bentham to Carlyle: Dickens' political development. J. Hist. Ideas **33**, 61–76 (1972)
27. Grossman, V.: Life and Fate. Vintage, London (1980 [2006])
28. Guidi, M.E.L.: Jeremy Bentham's quantitative analysis of happiness and its asymmetries. In: Bruni, L., Porta, P.L. (eds.) Handbook on the Economics of Happiness. Edward Elgar, Cheltenham (2007)
29. Harsanyi, J.: Cardinal welfare, individualistic ethics, and interpersonal comparisons of utility. J. Polit. Econ. **4**, 309–321 (1955)
30. Hutchison, T.W.: Bentham as an economist. Econ. J. **66**(262), 288–306 (1956)
31. Ingrao B.: Review of Pleasures of Benthamism. Victorian Literature, Utility, Political Economy, Kathleen Blake, 267 pp. Oxford University Press (2009). Econ. Philos. **27**(3), 346–352 (2011)
32. Ingrao B.: La pile instable d'assiettes. Contes économiques et récits romanesques. In: Baubeau, P., Péraud, A., Pignol, C., Reffait, C. (eds.) Récit romanesque et modèle économique, Romanesque, vol. 7, pp. 67–85. Classiques Garnier, Paris (2015)

33. Jennings, R.: Natural Elements of Political Economy. Longman, Brown, Green and Longmans, London (1855)
34. Jevons, W.S.: Brief account of a general mathematical theory of political economy. J. R. Stat. Soc. **XXIX**, 282–287 (1866)
35. Jevons, W.S.: The Theory of Political Economy, 2nd edn. Reprint in The Palgrave Archive edition of The Writings on Economics of W.S. Jevons, vol. II. Palgrave, New York (1879 [2001)
36. Kahneman, D.: Experienced utility and objective happiness: a moment-based approach. In: Kahneman, D., Tversky, A. (eds.) *Choices, Values and Frames*, chapter 37, pp. 673–692. Cambridge University Press and the Russell Sage Foundation, New York (2000)
37. Kahneman, D., Krueger, A.B.: Developments in the measurement of subjective well-being. J. Econ. Perspect. **20**, 2–24 (2006)
38. Kahneman, D., Tversky, A. (eds.): Choice, Values and Frames. Cambridge University Press, Cambridge; The Russel Sage Foundation, New York (2000)
39. Kahneman, D., Wakker, P.P., Sarin, R.: Back to Bentham? Explorations of experienced utility. Q. J. Econ. **112**(2), 375–405 (1997)
40. Lapidus, A., Sigot, N.: Individual utility in a context of asymmetric sensitivity to pleasure and pain: an interpretation of Bentham's felicific calculus. Eur. J. Hist. Econ. Thought **7**, 45–78 (2000)
41. Lucas, J.R.: My Keynesian education. Hist. Polit. Econ. **36**(Suppl. 1), 12–24 (2004)
42. Melville, H.: Bartleby, the scrivener. In: Coviello, P. (ed.) Billy Budd, Bartebly and Other Stories, pp. 17–54. Penguin Books, New York (1853 [2016])
43. Melville, H.: Billy Budd, Sailor. In: Coviello, P. (ed.) Billy Budd, Bartebly and Other Stories, pp. 243–328. Penguin Books, New York (1891 [2016])
44. Mill, J.S.: An essay on Jeremy Bentham. Lond. Westminster Rev. **29**, 467–506 (1838)
45. Mill, J.S.: Autobiography, 6th edn. Longmans, Green, and Co., London (1879)
46. Moscati, I.: Measuring Utility. From the Marginal Revolution to Behavioral Economics. Oxford University Press, New York (2019)
47. Moscati, I.: On the recent philosophy of decision theory. J. Econ. Methodol. **28**, 98–106 (2021)
48. Nussbaum, M.C.: Mill between Aristotle & Bentham. Daedalus **133**(2), 60–68 (2004)
49. Pareto, V.: Il compito della sociologia fra le scienze sociali. Rivista italiana di sociologia, 45–54. Reprint in Busino, G. (ed.) Œuvres complètes. Ecrits sociologiques mineurs, vol. XXII, pp. 130–139. Librairie Droz, Genève (1897 [1980])
50. Pareto, V.: I problemi della sociologia. Rivista italiana di sociologia, 145–157. Reprint in Busino, G. (ed.) Œuvres complètes, vol. XXII, pp. 165–177. Ecrits sociologiques mineurs. Librairie Droz, Genève (1899 [1980])
51. Pareto, V.: Manuale di economia politica (Montesano, A., Zanni, A., Bruni, L. (eds.)), Italian critical edn. Università Bocconi, Milano (1906 [2006])
52. Pareto, V.: Manuale di economia politica (revised by the author, transl. by A. Bonnet as Manuel d'économie politique), 2nd edn. V. Giarde & E. Briere, Paris (1909). Reprint in Busino, G. (ed.) Œuvres complètes, vol. VII. Manuel d'économie politique. Librairie Droz, Genève (1909 [1981])
53. Peart, S.J., Levy, D.M.: Denying human homogeneity: eugenics and the making of post-classical economics. Journal of the History of Economic Thought **25**(3), 261–288 (2003)
54. Pirandello, L.: Novelle per un anno. Mondadori, Segrate (1938)
55. Rabin, M.: Psychology and economics. J. Econ. Lit. **36**, 11–46 (1998)
56. Ranchetti, F.: Francis Ysidro edgeworth (Chap. 9). In: Ingrao, B., Ranchetti, F. (eds.) Il mercato nel pensiero economico: storia e analisi di un'idea dall'illuminismo alla teoria dei giochi, vol. XVII, pp. 379–422. Hoepli, Milano (1996)
57. Ranchetti, F.: Choice without utility? Some reflections on the loose foundations of standard consumer theory. In: Bianchi, M. (ed.) The Active Consumer: Novelty and Surprise in Consumer Choice, pp. 21–45. Routledge, London (1998)
58. Ranchetti, F.: Pareto e l'asino di Buridano: sui fondamenti della teoria della scelta razionale in economia. In: Economia, sociologia e politica nell'opera di Vilfredo Pareto, pp. 1000–1016. Olschki, Firenze (2000)

59. Schumpeter, J.A.: Note on the theory of utility (Appendix to Part III, Chap. 7). In: History of Economic Analysis, pp. 1019–1039. Allen & Unwin, London. Reprint Taylor and Francis, Abingdon (1954 [2006])
60. Scitovski, T.: The Joyless Economy An Inquiry into Human Satisfaction and Consumer Dissatisfaction. Oxford University Press, New York (1976)
61. Sigot, N.: "Be Quiet" mais modérement : le rôle de l'Etat dans la pensée économique de Jeremy Bentham. Revue Economique **44**, 23–50 (1993)
62. Sigot, N.: Bentham et l'économie: Une histoire d'utilité. Economica, Paris (2001)
63. Sigot, N.: John Stuart Mill versus Bentham: between the liberalism of freedom and the liberalism of happiness. In: Ege, R., Igersheim, H. (eds.) Freedom and Happiness in Economic Thought and Philosophy. From Clash to Reconcilation, pp. 67–84. Routledge, London (2011)
64. Stokes, P.M.: Bentham, Dickens, and the uses of the workhouse. Stud. English Lit. 1500–1900 **41**(4), 711–727 (2001)
65. Verri, P.: Discorso sull'Indole del Piacere e del Dolore. Carocci editore, Roma (1781 [2001])
66. Viner, J.: Bentham and J.S. Mill: The utilitarian background. Am. Econ. Rev. **39**, 2, 360–382 (1949)
67. von Neumann, J., Morgenstern, O.: Theory of Games and Economic Behavior, 3rd edn. Princeton University Press, Princeton. Reprint Wiley, New York (1944 [1967])
68. Walras, L.: Economique et mécanique. Metroeconomica **12**, 1, 3–11 (1909 [1960])
69. Wordsworth, W.: Preface to *Lyrical Ballads, with Pastoral and Other Poems* In: Gill, S. (ed.) Willam Wordsworth. pp. 595–615. Oxford University Press, Oxford (1802 [1990])

A Historical Perspective on the Theories Related to the Competitive Equilibrium Allocation

Masahiro Kawamata

Abstract Since the Marginal Revolution, mainstream economics has developed around the framework of general equilibrium theory and the fundamental theorems of welfare economics. However, we have to remember the prehistory of the marginal utility theory, in which ample pioneering works had been accumulated, which we call as a whole the theory of utility and scarcity. Tracing the development of the marginal utility theory since its prehistory, we can find out that there are three theoretical streams relating to the competitive equilibrium allocation. The theory of voluntary exchange was developed by Jevons and Edgeworth into the theory of core. The development of the theory of competitive markets has a theoretical jump from Walras' general equilibrium theory. Modern theory does not refer to the theory of social value developed by the Austrian school, but its pioneering contributions to the history of welfare economics should be appreciated.

1 Three Approaches to the Competitive Equilibrium Allocation

As is well known, in the 1870s, W. S. Jevons, C. Menger, and L. Walras constructed economic theories based on marginal analysis. Jevons and Walras considered their theories to be innovative, and made few references to previous research in their writings. Jevons' *Theory of Political Economy* was published in 1871, earlier than Walras' *Éléments d'économie politique pure* published in 1874–77. However, Walras, who believed in the superiority of the general equilibrium theory of his own, had a discussion with Jevons about the priority of the completion of the theory of competitive markets, and they decided to investigate the preceding works. In the preface to the second edition of Jevons [24], Jevons reported, for instance, the rediscovery of Gossen's marginal analysis.

This paper is a revised edition of my papers in Japanese [29, 30].

M. Kawamata (✉)
Keio University, Minato City, Japan
e-mail: mkawamata@keio.jp

T. Maruyama (ed.), *Marginal Revolution in Economics*, Monographs in Mathematical Economics 6, https://doi.org/10.1007/978-981-99-4342-5_6

Menger [39], on the other hand, studied a large amount of literature from the 16th century to the mid-19th century and constructed a value theory based upon marginal utility. The pioneering theory of marginal utility, inherited directly by Menger, is called "the theory of utility and scarcity".[1] The theory of utility and scarcity consists of the value theory for an individual based on which the theory of voluntary exchange is developed. Consider an economy consisting of a single individual characterized by a necessity function and resources she owns. The value of a good for the individual is defined by the necessity of the goods when the resources are exhausted. The necessity function may be interpreted as a marginal utility function. Menger interpreted the theory of utility and scarcity based on individual rationality, and suggested its extension to social value.

The basic model of voluntary exchange theory is the exchange theory of two persons and two goods. Individuals have a different necessity for every good, which is a function of the quantity of consumption less than the resources owned. Each individual can be more satisfied if they give the goods that are less needed by each other and get ones that are more needed. Therefore, they voluntarily negotiate the exchange rate, and execute an exchange at the rate provided that it is Pareto improving. Since the exchange will continue as long as there is room for Pareto improvement, the terminal allocation is Pareto efficient. Voluntary exchange theory was reformulated by Jevons [24] and developed by Edgeworth [10] as the limit theorem.

Marginal analysis is not explicitly used in the theory of utility and scarcity, but it certainly has the same theoretical structure as the marginal utility theory. The Marginal Revolution has contributed to the explicit use of marginal analysis in economic theories to construct the marginal utility theory.[2]

Jevons also adds the law of indifference to the two-person, two-goods exchange model to explain the determination of the exchange rate, but it is not easy to explain how a competitive equilibrium is determined in a general economic environment consisting of many economic agents and many goods. Walras [59] solved this problem based on the general equilibrium theory, in which a price mechanism, called *tâtonnement* with reference to transactions on actual stock exchanges, played a key role in systematically understanding the market function. He constructed the general equilibrium theory as an analogy to the system of astronomical mechanics with the help of his colleague mathematicians and scientists. After Walras, the general

[1] It is now known that there are many pioneering studies with similar theoretical structures to neoclassical economics since the Marginal Revolution, other than those mentioned by Menger [11, 18, 26].

[2] The term "Marginal Revolution" has become firmly established in the history of economics, but it is a very shallow expression. First, the Marginal Revolution is not a scientific revolution, which is a concept defined by Kuhn [32], As historians of economics agreed at the Bellaggio conference [3], the Marginal Revolution is quite different from the process of scientific revolutions and should not be called a revolution. Second, the term "Marginal" does not exactly describe the content of economic theories as they are actually constructed. "Marginal" in those theories merely express operations that mathematically characterize rationality and optimality. However, as pointed out by Hicks [17], Jaffé [22] and clearly formulated by Negishi [44], the theories of Jevons, Menger, and Walras are based on quite distinct approaches in addition to marginal analysis.

equilibrium theory was developed and reached at the stage of completion in the 1950s.

Thus, the purpose of this paper is to confirm the historical perspective such that there is a pioneering theory of utility and scarcity in the marginal utility theory, and there are three approaches to the competitive equilibrium allocation, namely, the theory of voluntary exchange, the general equilibrium theory with the fundamental theorems of welfare economics, and the theory of social value. We can recognize these approaches by tracking the theoretical development from the pioneering theory. We can see that each theoretical stream is developed by extending models or sub-models according to the preceding theoretical ideas.

This paper consists of the following sections. In Sect. 2, we confirm the development of voluntary exchange theory. Section 3 briefly follows the establishment of the general equilibrium theory and the basic theorem of welfare economics. Section 4 confirms the development of the value theory of the Austrian school. In Sect. 5, we make some remarks.

2 Theory of Utility and Scarcity

The pioneering theories of marginal utility to which Menger refers are called the theory of utility and scarcity, because they have in common that the value of a good is determined by the utility and scarcity of the good. For example, Galiani [14, pp. 58–59] states:

> Value, then, is a ratio; and this is compounded of two ratios, expressed by the names UTILITY and SCARCITY. I will explain what I mean by giving examples, so that there may be no dispute over the words. It is evident that air and water, which are elements very useful for human life, have no value, because they lack scarcity; and, on the other hand, a bag of sand from the shores of Japan would be a rare thing, but, assuming that it had no special utility, it would have no value.[3]

2.1 *The Theory of Individual Value in the Theory of Utility and Scarcity*

Ekelund and Hébert [11] suggest a hegemony of marginal analysis in economics before the Marginal Revolution. A value of a good is a subjective evaluation of the goods by an individual, and a price is a rate of the quantities of goods that are exchanged by individuals. Kawamata [27, 28] explains the theory of individual value in the theory of utility and scarcity such as Galiani [14, pp. 54–91, Book I, Chap. 2], Turgot [55, pp. 137–147], and Condillac [4, pp. 97–113]). Suppose an individual

[3] Quoted from p. 283 of an English translation of Galiani [14, pp. 54–91] by Monroe [41, pp. 281–299].

economy consisting of a single person and two goods indicated by $h = 1, 2$. The individual is characterized by the necessity function (which can be interpreted as a marginal utility function) $u_h(\cdot)$ and the resources of the goods she owns (ω_1, ω_2). The more resources of a good she owns, the more goods she can consume, and therefore the more her necessity is satisfied. Thus, the necessity for good h is a decreasing function of the quantity of good x_h consumed by the individual. In other words, it is the additional utility $u_h(x_h)$ obtained by consuming one more unit when the individual has just consumed x_h.

For every good h, the individual consumes $x_h = \omega_h \geqq 0$ if the quantity of resources she owns cannot satiate her necessity, where an individual satiates her necessity with ω_h if and only if $u_h(x_h) = 0$ at $x_h < \omega_h$. Then, the value of a good h, denoted by v_h, is the necessity of the good when the individual exhausts the quantity of resource ω_h. It means the v_h is the additional utility gained by consuming one more unit, when she consumed ω_h. The value of a good such as air and water may be zero because she has enough resources of the goods to satiate her necessity. Suppose a corner equilibrium condition such that $x_h = 0$ if marginal utility at $x_h = 0$ is less than marignal utility of other goods k at $x_k = \omega_k$, that is to say, $u_h(0) < u_k(\omega_k)$. Then, according to the Kuhn–Tucker equivalent theorem [31], the theory of individual value is derived from the individual utility maximization under the resource constraint, which Menger [39] constructed.

Let us translate the concept of individual value in the theory of utility and scarcity into the modern theory of consumer. In the theory of utility and scarcity, an individual's utility is supposed to be the sum of the utilities obtained separately from the consumption of each good. We denote a possible cunsumtion as (x_1, x_2) and a utility function as $U = U_1(x_1) + U_2(x_2)$, where an index i of the individual is omitted for the sake of simplicity.[4] An indifference curve of an individual is a set of her consumptions (x_1, x_2) at the same utility level $\bar{U} = U_1(x_1) + U_2(x_2)$. If we move the consumption on the indifference curve a little from a consumption (x_1, x_2), then

$$dU = 0 = \frac{dU_1}{dx_1}(x_1)\,dx_1 + \frac{dU_2}{dx_2}(x_2)\,dx_2, \quad \frac{dU_1}{dx_1}(\cdot) > 0, \quad \frac{dU_2}{dx_2}(\cdot) > 0,$$

and it means that the utility level remains the same if the quantity consumed of good 1 decreases, $dx_1 < 0$ and that of 2 increases, $dx_2 > 0$, or vice versa, at the rate $-dx_2/dx_1(x_1, x_2) > 0$, which is called the marginal rate of substitution at the consumption (x_1, x_2) and abbreviated as MRS. An additive utility function with diminishing marginal utility implies that the utility function is concave and therefore the indifference curve is convex to the origin.

If the goods are not free, an individual exhausts the resources of the goods $x_h = \omega_h$, and the individual value of good h is equal to $dU_h/dx_h(\omega_h)$ for every $h = 1, 2$. The MRS is the rate of individual value of good 1 to that of good 2. The line representing MRS is tangent to the indifference curve at (ω_1, ω_2), as is shown in Fig. 1.

[4] We suppose that if $(x_1, x_2) \geqq 0$ it is possible for the individual.

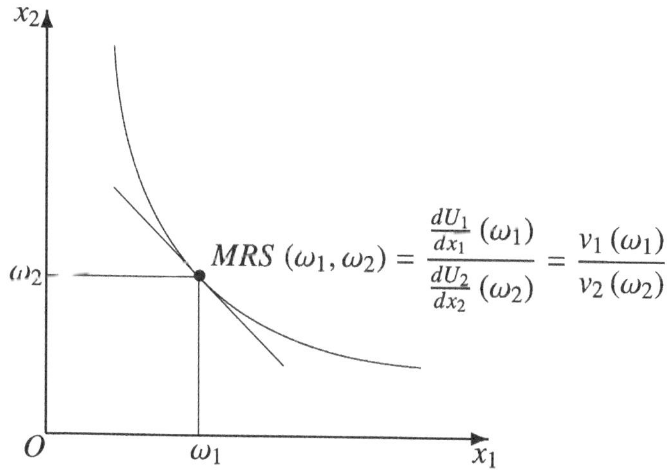

Fig. 1 Indifference curve and marginal rate of substitution

2.2 *Voluntary Exchange Equilibrium in an Exchange Economy*

In the theory of utility and scarcity, Turgot [55] constructed a clear theory of voluntary exchange [28]. Since the necessity to individuals of their goods and the quantity of resources they own are different, the values of goods to individuals are also different.[5] Therefore, the economic states of all individuals will be better if each individual gives the other a certain quantity of a good that are relatively unnecessary and receives the quantity of a good that is relatively necessary in exchange for it. It implies such an exchange is Pareto improving.

Every individual compares their individual values at the initial distributions of goods, or initial allocations, and a new allocation is proposed by some individual. The individuals discuss the proposal among themselves. If some group of individuals prefers the initial allocation to the allocation proposed, it objects to the proposal and the proposal is withdrawn or blocked. Then a new proposal is made by someone. This process, which we call voluntary exchange, continues until an allocation is found that no group objects to.[6] Turgot [55] showed that through a voluntary Pareto improving exchange, the individual values of goods, which differ between individuals, are equalized and equal to the exchange rate. This values of goods are called the exchange values.

Let us follow Turgot's discussions, and interpret them in modern economic theory, in particular, in an Edgeworth box diagram.[7] Suppose an exchange economy

[5] Turgot [55] calls the individual value "esteem value".

[6] Feldman and Serrano [12, pp. 33–49] calls it barter exchange.

[7] See Mas-Colell et al. [38, pp. 515–686] and Feldman and Serrano [12, pp. 33–77].

consisting of two individuals, A and B, and two goods, 1 and 2. For every individual $i = A, B$, we denote individual i's consumption of goods by $x_i = (x_{i1}, x_{i2})$. Individual i is characterized by the utility function $U_i(x_{i1}, x_{i2}) = U_{i1}(x_{i1}) + U_{i2}(x_{i2})$ and i's resources $\omega_i = (\omega_{i1}, \omega_{i2})$. The total quantity of good 1 is given $\omega_1 = \omega_{A1} + \omega_{B1}$ and that of good 2 is given $\omega_2 = \omega_{A2} + \omega_{B2}$. The rectangle has dimensions ω_1 by ω_2. An allocation $(x_A, x_B) = ((x_{A1}, x_{A2}), (x_{B1}, x_{B2}))$ represented by a set of points contained in the Edgeworth box is feasible, because it satisfies $x_{A1} + x_{B1} \leqq \omega_1$ and $x_{A2} + x_{B2} \leqq \omega_2$.

Suppose all the goods are not free goods. By the theory of individual value, the individual values of A and B at an allocation (x_A, x_B) are $dU_{ih}/dx_{ih}(x_{ih})$ for every $i = A, B$ and for every $h = 1, 2$. For example, the individual values of A and B at the initial allocation (ω_A, ω_B) are

$$v_{A1}(\omega_{A1}) = \frac{dU_{A1}}{dx_{A1}}(\omega_{A1}), \; v_{A2}(\omega_{A2}) = \frac{dU_{A2}}{dx_{A2}}(\omega_{A2}),$$
$$v_{B1}(\omega_{B1}) = \frac{dU_{B1}}{dx_{B1}}(\omega_{B1}), \; v_{B2}(\omega_{B2}) = \frac{dU_{B2}}{dx_{B2}}(\omega_{B2})$$

and in the Edgeworth box, they are represented by the marginal rates of substitution at the point of initial allocation, namely

$$MRS_A(\omega_{A1}, \omega_{A2}) = \frac{v_{A1}(\omega_{A1})}{v_{A2}(\omega_{A2})} < MRS_B(\omega_{B1}, \omega_{B2}) = \frac{v_{B1}(\omega_{B1})}{v_{B2}(\omega_{B2})},$$

which are represented by tangent lines of the indifference curves at the initial allocation, as is shown in Fig. 2.

Any allocation represented by an interior point of the set shaped like a lens enclosed with the indifference curves of two individuals passing through a certain allocation is Pareto superior to the allocation represented by the end point of the lens.[8] Therefore, the exchange preceding the end point to the interior of the lens is Pareto improving. Turgot also explains that an exchange takes place because it increases the satisfaction of both individuals.

> At the moment of exchange the one who, for example, gives four measures of corn for five armfuls of wood, without doubt prefers these five armfuls of wood to the four measures of corn; he will give them a higher esteem value. But for his part, the one who receives the four measures of corn also prefers them to the five armfuls of wood. This superiority of the esteem value attributed by the acquirer to the thing he acquires over the thing he gives up is essential to the exchange, for it is the sole motive for it. Each would remain as he was, if he did not find an interest, a individual profit, in exchange; if, in his own mind, he did not consider what he receives worth more than what he gives. [55, p. 142]

[8] If there are an allocation $x = (x_A, x_B)$ and another allocation $z = (z_A, z_B)$ such that $U_i(z_i) \geq U_i(x_i)$ for all $i = A, B$, and $U_i(z_i) > U_i(x_i)$ for at least one $i = A, B$, then z is called Pareto superior to x or Pareto improving of x, namely a Pareto improvement of x. An allocation x is Pareto efficient if there is no Pareto improvement of x.

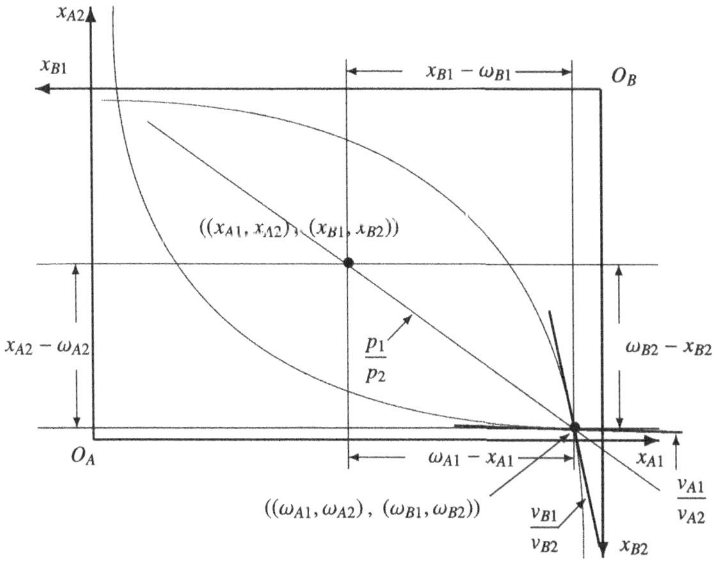

Fig. 2 Initial allocation and transactions in the exchange economy

Turgot defines the exchange value in terms of the individual value (which Turgot calls "the esteem value"), and explains the distinction between them as follows:

> This *exchange value* is not exactly the same as the esteem value, or, in other words, the interest which each of the two separately attached to the two desired objects, whose possession he compares in order to determine what he must give up of the one to acquire some of the other, since the result of the comparison might not be the same in the minds of the two contracting parties. This first value, to which we have given the name *esteem value*, is arrived at through the comparison which each separately makes between the two interests that contend with one another in his case; it exists only in the interest of each of the two taken separately. *Exchange* value on the contrary, is adopted by both the contracting parties, who recognise its equality and who make it the condition of the exchange. In the determination of *esteem* value, each man, taken separately, has compared only two interests, attached by him to the object he has and to the one he desires to have. In the determination of *exchange value*, there are two men who compare, and four interests (which are) compared by them; but the two individual interests of each of the two contracting parties have first been compared between themselves separately, and it is the two results which are then compared together, or rather, discussed by the two contracting parties, in order to form one *average esteem value*, which becomes precisely the *exchange value*, to which we believe we must give the name of *appreciative value* (valeur appreciative) because it determines the *price* or the condition of exchange. [55, p. 143]

If the exchange is carried out at a rate between the individual values of the goods, the allocation after the exchange is a Pareto improvement of that before the exchange. Also, no exchange will take place unless the exchange results in a Pareto improving allocation.

Suppose an individual gives $dx_1 < 0$ of good 1 and receives $dx_2 > 0$ of good 2 from a consumption $x = (x_1, x_2)$, then she consumes $(x_1 + dx_1, x_2 + dx_2)$. By getting the total derivative of the utility function $U = U_1(x_1) + U_2(x_2)$, we get the variation of utility in this exchange,

$$dU = \frac{dU_1}{dx_1}(x_1)\,dx_1 + \frac{dU_2}{dx_2}(x_2)\,dx_2, \text{ where } \frac{dU_1}{dx_1} > 0, \frac{dU_2}{dx_2} > 0.$$

Exchanges are made between goods of equal amount in terms of proposed prices of goods. Denote proposed prices of goods p_1, p_2 defined by the equation:

$$p_1 dx_1 + p_2 dx_2 = 0 \quad \text{that is} \quad -\frac{dx_2}{dx_1} = \frac{p_1}{p_2}.$$

Since, also, $dU_1/dx_1\,(x_1)$ and $dU_2/dx_2\,(x_2)$ are respectively equal to the individual value of goods 1 and 2 at (x_1, x_2),

$$\frac{v_1}{v_2} = \frac{dU_1}{dx_1}(x_1) \Big/ \frac{dU_2}{dx_2}(x_2).$$

Therefore, we have the following relationship between individual values and an exchange rate:

$$\begin{aligned} dU &= \frac{dU_1}{dx_1}(x_1)\,dx_1 + \frac{dU_2}{dx_2}(x_2)\,dx_2 \\ &= \frac{dU_2}{dx_2}(x_2)\,dx_1 \left\{ \left(\frac{dU_1}{dx_1}(x_1) \Big/ \frac{dU_2}{dx_2}(x_2) \right) - \left(-\frac{dx_2}{dx_1} \right) \right\} \\ &= \frac{dU_2}{dx_2}(x_2)\,dx_1 \left\{ \frac{v_1}{v_2} - \frac{p_1}{p_2} \right\} > 0 \iff dx_1 \left\{ \frac{v_1}{v_2} - \frac{p_1}{p_2} \right\} > 0. \end{aligned}$$

Since $dU_2/dx_2 > 0$, $dU > 0$ follows:

1. if $v_1/v_2 > p_1/p_2$, then the individual gives good 2, $dx_2 < 0$, and receives good 1, $dx_1 > 0$,
2. if $v_1/v_2 < p_1/p_2$, then the individual gives good 1, $dx_1 < 0$, and receives good 2, $dx_2 > 0$.

In the situation of Fig. 2, the exchange rate p_1/p_2 would be proposed in the following range:

$$MRS_A(\omega_{A1}, \omega_{A2}) = \frac{v_{A1}(\omega_{A1})}{v_{A2}(\omega_{A2})} < \frac{p_1}{p_2} < \frac{v_{B1}(\omega_{B1})}{v_{B2}(\omega_{B2})} = MRS_B(\omega_{B1}, \omega_{B2}).$$

Turgot explains the adjustment process of the exchange rate as follows:

> Independently of this kind of mental evaluation by which each of them compares the interest he has of keeping to that of acquiring, both are also animated by a general interest independent of all comparisons, that is the interest of keeping as much of their own goods as they can, and of acquiring as much as they can of that of the other. With this in mind, each will keep secret

> the mental comparison which he has made of his two interests, of the two values which he attached to the goods for exchange, and he will sound out the owner of the goods he desires by lower offers and higher demands. Since the other will follow the same procedure, they will discuss the conditions of the exchange, and as they both have a great interest in coming to an agreement, they will finally do so. Turgot [55, p. 142]

That is, in Fig. 2, individual A negotiates to set p_1/p_2 higher and B negotiates to set it lower. However, the trading partners behave in the same way, and they will know each other, so in the end, the transactions are conducted so that the individual values are equal to the exchange rate. It should be noted that an exchange value is not always a competitive equilibrium price with respect to (ω_A, ω_B).

3 Theory of Voluntary Exchange[9]

The theory of voluntary exchange began with the theory of utility and scarcity and was formulated in terms of marginal analysis by Jevons [24].[10]

3.1 *Jevons' Theory of Voluntary Exchange*

In his theory of exchange, Jevons [24, pp. 94–95]assumes the two-person, two-goods exchange economy which is the same as that of Turgot, and defines two types of exchange rate. One is the exchange rate of goods in the Pareto efficient allocation reached by voluntary Pareto improving exchange. Jevons denotes it as dy/dx. The other is the exchange rate of goods that stands between the initial allocation and another allocation realized by exchange from the initial allocation. Jevons denotes it as y/x. Then, Jevons shows that according to the law of indifference, they have to be equal at the equilibrium of freely competitive markets.

The exchange rate dy/dx is realized through the process of voluntary Pareto improving exchange. Jevons states, "imagine that there are now exchange groups that own only grain and exchange groups that own only beef. Under these circumstances, it is certain that giving a portion of the grain and receiving a portion of the beef will significantly increase the utility. In what respect would the exchange be unfavorable? This issue must include both exchange rate and utility." Jevons' exposition is illustrated by Fig. 3.

The horizontal axis represents individual i's consumptions of good 1 and good 2, the vertical axis represents the magnitude of the marginal utility of an individual

---

[9] The results of this section is indebted to Negishi [44, pp. 324–338], but is necessary for us to complete the demonstration of this paper.

[10] It is clear that there is no inheritance between the theory of utility and scarcity and the theory of Jevons. After investigating his pioneering work on the priorities of marginal utility theory, he first mentioned Condillac [4] as a good value theory without mathematics, and highly appreciated his theory. Jevons did not know the manuscript of Turgot [55] because it had not been published.

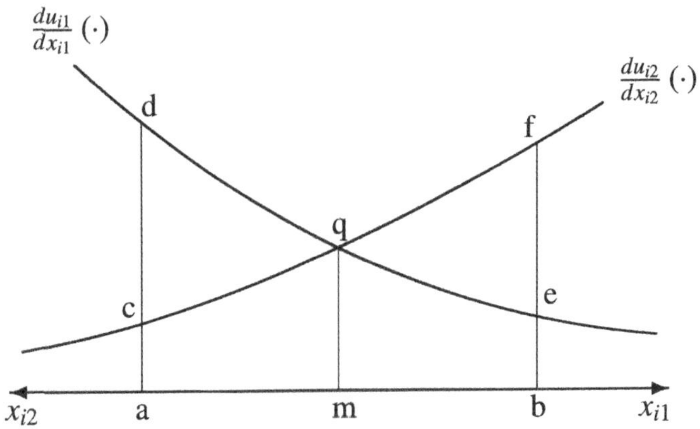

Fig. 3 Voluntary equilibrium in an exchange economy

i, the curve downward to the right represents the marginal utility curve of good 1, and the curve downward to the left represents the marginal utility curve of good 2. It makes sense that the two marginal utility curves are randomly superimposed. Given prices of goods 1 and 2, v_1, v_2, the quantity of good 1 x_{i1} is exchanged with quantity of good 2 x_{i2} to satisfy $v_1 x_{i1} = v_2 x_{i2}$. The slope of the curve differs depending on the exchange rate v_1/v_2 between good 1 and good 2.

Given the marginal utility curve of good 1, the relative position of the marginal utility curve of good 2 varies due to variations in the initial allocation prior to exchange. This figure shows that any final allocation reached by the repetition of voluntary exchanges is Pareto efficient, regardless of the initital allocation. In other words, it only explains that the equilibrium reached by voluntary exchange is a Pareto efficient allocation. This argument holds for both individuals A and B.

The voluntary exchange equilibrium is an allocation that is finally reached through the process of voluntary exchange, which is the Pareto improving reallocation of resources. Thus, the following relationship that holds:

$$\frac{dU_{A1}}{dx_{A1}}\bigg/\frac{dU_{A2}}{dx_{A2}} = -\frac{dx_{A2}}{dx_{A1}} = \frac{v_1}{v_2} = -\frac{dx_{B2}}{dx_{B1}} = \frac{dU_{B1}}{dx_{B1}}\bigg/\frac{dU_{B2}}{dx_{B2}}.$$

This equation defines the values of the goods, namely, their relative prices in the voluntary exchange equilibrium.

3.2 *Edgeworth's Limit Theorem*

Edgeworth [10] assumes an exchange economy similar to that of Jevons, generalizing the form of utility function from an additive one to $U = U(x_1, x_2)$; he introduced the concept of indifference curve and devised the apparatus of the so-called Edgeworth box.[11] Each indifference curve is assumed to be convex to the origin [10, p. 36]. For instance, an additive utility function with diminishing marginal utility satisfies this assumption. Given resources of goods $\omega = (\omega_1, \omega_2)$, an allocation (x_A, x_B) such that $x_A + x_B \leqq \omega$ is called feasible.

Any allocation represented by points in the Edgeworth box (Fig. 4) is feasible, that is, the interior of the box is the set of feasible allocations in this exchange economy. If individual $i \in \{A, B\}$ does not execute any exchange, she can secure the utility level of $U_i(\omega_{i1}, \omega_{i2})$. Therefore, the exchange contract must guarantee at least the utility level achieved by the initial allocation. The exchange takes place because it results in a Pareto improving reallocation of resources. It implies that the voluntary exchange eventually leads to a Pareto efficient allocation. We suppose that Turgot [55] expected voluntary exchanges to reach the allocation on the zone DF, namely, the core of this exchange economy.[12,13]

Jevons assumed the law of indifference to show that in the two-person, two-goods exchange model, voluntary negotiations ultimately reach the Walras equilibrium. However, the law of indifference holds under perfect competition. In fact, Jevons supposed competitive situations among trading bodies consisting of a large number of individuals.[14] Reflecting Jevons' ideas, Edgeworth considers the so-called replica economy consisting of trading bodies, in which economic agents are identical in utility function and initial allocation.

In Fig. 4, suppose that there are two individuals whose utility functions (indifference cureves) and resources of goods are identical in each trading body, i.e. trader A_1 and trader A_2 of type A, and trader B_1 and trader B_2 of type B. The curve connecting the origin O_A and O_B is a contract curve that is a set of Pareto efficient allocations. Edgeworth [10, p. 35] points out that (1) individual traders of the same type have the same equilibrium allocation, and therefore the equilibrium

[11] Full-fledged Edgeworth boxes do not appear in Edgeworth's own writings. As Jaffé [21] points out, perhaps it is Pareto [46] who drew the Edgeworth box for the first time.

[12] Consider a group of persons S, called a coalition. S might be one person, several persons, or all. Let $\{s_i\}_{i \in S}$ represent a set of bundles of goods, one bundle for each member of S. We say that $\{s_i\}_{i \in S}$ is feasible for S if

$$\sum_{i \in S} s_{ij} = \sum_{i \in S} \omega_{ij}, \text{ for every good } j.$$

If x is a proposed allocation and S is a coalition, we say that S can block x if there is a feasible set of bundles $\{s_i\}_{i \in S}$ such that:

$$U_i(s_i) \geq U_i(x_i) \text{ for all } i \text{ in } S,$$
$$U_i(s_i) > U_i(x_i) \text{ for at least one } i \text{ in } S.$$

The core is that set of allocations which cannot be blocked by any coalition. See Feldman and Serrano [12, pp. 37–40] and Mas-Colell et al. [38, Chap. 18].

[13] Desai [8] investigates a pioneering analysis of the core by Turgot [55].

[14] Jevons uses the concept of a trading body as a representative or average individual, because "questions which appear, and perhaps are, quite indeterminate as regards individuals, may be capable of exact investigation and solution in regard to great masses and wide averages" Jevons [24, p. 16].

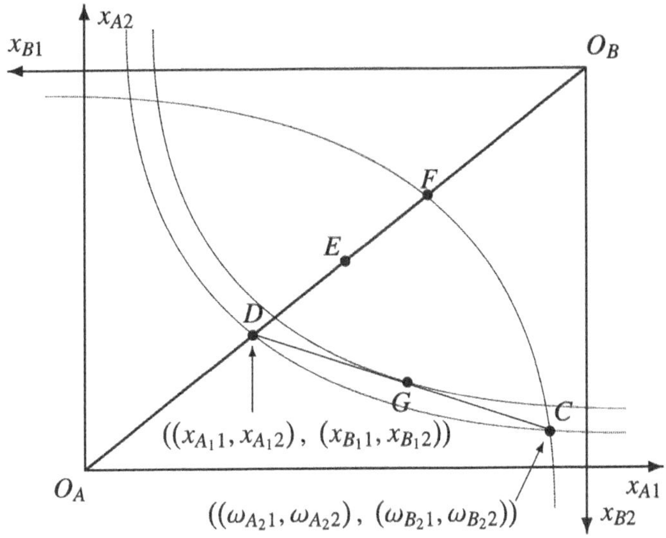

Fig. 4 Edgeworth's limit theorem

allocation may be represented by single point in the Edgeworth box, (2) the point is on the zone DF.

> This being premised, let us now introduce a second X and a second Y; so that the field of competition consists of two Xs and two Ys. And for the sake of illustration (not of the argument) let us suppose that the new X has the same requirements, the same nature as the old X; and similarly that the new Y is equal-natured with the old.
>
> Then it is evident that there cannot be equilibrium unless (1) all the field is collected at one point; (2) that point is on the contract-curve. For (1) if possible let one couple be at one point, and another couple at another point, it will generally be the interest of the X of one couple and the Y of the other to rush together, leaving their partners in the lurch. And (2) if the common point is not on the contract-curve, it will be the interest of all parties to descend to the contract-curve. Edgeworth [10, p. 35]

First, suppose the allocations of traders of the same type A_1 and A_2 or B_1 and B_2 are different in the core. For example, suppose A_1 and B_2 contract at D, and A_2 and B_1 contract at F. In this case, A_1 and B_1 have a disadvantageous contract, but when they divide their initial allocation into two equal parts, they get the allocation

$$H = \left(\left(\frac{\omega_{A_1 1} + \omega_{B_1 1}}{2}, \frac{\omega_{A_1 2} + \omega_{B_1 2}}{2}\right), \left(\frac{\omega_{A_1 1} + \omega_{B_1 1}}{2}, \frac{\omega_{A_1 2} + \omega_{B_1 2}}{2}\right)\right)$$

represented by the center of the Edgeworth box, which is better for A_1 than D and better for B_1 than F, and therefore D and F are blocked by H. Thus, individual traders of the same type have the same equilibrium allocation. This is known as the equal-treatment property [38, p. 656, Proposition 18.B.2], by which we can still use

Fig. 4 by interpreting it as describing the trade between the representative traders of each trading body.

The second proposition is obvious. Any final settlement of voluntary exchange between A and B is an allocation from D to F. Here comes an indeterminacy. However, in Fig. 4 no allocation close to D or F is executed. Now, suppose that A_1 and B_1 have negotiated and the allocation $D = \left(\left(x_{A_1 1}, x_{A_1 2}\right), \left(x_{B_1 1}, x_{B_1 2}\right)\right)$ is established. A_2 and B_2 stay at C, and A_2 have $\left(\omega_{A_2 1}, \omega_{A_2 2}\right)$. Then, if A_1 and A_2 divide the goods they hold into two equal parts, their allocation becomes that of point G as follows:

$$G = \left(\left(\frac{x_{A_1 1} + \omega_{A_2 1}}{2}, \frac{x_{A_1 2} + \omega_{A_2 2}}{2}\right), \left(\frac{x_{A_1 1} + \omega_{A_2 1}}{2}, \frac{x_{A_1 2} + \omega_{A_2 2}}{2}\right)\right).$$

Since the indifference curves of A_1 and A_2 passing through G are on the upper right of the indifference curve passing through C, D, for A_1 and A_2, G is Pareto superior to C, D. Therefore, if A_1 and A_2 give their allocations in G little by little to B_2, everyone will have a Pareto improvement of C, D. Since such an allocation can be realized, D will not be executed. Similarly, F is not executed either. After showing such a proof policy, Edgeworth [10, pp. 37–40] points out that the core converges to the Walras allocation when the number of members of the A and B types becomes infinite.

The concept of a core was defined by Gillies [15]. Shubik [52] applied it to the market economy. Debreu and Scarf [7] proved Edgeworth's limit theorem in a replica economy consisting of many types of individuals. This theorem generally proves that if the number of players increases infinitely, the core will converge to the set of Walras allocations.

In an economy consisting of a large number of individuals, it is usually not possible for individuals to have sufficient information about each other's characteristics of preference ordering and resources. Then, even if individuals who are dissatisfied with the current situation try to form a coalition that produces mutual benefits, it is not easy to find a partner to form a coalition with or to negotiate how to distribute the benefits of the coalition, and it requires a large transaction cost. However, if there were no transaction cost, Edgeworth's limit theorem could be interpreted as supporting Walras' assumption of perfectly competitive markets [44, pp. 338–340].

4 Competitive Equilibrium and Pareto Efficiency

Adam Smith [53, pp. 72–81, Book I, Chap. 7] had a notion of price mechanism such that (1) the demand for a good depends on the market price, (2) the supply of a good also depends on the market price, and (3) the market price is determined by the equilibrium of supply and demand for that good, but the behavior of individual economic agents and the transactions of the market as a whole were not understood consistently. Since Walras was the first to systematize a theory of the perfectly competitive market after introducing the price mechanism (*tâtonnement*), there was no systematic theory of competitive markets before Walras [2, pp. 2–5]. However, in addition to Smith, the source of the theory of competitive markets can be found in various studies before Walras.

The origin of the theory of competitive markets lies in the recognition such that the prices of goods are determined by the equilibrium between supply and demand of their markets. It starts, for example, with the early idea that if the demand for a good increases due to an increase in the population or changes in people's tastes, the price of that good will rise, or if the supply of a good is strengthened due to the availability of raw materials or the innovation of technology, the price of that product will fall. The price theories of Smith [53], Malthus [36] and Mill [40] are considered to belong to this stream.[15] The price of a good reflects the demand factor in the short term and is determined by the production cost on the supply side in the long term.

Primitive concepts of the demand function that explain the law of demand based on utility can be found in the theory of utility and scarcity. For every person, the more a person owns the resource of a goods, the more she consumes and the more she fills the utility or necessity of the goods. Thus, as the consumption increases, the utility or necessity is gradually filled, and eventually it becomes satiated. This means that utility is essentially a function of consumption and satisfies the law of diminishing marginal utility. The marginal utility function is a function that indicates the marginal utility for the consumption of a good, but it can also be considered to represent the quantity of demand for the good that can obtain the marginal utility that offsets the price paid in the market of the good. This means that a demand function is an inverse of a marginal utility function and provides a reason to equate the demand function with the marginal utility function.[16] Typical examples of this approach are the theories of Galiani [14, pp. 86–91], J. B. Say [50], and Dupuit [9].

Since, prior to Edgeworth, the utility function was usually assumed to be separable and additive, it was possible to equate the marginal utility curve and the demand curve separately for every good. However, in order to derive the supply function of a product which is produced by combining multiple productive factors, one has to solve the mathematical problem of profit maximization under the constraints of the production function. It was not possible to relate production technology to the supply curve until Marshall [37]. In the theories of Smith, Malthus, and J. S. Mill, the supply in the short run was supposed to be constant, and the supply price in the long run was supposed to be equal to the constant average cost.

Inheriting J. S. Mill's supply–demand equilibrium theory, A. Marshall constructed a partial equilibrium theory based on marginal analysis. Marshall [37] provides a rationale for measuring utility based on Dupuit's ideas. He showed that the law of constant marginal utility of money is necessary to measure utility in terms of money.

---

[15] It is not difficult to find the germ of this theory even in very early economic theory such as Pietro Verri [57, pp. 14–22].

[16] If we interpret this idea in the general equilibrium theory, we must refer to the law of constant marginal utility of money. However, before Walras [59], the demand function was still a primitive notion not integrated into a systematic theory of competitive markets.

This means that Marshall implicitly assumed a quasi-linear utility function [25, pp. 91–95], which may reduce the framework of general equilibrium analysis to that of partial equilibrium analysis. Marshall also developed the cost–benefit analysis, because scientific analysis of economic problems becomes possible only if the concepts that are the subject of economic analysis are measurable, according to his own scientific philosophy [37, pp. 12–23]. He derives the demand function of the goods as a consequence of maximizing utility and the supply function as a consequence of maximizing profits, and he laid the foundation for supply–demand equilibrium analysis.

4.1 Jevons Equilibrium and Its Limits

Within the framework of the two-person, two-goods exchange economy, Jevons [24] explained how an exchange equilibrium is determined based on the voluntary exchange theory, which is independent of the supply–demand equilibrium theory. As is shown above, Jevons demonstrates that voluntary exchange leads to a Pareto efficient allocation, and defines an exchange rate that holds for Pareto efficient allocations reached by voluntary exchange. Then, he defines another exchange rate that holds between the initial allocation and another allocation realized by exchanges. Assuming that the law of indifference holds in competitive markets, Jevons determines the exchange equilibrium based on the equality of exchange rates defined above. Jevons' theory of exchange is certainly a theory of competitive markets, but his set of apparatuses may not be robust enough to analyze market mechanisms.

Now, in Fig. 2, suppose that the exchange will be carried out as follows: A gives good 1, $\omega_{A1} - x_{A1}$, to B and obtain good 2, $x_{A2} - \omega_{A2}$, from B, and B gives good 2, $\omega_{B2} - x_{B2}$, to A and obtain good 1, $x_{B1} - \omega_{B1}$, from A. Denoting prices of goods 1, 2 as p_1, p_2, we can define the relative price p_1 / p_2 as an exchange rate of good 2 to good 1 in any exchange. Thus, the relative price p_1 / p_2 holds between the equilibrium allocation $\left((x^*_{A1}, x^*_{A2}), (x^*_{B1}, x^*_{B2})\right)$ reached by the process of voluntary exchange and the initial allocation $((\omega_{A1}, \omega_{A2}), (\omega_{B1}, \omega_{B2}))$, and therefore the following equation holds:

$$-\frac{\omega_{A2} - x^*_{A2}}{\omega_{A1} - x^*_{A1}} = \frac{p^*_1}{p^*_2} = -\frac{\omega_{B2} - x^*_{B2}}{\omega_{B1} - x^*_{B1}}.$$

This condition is the same as the budget constraint equation that holds for every individual.

Then, because the prices that are established in competitive markets are unique by the law of indifference,

$$\frac{v_{A1}\left(x^*_{A1}\right)}{v_{A2}\left(x^*_{A2}\right)} = \frac{p^*_1}{p^*_2} = \frac{v_{B1}\left(x^*_{B1}\right)}{v_{B2}\left(x^*_{B2}\right)}.$$

The set of equilibrium conditions in competitive markets implies that each individual maximizes her utility under the budget constraint. It is clear that the Jevons' exchange equilibrium is equivalent to the Walras equilibrium of the above exchange economy.[17] Therefore, we can be sure that Jevons solved the problem of exchange theory within the framework of two-person, two-goods exchange economy.

In the case of a two-person, two-goods economy, the exchange rate is unique, and an agreement on what exchange rate to be carried out is easily established between the two persons, because each individual can easily reject the other's proposal if the other's proposal does not benefit for her. However, Jevons' model is not extendable to the general economic environment consisting of many persons and many goods. The two-person, two-goods model is a typical simple model for analyzing economic activity. Due to its simplicity, it has specific attributes not extendable to complex models. For example, suppose the model in the case of a 3-person, 3-goods exchange economy is as follows Feldman and Serrano [12, p. 47]:

$$\begin{array}{lll} A & U_A = 2x_{A1} + x_{A2} + 3x_{A3} & \omega_A = (1, 0, 0)\,, \\ B & U_B = 3x_{B1} + 2x_{B2} + x_{B3} & \omega_B = (0, 1, 0)\,, \\ C & U_C = x_{C1} + 3x_{C2} + 2x_{C3} & \omega_C = (0, 0, 1)\,. \end{array}$$

A tries to give good 1 and obtain good 3, the owner of which, C, tries to give good 3 and obtain good 2, the owner of which, B, tries to give good 2 and obtain good 1, and therefore there is no exchange in the Jevons equilibrium. No direct exchange occurs because there is no Pareto-improving allocation between any two individuals. However, a Walras equilibrium clearly exists, and it is the price system $p^* = (1, 1, 1)$ and equilibrium allocation $x_A^* = (0, 0, 1)$, $x_B^* = (1, 0, 0)$, $x_C^* = (0, 1, 0)$.

As far as the two-person, two-goods exchange model is concerned, it is clear that Jevons solves the problem of exchange equilibrium. However, his approach relies on the specificities of the two-person, two-goods exchange economy. It was Walras who extended a theory of competitive markets to a general economic environment consisting of many persons and many goods.

4.2 General Equilibrium Theory

In the case of Walras [59], as can be recognized from the writing circumstances of his main work [23] and his own explanation [60], the idea of a supply–demand equilibrium equation system in general equilibrium theory was not developed from earlier economic theories, but imported from Physics. Walras in his youth had a strong interest in natural science and his idea of the general equilibrium was inspired by Louis Poinsot's astronomical system (*Élémens de Statique*, 1803). Walras also refers

---

[17] A Walras equilibrium of an exchange economy is an allocation and a price system in which each individual chooses a consumption so as to maximize her utility under the budget constraint and the total consumptions equals the total resources for each good.

to Cournot [5] for Walras' Law and the idea of perfect competition [44, pp. 241–246]. Walras introduced a price mechanism called *tâtonnement*, inspired by procedures on the Paris and London stock exchanges. *Tâtonnement* would be the first game rule introduced in economics, which made it possible to construct a general equilibrium theory.

Walras considered the general economy consisting of many consumers, many producers and many goods. According to Walras' *tâtonnement*, there is an auctioneer who provisionally set prices for all goods. Each consumer chooses a consumption so as to maximize her utility under the budget constraint and informs of this consumption the auctioneer. Each producer chooses a production so as to minimize cost under the constraint of the production coefficient, and informs the auctioneer of this production. Based on the consumption and production information, the auctioneer adjusts each price according to the sign of excess demand, and exchanges are carried out only when an equilibrium is attained.

However, Walras was unable to solve the mathematical problems of consumer and producer behavior, so he turned to his colleague for help. Walras asked Picard, a professor of mechanics at Lausanne Academy, around 1872 how to derive the demand function. He received a polite, illustrated explanation [20, pp. 308–311, L211] and made full use of that explanation in his *Élémens* [13]. For the problem of minimizing cost under the constraint of production coefficient, in January 1877, he sought advice from Amstein, a professor of mathematics at Lausanne Institute of Technology, and received an appropriate commentary [20, pp. 516–520, L364]. However, Walras could not digest Amstein's explanation using Lagrangian multiplier method, so he did not use Amstein's explanation in his *Élémens* [19]. There were still problems to be clarified in Walras' theory, which were mostly corrected by Pareto [45, 46].

Walras' construction of the general equilibrium theory has been largely praised by Schumpeter [51] for its contribution to modern economics, and the introduction of a price mechanism was crucial to his economic theory. *Tâtonnement* was applied to the socialist economy by Lange [33] during the controversy over socialist economic calculation. It can be applied to the resource allocation mechanism of various planned economies, and mechanisms with higher performance have been developed. *Tâtonnement* would be the origin of the theoretical development that leads to the theory of mechanism design under modern asymmetric information [28, pp. 41–43].

4.3 Pareto Efficiency of a Competitive Equilibrium

The function of the competitive market was already recognized in mercantilism [34, pp. 211–216], but the mercantile policy could not achieve its target because the mercantile idea of the wealth of nations was not appropriate. Based on the reflection of the failure of mercantile policy, physiocrats promoted policy management entrusted to the natural law that governs economic activities of the freely competitive market.

Smith also argues that free competition maximizes the wealth of nations.[18] The efficiency of the competitive market was also suggested in the theory of utility and scarcity.[19]

Thus, the competitive market was expected to be an efficient resource allocation mechanism that would bring about social benefits (maximization of GDP or social welfare). Gossen [16, pp. 115–116] tried to prove that free competition attained the optimal allocation, but the proof was not successful because his criterion of the optimality allocation was given by maximization of the utilitarian social welfare function [28, p. 39]. The fundamental theorems of welfare economics cannot be proved unless the framework of general equilibrium theory is established and ordinalism is introduced into economics. The general equilibrium theory was constructed by Walras and modified by Pareto, and ordinalism was introduced by Pareto. Pareto [47, pp. 287–292] wrote a letter to Pantaleoni in 1899, in which he suggested that an indifference curve can be derived from a preference ordering that satisfies appropriate properties, He himself seems to assume that the preference ordering should satisfy the completeness, monotonicity, convexity, and transitivity that would be implicitly assumed. He further argues that an ordinal utility function can be inferred from the map of the indifference curve.[20] Pareto [46] defined Pareto efficiency of allocations and tried to show that any competitive equilibrium of a market economy is Pareto efficient.[21]

The idea that the optimum state of the market economy is achieved through perfect and free competition has been inherited since the establishment of political economy. The notions of competitive market and optimality are like identical twin sisters and can be viewed as a single economic thought of an "invisible hand" or "laissez-faire".

5 Theory of Social Value

The fundamental theorems of welfare economics characterize Pareto efficiency as a necessary and sufficient norm for competitive equilibrium, but it is insufficient as norms for judging social states. In order to make appropriate policy decisions, it is necessary to define the norm of optimality that is stronger than Pareto efficiency.

---

[18] Smith's claim [53, pp. 454–456] does not mean the fundamental theorems of welfare economics [38, p. 549, Proposition 16.C.1], but a proposition that means the efficiency of decentralized production [38, p. 148, Proposition 5.E.1].

[19] Galiani [14, pp. 89–90] states that " this equilibrium fits in wonderfully with the just abundance of the conveniences of life and earthly happiness, although it results, not from human prudence or virtue, but from the base incentive of sordid gain: Providence, out of infinite love for men, having so ordered the relations of things that even our base passions, as if in spite of us, are often arranged for the good of the whole" [41, p. 298]. See Kawamata [28, pp. 38–39] about Turgot's idea.

[20] Debreu [6, pp. 50–59] shows that under a certain set of assumptions, if the preference preordering is complete, transitive, and continuous, then there exists a continuous utility function.

[21] Pareto assumes ordinalism in economics, but he clearly reveals a cardinalist position in sociology that includes the field of welfare economics [54].

It was the Austrian School that developed the theory of optimal allocation in the process of the Marginal Revolution.

5.1 Menger's Theory of Individual Value

Carl Menger [39, pp. 51–60] investigated pioneering theories of utility and scarcity including Montanari [42], Galiani [14], Turgot [55] and Condillac [4] and so on, and constructed the theory of value based on the individual rationality.

Menger [39, pp. 51–60] essentially assumes an individual who has a decreasing marginal utility function such as

$$\frac{dU_h}{dx_h}(x_h) = a_h - b_h x_h,$$

where a_h and b_h are constants, and satisfy $a_h > b_h > 0$, for every good h. The individual has the disposable resource ω_h of all goods h. He supposes there is a satiation consumption x_h^s where the marginal utility is zero, and therefore $x_h^s = a_h/b_h$. Good h is called an economic good if $a_h/b_h > \omega_h$, and a non-economic good if $a_h/b_h \leqq \omega_h$. The value of an economic good is equal to the marginal utility at the disposable resources $dU_h/dx_h\,(\omega_h)$, and that of a non-economic good zero. Menger's theory is an extension of the theory of utility and scarcity based on the individual's utility maximization principle [27, 28].

Menger [39] merely presented the theory of value in the Robinson Crusoe economy in his *Gruntsätze für Volkswirtschaftslehre*, but suggested that it characterizes the nature of value in a socialist economy. Wieser characterize the significance of Menger's theory of value as follows:

> Value generally and in every form, even in that of use, and even where there is no exchange — as e.g. in a community organised on a socialist basis — must be the same for all similar goods, and must obey a marginal law. Jevons, Gossen, and Walras have not gone so far as to assert this. ... If a socialist community were to give up exchange — the payment of buyer to seller — it would not on that account require to give up this measuring scale for the valuation of goods. It could continue to value similar goods at the same figure, and to bring them all under a marginal law. ... Menger's theory of value differs essentially from its rivals on this point. He asserts that the law of equality and the marginal law refer not only to price but to value. In my opinion this places his theory in advance of all the others, and wins for him the fame of being the first to lay a perfect foundation for the theory of value. The other authors we have named examine only the laws of want and the laws of price. Menger alone includes the laws of value. Wieser [58, pp. 26–27]

Austrian economists who inherited Menger's economics were characterized by insights into the concept of value in the economy as a whole through consideration of value determination in the Robinson Crusoe economy. For example, one of the successors of the Austrian School, J. A. Schumpeter suggests the following:

> Now the Austrians were in the habit of using the model of Robinson Crusoe economy for the purpose of explaining certain fundamental properties of economic behavior. Therefore,

> it was particularly easy for them to realize that there was nothing specifically capitalist about their basic concept of value and its derivates such as cost and imputed returns. Schumpeter [51, pp. 986–987]

5.2 *Wieser's Theory of Natural Value*

Wieser interpreted Menger's theory as the theory of value determination in the socialist economy where the resource of goods is not privately owned but owned by the society. Wieser [58, p. 61] explains:

> Even in a community or state whose economic affairs were ordered on communistic principles, goods would not cease to have value. Wants there would still be there as elsewhere; the available means would still be insufficient for their full satisfaction; and the human heart would still cling to its possessions. All goods which were not free would be recognised as not only useful but valuable; they would rank in value according to the relation in which the available stocks stood to the demand; and that relation would express itself finally in the marginal utility. Social supply and demand, or amount of goods and utility socially compared with one another, would decide value. The elementary laws of valuation, as we have explained them, would be entirely and unlimitedly effective for the whole community.
>
> That value which arises from the social relation between amount of goods and utility, or value as it would exist in the communist state, we shall henceforth call "Natural Values."

This implies that the value of a good is independent of the private ownership of resources.

Menger's theory of individual value is easily extendable to the theory of social value in an economy consisting of many individuals by interpreting the individual utility function as a social welfare function defined as the aggregation of all the individual utility functions. In a socialist economy, the social value of a good is determined so as to maximize the social welfare function subject to the constraints of resources.[22] Thus, the essence of the problem is in the aggregation of individual utility to social welfare. The difficulty of the problem would be shown later by Arrow's General Impossibility Theorem [1].

Wieser avoids answering this difficult problem directly, but he defines natural value by characterizing the norms that the allocation of natural value satisfies. He characterizes natural value as an exchange value independent of purchasing power. Wieser [58, p. 62] states that:

> Natural value is one element in the formation of exchange value. It does not, however, enter simply and thoroughly into exchange value. On the one side, it is disturbed by human imperfection, by error, fraud, force, chance; and on the other, by the present order of society, by the existence of private property, and by the differences between rich and poor, — as a consequence of which latter a second element mingles itself in the formation of exchange value, namely, purchasing power. In natural value goods are estimated simply according to their marginal utility; in exchange value, according to a combination of marginal utility and purchasing power.

[22] I confine my demonstration within the framework of the exchange economy. For more general cases including production, see Negishi [44, pp. 289–294].

As shown above, the theories of Jevons and Walras are distinct from each other, but the concepts of exchange value in their theories means competitive equilibrium prices. Since purchasing power implies income, in natural value, equal incomes are distributed to all individuals. Thus, we can define natural value as (1) the value of goods in a socialist economy and (2) the exchange value when the incomes (purchasing power) of all individuals are equal to one another.

There are two possible interpretations of what the statement that incomes are equal means. One is based on interpreting the income as a real income, the other as a nominal income. The former interpretation is given by Negishi [44, pp. 289–297]. Negishi formulates Wieser's theory of natural value in the socialist economy. Suppose a socialist economy consisting of two persons and two goods, where the resources of the goods (ω_1, ω_2) are owned by the society. Individual $i = A, B$ consumes (x_{i1}, x_{i2}) and $x_{A1} + x_{B1} \leqq \omega_1, x_{A2} + x_{B2} \leqq \omega_2$. The welfare to be maximized in the socialist economy is a function of individual utilities denoted by $W(U_A, U_B)$. Since the marginal use of a good is equalized everywhere in the socialist economy and stratified marginal utility signifies the natural values, we have

$$\frac{\partial W}{\partial U_A}\frac{\partial U_A}{\partial x_{Ah}} = \frac{\partial U_A}{\partial x_{Ah}} = v_h = \frac{\partial U_B}{\partial x_{Bh}} = \frac{\partial W}{\partial U_B}\frac{\partial U_B}{\partial x_{Bh}}, \quad h = 1, 2, \tag{1}$$

where v_h is the natural value of the good h. We can integrate (1) into

$$W(U_A, U_B) = U_A(x_{A1}, x_{A2}) + U_B(x_{B1}, x_{B2}), \tag{2}$$

which is known as the utilitarian social welfare function (USWF). The consumptions of individuals must satisfy the resource constraint such that

$$(x_{A1} + x_{B1}, x_{A2} + x_{B2}) \leqq (\omega_1, \omega_2). \tag{3}$$

Consider the maximization of (2) under the constraint of (3). If we denote the Lagrangian multipliers corresponding to (3) v_h, condition (1) is satisfied by the condition for the constrained maximization. That is to say, natural values are derived as Lagrangian multipliers from the maximization of the maximization of the USWF.

Negishi [43, 44] also shows that exchange values are derived as Lagrangian multipliers from the maximization of the maximization of the following social welfare function:

$$W(U_A, U_B) = a_A U_A(x_{A1}, x_{A2}) + a_B U_B(x_{B1}, x_{B2}), \tag{4}$$

where a_i is the inverse of the marginal utility of income for $i = A, B$.

Another interpretation of natural value is based on the prices of distribution economy by Malinvaud [35, pp. 111–114]. Consider a distribution economy consisting of two persons A, B and two goods 1, 2, where the resources of goods are not owned privately but by the society, and a nominal income R_i is distributed to individual $i = A, B$. We can denote an equilibrium of a distribution economy such as $((U_A, U_B), (R_A, R_B), (\omega_1, \omega_2))$ by:

(α) individual i chooses her consumption $x_i = (x_{i1}, x_{i2})$ so as to maximize her utility $U_i(x_{i1}, x_{i2})$ subject to the income constraint $p_1 x_{i1} + p_2 x_{i2} \leqq R_i$, for $i = A, B$,
(β) $(x_{A1} + x_{B1}, x_{A2} + x_{B2}) \leqq (\omega_1, \omega_2)$.

Then, we have an equilibrium $\left(\left(x_A^*, x_B^*\right), \left(p_1^*, p_2^*\right)\right)$, in which the equilibrium prices are interpreted as exchange values of the distribution economy, because the equilibrium prices of distribution economy corresponds to competitive prices of a private ownership economy in which private ownership of the resources is specified to satisfy the equation $R_i = p^*\omega_i$, as Malinvaud [35, pp. 114–119] shows. Thus we can interpret natural values of goods of a socialist economy as prices of goods of a distribution economy in which the nominal incomes are the same for all individuals.

In this case, an allocation of the natural value is Pareto efficient, because it corresponds to a competitive equilibrium allocation [12, pp. 218–221]. Since the nominal incomes of all individuals are equal to one another, all the individuals i choose the best consumptions under the same income constraints with one another. Therefore, $U_i(x_i) \geq U_i\left(x_j\right)$ for every pair of people $\{i, j\} \in \{A, B\}$. An allocation (x_A, x_B) is called envy-free or equitable if $U_i(x_i) \geq U_i\left(x_j\right)$ for every pair of people $\{i, j\} \in \{A, B\}$. A Pareto efficient and equitable allocation is called a fair allocation [56]. It means that an allocation of the natural value is a fair allocation.

6 Concluding Remarks

By tracing the major contributions in the Marginal Revolution to the theories of utility and scarcity, we can confirm that there are three streams of theories related to the competitive equilibrium allocation. The first stream is the voluntary exchange theory leading to those of Jevons and Edgeworth. The theory of core in the cooperative game theory was born in this stream. The second is the stream of the theory of social value due to the Austrian school. The third is the "invisible hand" or "laissez-faire" stream characterized by the theory of competitive markets and the fundamental theorems of welfare economics.

The theories of voluntary exchange and of social value were developed by extending the models from the theory of utility and scarcity, but the development of the theory of competitive markets has a theoretical jump by Walras' general equilibrium theory. The Marginal Revolution was promoted not only by the introduction of marginal analysis, but also by the establishment of the general equilibrium theory as well as the theories of voluntary exchange and of social value.

It is well known that the core theory evolved from the theory of voluntary exchange by Jevons and Edgeworth, and that the general equilibrium theory and welfare economics were constructed by Walras and Pareto. However, the Austrian theory of social value has not been so familiar. A reason seems to be the rise of ordinalism, which rejects the interpersonal comparison of utility. From the end of the 19th century, economic theory was in the process of "scientification," in which value judgment was excluded from economic theory as far as possible. Subsequently, the ordinalism introduced by Pareto has become a paradigm of microeconomics since Robbins [48] and Samuelson [49]. Wieser's theory of natural value is incompatible with ordinalism because it presupposes a value judgment regarding the fairness of income distribution in addition to Pareto efficiency. In the theory of cooperative games, optimal allocation is characterized as a solution agreed upon by the bargaining of social members. The Austrian theory of social value was revived by the theory of bargaining games.

References

1. Arrow, K.J.: Social Choice and Individual Values, 2nd edn. Yale University Press, New Haven (1951/1963)
2. Arrow, K.J., Hahn, F.: General Competitive Analysis. Holden Day, San Francisco (1971)
3. Black, R.D.C., Coats, A.W., Goodwin, C.D.W. (eds.): The Marginal Revolution in Economics. Duke University Press, Durham (1973)
4. Bonnot, E., de Condillac, A.: Commerce and Government Considered in Their Mutual Relationship. Liberty Fund, trans. Shelagh Eltis, Indianapolis (1776/2008). Translated from Le commerce et le gouvernement considérés relativement l'un a 1'autre, Houel, Paris (1776)
5. Cournot, A.A.: Recherches sur les principes mathématiques de la théorie des richesses, Vol. 8 of A. A. Cournot OEuvres compètes. Librairie J. Vrin, Paris (1838/1980)
6. Debreu, G.: Theory of Value, An Axiomatic Analysis of Economic Equilibrium. Wiley, New York (1959)
7. Debreu, G., Scarf, H.: A limit theorem on the core of an economy. Int. Econ. Rev. **4**, 235–46 (1963)
8. Desai, M.: A pioneering analysis of the core: Turgot's essay on value. Rech. Econ. de Louvain/Louvain Econ. Rev. **53**(2), 191–198 (1987)
9. Dupuit, J.: De la eesure de l'utilité des travaux publiques. Ann. Ponts Chaussées **8**(2), 332–375 (1844)
10. Edgeworth, F.Y.: Mathematical Psychics. Kegan Paul, London (1881/1967). Reprints of Economic Classics. Augustus M, Kelley (1967)
11. Ekelund, R.B.J., Hébert, R.F.: Retrospectives: the origins of neoclassical microeconomics. J. Econ. Perspect. **16**, 197–215 (2002)
12. Feldman, A.M., Serrano, R.: Welfare Economics and Social Choice Theory, 2nd edn. Springer, London (2006)
13. Fukuoka, M., Maruyama, T.: Walras in the history of utility theory in commemoration of the 100th anniversary of the publication of "Éléments.". Mita J. Econ. **69**(5), 231–250 (1976). (In Japanese)
14. Galiani, F.: Della moneta, vol. III of Scrittori classici italiani di economia politica, Parte moderna edited by P. Custodi. Destefanis, Milano (1750/1803)
15. Gillies, D.G.: Solutions to general non-zero-sum games. In: Tucker, A.W., Luce, R.D. (eds.) Contributions to the Theory of Games IV, pp. 47–85. Princeton University Press, Princeton (1959)
16. Gossen, H.H.: The Laws of Relations, and the Rules of Human Action Derived Therefrom. MIT Press, Massachusetts (1854/1983). Trans. R. C. Blitz with an introduction by N. Georgescu-Roegen; Translated from Entwicklung der Gesetze des menschlichen Verkehrs, und der daraus flissenden Regeln für menschliches Handeln. Friedrich Vieweg & Sohn, Braunschweig (1854)
17. Hicks, J.R.: The scope and status of welfare economics. Oxf. Econ. Pap. **27**, 307–326 (1975)
18. Hutchison, T.W.: Before Adam Smith, The Emergence of Political Economy 1662–1776. Blackwell, Oxford (1988)
19. Jaffé, W.: New light on an old quarrel - Barone's unpublished review of Wicksteed's Essay on the coordination of the laws of distribution and related documents. Cahiers Vilfredo Pareto **3**, 61–102 (1964)

20. Jaffé, W.: Correspondence of Léon Walras and Related Papers, vol. I, 1857–1883. North-Holland, Amsterdam (1965)
21. Jaffé, W.: Edgeworth's contract curve. Hist. Polit. Econ. **6**, 343–359, 381–404 (1974)
22. Jaffé, W.: Menger, Jevons, and Walras De-Homonized. Econ. Inq. **14**, 511–524 (1976)
23. Jaffé, W.: The birth of Léon Walras' Éléments. Hist. Polit. Econ. **9**, 198–214 (1977)
24. Jevons, W.S.: The Theory of Political Economy, 5th edn. Reprints of Economic Classics, Augustus M. Kelley, New York (1871/1965)
25. Katzner, D.W.: Static Demand Theory. Macmillan, New York (1970)
26. Kauder, E.: A History of Marginal Utility Theory. Princeton University Press, Princeton (1965)
27. Kawamata, M.: The Negishi method in the history of general equilibrium theory. In: Ikeo, A., Kurz, H.D. (eds.) A History of Economic Theory: Essays in Honour of Takashi Negishi, pp. 120–136. Routledge, London and New York (2009)
28. Kawamata, M.: Individual rationality and mechanism in the history of microeconomic theory. In: Yagi, K., Ikeda, Y. (eds.) Subjectivism and Objectivism in the History of Economic Thought, pp. 29–47. Routledge, London and New York (2012)
29. Kawamata, M.: A reconsideration of the marginal revolution. Mita J. Econ. **111**(3), 111–145 (2018). (In Japanese)
30. Kawamata, M.: Three approaches to an allocation of competitive equilibrium. Mita J. Econ. **113**(4), 49–79 (2021). (In Japanese)
31. Kuhn, H.W., Tucker, A.W.: Nonlinear programming. In: Neyman, J. (ed.) Proceedings of the Second Berkeley Symposium on Mathematical Statistics and Probability, pp. 481–492. University of California Press, Berkeley (1951)
32. Kuhn, T.S.: The Structure of Scientific Revolution, 2nd edn. The University of Chicago Press, Chicago (1970)
33. Lange, O.R.: On the economic theory of socialism. Rev. Econ. Stud. **4**, 53–71, 123–142 (1936–37)
34. Magnusson, L.: Mercantilism: The Shaping of an Economic Language. Routledge, London (1994)
35. Malinvaud, E.: Lectures on Microeconomic Theory, revised Elsevier Science Publisher, New York (1985)
36. Malthus, T.R.: Principles of Political Economy. Cambridge University Press, Cambridge (1820/1989)
37. Marshall, A.: Principles of Economics, 8th edn. Macmillan, London (1890/1920)
38. Mas-Colell, A., Whinston, M.D., Green, J.R.: Microeconomic Theory. Oxford University Press, New York (1995)
39. Menger, C.: Principles of Economics. Free Press, Glencoe. Translated and edited by J. Dingwall and B. F. Hoselitz with an introduction by F. A, Hayek (1950)
40. Mill, J.S.: Principles of Political Economy with Some of Their Applications to Social Philosophy. John W. Parker, London (1848/1871)
41. Monroe, A.E. (ed.): Early Economic Thought: Selections from Economic Literature prior to Adam Smith. Harvard University Press, Cambridge (1924)
42. Montanari, G.: Trattato mercantile delle monete, vol. III of Scrittori classici italiani di economia politica. Destefanis, Parte antica, Milano (1687/1803)
43. Negishi, T.: Welfare economics and existence of an equilibrium for a competitive economy. Metroeconomica **14**, 92–97 (1960)
44. Negishi, T.: History of Economic Theory. North-Holland, Amsterdam (1989)
45. Pareto, V.: Cours d'éoconomie politique, new edition. Librairie Droz, Genéve (1896–97/1964)
46. Pareto, V.: Manuel d'économie politique. Giard et Briére, Paris. Reprinted in Œuvres complètes, ed. G. Busino, Tome 7. Librairie Droz, Genève (1909/1966)
47. Pareto, V.: Lettere a Maffeo Pantaleoni 1890–1923, vol. 2, 1897–1906. Banca Nazionale del Lavoro, Roma (1960). Reprinted in Œuvres Complètes, ed. G. Busino, Tome 28-2. Librairie Droz, Genève (1984)
48. Robbins, L.C.: An Essay on the Nature and Significance of Economic Science, 2nd edn. Macmillan, London (1935)

49. Samuelson, P.A.: Foundations of Economic Analysis. Harvard University Press, Cambridge (1947)
50. Say, J.B.: Traité d'économie politique ou simple exposition de la maniére dont se forment, se distribuent et se composent les richesses, 2nd edn. Antoine Augustin Renouard, Paris (1803/1814)
51. Schumpeter, J.A.: History of Economic Analysis. Oxford University Press, New York (1954)
52. Shubik, M.: Edgeworth market games. In: Tucker, A.W., Luce, R.D. (eds.) Contributions to the Theory of Games IV, pp. 267–278. Princeton University Press, Princeton (1959)
53. Smith, A.: An Inquiry into the Nature and Causes of the Wealth of Nations, vol. I, 4th edn. Clarendon Press, Oxford (1776/1976)
54. Tarascio, V.J.: Paretian welfare theory: some neglected aspects. J. Polit. Econ. **77**, 1–20 (1969)
55. Turgot, A.R.J.: Value and money (1769). In: Groenewegen, P.D. (ed.) The Economics of A. R. J. Turgot, pp. 133–148. Springer, Dordrecht (1977)
56. Varian, H.R.: Equity, envy, and efficiency. J. Econ. Theory **9**, 63–91 (1974)
57. Verri, P.: Reflections on Political Economy (1781/1993). Reprints of Economic Classics. Augustus M. Kelley, Fairfield, NJ. Translated by Barbara McGilvray in collaboration with Peter D. Groenewegen from Meditazioni sull'economia politica, definitive edition (1781)
58. von Wieser, F: Natural Value, Reprints of Economic Classics. Augustus M.Kelley New York (1889/1971). Translated by C. A. Malloch. Translated from Der natürliche Werth. Höolder, Wien (1889)
59. Walras, L.: Éléments d'économie politique pure, ou théorie de la richesse sociale. Corbaz, Lausanne/R. Pichon et R. Durand-Auzias, Paris (1874–77/1926). Translated by W. Jaffé as Elements of Pure Economics, Allen and Unwin, London (1954)
60. Walras, L.: Économique et mécanique. Metroeconomica **12**, 3–13 (1909/1960). First published in 1909

Behavioral Heterogeneity Pareto Distributions of Homothetic Preference Scales and Aggregate Expenditures Income Elasticities

Jean-Michel Grandmont

Abstract We evaluate the income elasticity of the aggregate budget share spent on a *sub-group* of commodities, in a competitive framework, by a continuum of agents having the same income, but heterogeneous behavior described by an "homothetic preferences scaling factor" having a bounded Pareto distribution in the population. If individual budget share increases *globally* significantly in the limit from low to large incomes, aggregate budget share is *locally* increasing with medium range incomes when the logarithm of the heterogeneity factor has an increasing (exponential) density with a large support. Aggregate income elasticity converges to that exponential density parameter when its support becomes infinitely large. Symmetric results hold in the decreasing case. Applications are made to commodity market expenditures, wealth effects on portfolio choice, concave expenditures, that are compatible with standard (expected) utility maximization or other "behavioral" decision making processes. In particular, we show that the *aggregate* income share invested in risky assets is increasing with wealth under weak conditions, even in the case of several risky assets, an outcome that was impossible to obtain at the *individual* level, except under extremely specific assumptions.

1 Introduction

This work seeks to make theoretical progress on the analysis of macroeconomic aggregates by relying significantly on a *distributional* approach, i.e. on specified properties of distributions of heterogeneous individual *preferences* characteristics.

I have benefited from conversations with Frédérique Bec, Xavier d'Haultfoeuille, Francis Kramarz and Laurent Linnemer. I am very grateful to Fanda Traore for her efficient typing. As usual, I am sole responsible for any remaining shortcomings of the paper.

J.-M. Grandmont (✉)
Department of Economics,ICEF, University Ca' Foscari di Venezia at San Giobbe, Venice, Italy
e-mail: Jean-Michel.Grandmont@ensae.fr

CREST, Institut Polytechnique of Paris, Palaiseau, France

RIEB Fellow, University of Kobe, Kobe, Japan

T. Maruyama (ed.), *Marginal Revolution in Economics*, Monographs in Mathematical Economics 6, https://doi.org/10.1007/978-981-99-4342-5_7

Such an approach has been initiated in particular by W. Hildenbrand [23, 24] and the author (1987, 1992). It aims at coping with the observed behavioral diversity of economic agents, which has been well documented empirically for long in households demand analysis (see Barten [2, 3], Calvet and Comon [9], Deaton and Muellbauer [16], Jorgenson and Selsnik [28], Lewbel [31], Lewbel and Pendakur [34], Matzkin [36], McFadden [37], Muellbauer [38, 39], Prais and Houthakker [40]), or in households heterogeneous portfolio choices (see Calvet et al. [7, 8, 10], Chan and Kogan [14], Curcuru et al. [15]): for useful recent surveys, see Blundell and Stoker [6], Lewbel [32]. A distinguishing feature of the approach adopted here is that we take as a requirement that the introduced heterogeneity should be compatible with individual agents behaving "rationally", i.e. through (expected) utility maximization, as well as with various kinds of "behavioral" biases. This is of course to be contrasted with a large part of (micro or macro) economic theorizing, that is often based more or less implicitly (and misleadingly) on a "representative agent" approach, in which all agents are supposed to behave similarly, whether in markets for goods and services, or in their portfolio choices in finance.

The distributional approach employed in this work, as in previous contributions of the author on this topic (1987, 1992, 1993), seeks to account for behavioral heterogeneity through "*preference scaling factors*" (the word "preference" being there as a reminder of our requirement of individual behaviors being compatible with "rationality"). These are essentially (possible commodity specific) versions of the well known "*households equivalence scales*", due to Prais and Houthakker [40], Barten [2, 3], that have been subsequently much used in econometric studies (Blundell and Lewbel [5], Deaton and Muellbauer [16], Jorgenson and Selsnik [28], Lewbel [31], Lewbel and Pendakur [33], Muellbauer [38, 39]). Such a structure was also used long ago in general equilibrium analysis to show that suitably dispersed distributions on the space of consumers' characteristics led to a nice "smoothing" of competitive aggregate demand (Mas-Colell and Neuefind [35], Dierker et al. [17]). Specifically, one considers a collection (continuum) of agents who have the same income and the same individual (possibly (expected) utility maximizing) "base" behavior, but who behave actually differently, i.e. as if the prices they face were multiplied by these (possibly commodity specific) heterogeneity scaling factors. Previous analysis showed that when the distributions of such commodity specific scaling factors (actually, of their logarithms), were "flat enough", aggregate expenditures became less sensitive to prices (Grandmont [21], Quah [42, 43], Giraud and Maret [19]). Such aggregate insensitivity property has been further shown to occur as well for non-parametric definitions of "behavioral heterogeneity" (see Kneip [30], Jerison [27], Hildenbrand and Kniep [25] and their references). These "behavioral heterogeneity" concepts were however mostly defined directly as properties of the distributions of individual demand functions themselves, without any explicit guarantee that there were compatible with standard individual "rational" utility maximization, in the spirit of the early "irrational" approach of Becker [4].

The aim of this paper is to show that one can actually get sharper aggregate properties by exploiting further the convenient parametric nature of the distributions of such heterogeneous preference scaling factors. We focus here on the simple case where such scaling factors are "homothetic", i.e. the same for all commodities exchanged in the market, as in Grandmont [20], Quah [42]. Such a formulation is indeed particularly adapted to an application we wish to make, i.e. portfolio choices in financial markets, if only to be compatible with standard individual decision making specifications, including VNM expected utility maximization, ambiguity, loss aversion, or whatever. We shall show that focusing, for the distributions of such homothetic scaling factors, on *power laws*, which do have important applications in economics and finance (Gabaix [18]), generates fruitful implications also here. We consider accordingly a continuum of agents who have the same income and the same "base" behavior, but whose actual behavior is indexed by such a multiplicative heterogeneity scaling factor, that is distributed among the agents following a *bounded Pareto distribution*. We focus on the income elasticity of the aggregate budget share spent by these agents on a *sub-group* of commodities. We derive then sharp results on the aggregate income elasticity from rather weak limiting properties of individual "base" behavior for quite low and large income levels, and by relying significantly on the specification of a Pareto distribution for the scaling factor. While our analysis is compatible with standard "rational" (expected) utility maximization over the entire income range, as mentioned earlier, it is compatible as well with any other "behavioral", more or less "irrational" departures in the medium income range, as we do not require any specific properties of the individual "base" behavior there. In the "increasing" case, one requires that the individual "base" budget share spent on the subgroup of commodities, or assets, under consideration, is very small for low incomes, and quite large for high incomes. Then if the logarithm of the homothetic scaling factor has an increasing (exponential) density with a large support, the aggregate budget share is locally increasing for medium range incomes. Further, the aggregate income elasticity converges to the (positive) parameter of this exponential density when its support becomes infinitely large. Symmetric results are valid in the "decreasing" case.

A detailed application is made to expenditures on a subgroup of commodities in a competitive framework. There, low budget shares allocated to this subgroup for low incomes can be viewed as resulting from low marginal rates of substitution for these commodities at low consumption levels. And conversely, large budget shares spent on the subgroup for large incomes, are resulting from large marginal rates of substitution in favor of these commodities at large consumption levels. Another detailed application is made to aggregate portfolio choice in financial markets involving several risky securities and a single riskfree asset. There, a low budget share invested in the riskless asset for low incomes, at the individual level, is then the consequence of very large degrees of relative risk aversion (or of loss aversion) in that income range. Symmetrically, a very low degree of risk aversion (i.e. risk neutrality) for large incomes implies a vanishing investment, at the individual level, in the risk free asset. Our approach enables us then to get the *aggregate* budget share invested in the riskfree asset to be locally decreasing with medium range wealth levels. Such a property has been known to be valid at the individual level since the contribution of Arrow [1], when individual relative degrees of risk aversion are decreasing everywhere with income, provided however that there is a single risky asset to be invested in. By contrast, when there are several risky securities as here, such a property has been well known to be impossible to obtain at the individual level, without imposing extremely particular specifications, e.g. of the agent's VNM expected utility (Cass and Stiglitz [12, 13], Hart [26]). Our results show that going to the aggregate level does improve significantly the perspective. Detailed applications to higher order derivatives, i.e. concavity or convexity of aggregate expenditures, are also presented, a topic that has attracted a significant attenion in the literature (see Caroll and Kimball [11]).

The paper is organized as follows. We present the basic framework and assumptions in Sect. 2. We state the core formal results on aggregate income elasticities in Sect. 3 when the homothetic heterogeneity scaling factor has a bounded Pareto distribution, i.e. when its logarithm has an exponential density. The case where that density is uniform is considered in Sect. 3.1, and monotonically increasing or decreasing in Sect. 3.2. Detailed applications to market demand are given in Sect. 4.1, to aggregate portfolio choice in Sect. 4.2, to concavity of aggregate expenditures in Sect. 4.3. We conclude briefly in Sect. 5. Proofs are gathered in an appendix.

2 Framework

We study aggregate expenditures for a particular *group* of commodities (consumption goods, services, physical or financial assets) of a collection (continuum) of agents in a given period, in a competitive set up. Their common "base" behavior is described by an *individual expenditure function* $w^*(p, \beta)$, where $p = (p_1, \ldots, p_n) \gg 0$ is a vector of positive market prices for all commodities exchanged in the period under consideration and $\beta > 0$ is income. It satisfies $0 \leqq w^*(p, \beta) \leqq \beta$ and is homogenous of degree 1. The corresponding *individual budget share function* $s^*(p, \beta) = w^*(p, \beta)/\beta$ lies accordingly in [0,1] and is homogenous of degree 0. Both expenditure functions are continuously differentiable in $(p, \beta) \gg 0$ up to any required order.

The agents' actual behavior is governed by these expenditure functions up to a *"behavioral heterogeneity" scaling factor* λ, an arbitrary positive real number. An agent corresponding to the scaling factor λ has "tastes" for all commodities that are homothetically divided by λ: she behaves accordingly "as if" the market price vector were λp instead of p. We shall work for analytical convenience with the heterogeneity parameter $\alpha = \log \lambda$. If the agent's income is $b > 0$, the resulting expenditures functions are given by

$$w(\alpha, p, b) = w^*(e^{\alpha} p, b) = e^{\alpha} w^*(p, e^{-\alpha} b), \tag{2.1}$$

$$s(\alpha, p, b) = s^*(e^{\alpha} p, b) = s^*(p, e^{-\alpha} b). \tag{2.2}$$

There is a continuum of such individual agents, normalized to 1, that is described by a probability distribution over the behavioral heterogeneity parameter, generated by a density function $g(\alpha) \geqq 0$. It is concentrated on an interval $[\alpha_0, \alpha_1]$ with $\alpha_0 < \alpha_1$, i.e. $g(\alpha) = 0$ for $\alpha \notin [\alpha_0, \alpha_1]$, and continuously differentiable up to any required order on that interval. Under the assumption that all these agents face the same current price system $p \gg 0$ and have the same current income $b > 0$, aggregate expenditures and budget shares for the group of commodities under consideration, are given by

$$W(p, b) = \int_{\alpha_0}^{\alpha_1} w(\alpha, p, b) g(\alpha) d\alpha, \tag{2.3}$$

$$S(p, b) = W(p, b)/b = \int_{\alpha_0}^{\alpha_1} s(\alpha, p, b) g(\alpha) d\alpha. \tag{2.4}$$

Our aim is to analyze how these aggregate expenditures vary with income, namely to evaluate $\partial W(p, b)/\partial b$ and $\partial S(p, b)/\partial b$, in relation with the properties of individual expenditure functions and of the behavioral heterogeneity scaling factor density distribution $g(\alpha)$. It is clear from (2.2) that $b\dfrac{\partial s}{\partial b}(\alpha, p, b) = -\dfrac{\partial s}{\partial \alpha}(\alpha, p, b)$, so that by differentiation of (2.4)

$$b\frac{\partial S}{\partial b}(p, b) = -\int_{\alpha_0}^{\alpha_1} \frac{\partial s}{\partial \alpha}(\alpha, p, b) g(\alpha) d\alpha. \tag{2.5}$$

An elementary integration by parts yields accordingly

Lemma 1 *Aggregate expenditures' income derivatives satisfy*

$$b\frac{\partial S}{\partial b}(p, b) = \int_{\alpha_0}^{\alpha_1} s(\alpha, p, b) g'(\alpha) d\alpha - [g(\alpha) s(\alpha, p, b)]_{\alpha_0}^{\alpha_1} \tag{2.6}$$

where $[g(\alpha) s(\alpha, p, b)]_{\alpha_0}^{\alpha_1} = g(\alpha_1) s^*(p, e^{-\alpha_1} b) - g(\alpha_0) s^*(p, e^{-\alpha_0} b)$,

$$\frac{\partial W}{\partial b}(p, b) = S(p, b) + b\frac{\partial S}{\partial b}(p, b). \tag{2.7}$$

The above suggests a potentially fruitful research strategy. Aggregate expenditures' income derivatives, or elasticities, depend on two interacting features: *a)* the relative sizes of individual "base" budget shares for low and large incomes at the endpoints of the distribution, namely $s^*(p, e^{-\alpha_1} b)$ compared to $s^*(p, e^{-\alpha_0} b)$, together with *b)* the shape of the scaling parameter density, through its derivative $g'(\alpha)$ and its values at the endpoints of the distribution, $g(\alpha_0)$ compared to $g(\alpha_1)$. As a quick simple example, one is sure to get a locally increasing aggregate budget share, i.e. $b\partial S(p, b)/\partial b > 0$, if the scaling parameter density is non-decreasing, $g'(\alpha) \geqq 0$ (in which case $g(\alpha_1) \geqq g(\alpha_0)$), whenever the individual budget share $s^*(p, \beta)$ is much higher for large incomes β than for low ones, so that,

$$g(\alpha_0) s^*(p, e^{-\alpha_0} b) - g(\alpha_1) s^*(p, e^{-\alpha_1} b) > 0. \tag{2.8}$$

In such a configuration, a significant increase of individual base behavior "in the large" (for large incomes β compared to low ones) translates into a locally increasing behavior for intermediate incomes b through a smoothing effect of the distribution of the behavioral heterogeneity scaling parameter. In the sequel, we shall rely on the two following alternative configurations where base individual budget shares increase (assumption *(H.0)*), or decrease (assumption *(H.1)*), more or less significantly when comparing large incomes β to low ones.

(H.0) *(Increasing individual budget share "in the large") Given* $p \gg 0$, *there exist* $0 < \beta_0^* < \beta_1^*$ *and* $0 \leqq s_0^* < s_1^* \leqq 1$ *such that* $s^*(p, \beta) \leqq s_0^*$ *for* $0 < \beta \leqq \beta_0^*$ *and* $s_1^* \leqq s^*(p, \beta)$ *for* $\beta_1^* \leqq \beta$.

(H.1) *(Decreasing individual budget share "in the large") Given* $p \gg 0$, *there exist* $0 < \beta_0^* < \beta_1^*$ *and* $1 \geqq s_0^* > s_1^* \geqq 0$ *such that* $s^*(p, \beta) \geqq s_0^*$ for $0 < \beta \leqq \beta_0^*$ *and* $s_1^* \geqq s^*(p, \beta)$ *for* $\beta_1^* \leqq \beta$.

In order to ensure the possibility that these increasing or decreasing properties of individual behavior, "in the large", may induce corresponding local monotonicity properties in the aggregate, we consider distributions of the scaling parameter that have a large enough support. In addition to either *(H.0)* or *(H.1)*, we shall assume throughout

(H) *Given* β_0^* *and* β_1^* *in either (H.0) or (H.1), the support of the distribution of the heterogeneity scaling parameter* α *is large, so that* $e^{\alpha_0}\beta_1^* < e^{\alpha_1}\beta_0^*$. *The agents' common income b lies in the open interval* $(e^{\alpha_0}\beta_1^*, e^{\alpha_1}\beta_0^*)$, *so that the scaled incomes* $e^{-\alpha}b$ *at the endpoints of the density distribution* $g(\alpha)$ *satisfy*

$$e^{-\alpha_1}b < \beta_0^* \text{ and } \beta_1^* < e^{-\alpha_0}b. \tag{2.9}$$

This strategy will enable us to get approximate relative evaluations of the individual "base" budget shares $s^*(p, e^{-\alpha}b)$ at the endpoints of the density distribution α_0, α_1, as for instance in (2.8), by direct comparison of $s^*(p, \beta_0^*)$ and $s^*(p, \beta_1^*)$. As for the distribution of the behavioral heterogeneity parameter itself, we shall focus on monotone distributions, and actually *bounded exponential distributions for the indexing parameter* $\alpha = \log\lambda$, which corresponds to the case of *bounded Pareto distributions for the heterogeneity scaling factor* λ itself. That will enable us to derive sharp results on the income elasticities of aggregate expenditures for large heterogeneity distributions, i.e. when $|\alpha_1 - \alpha_0|$ tends to $+\infty$.

3 Exponential Distributions of $\alpha = \log\lambda$

We consider bounded exponential distributions for the indexing parameter α, i.e. $g'(\alpha) = \varepsilon g(\alpha)$ for α in $[\alpha_0, \alpha_1]$ in which case

$$g(\alpha) = \gamma_0 e^{\alpha\varepsilon} \text{ for } \alpha \in [\alpha_0, \alpha_1], \; g(\alpha) = 0 \text{ for } \alpha \notin [\alpha_0, \alpha_1], \text{ with}$$
$$\gamma_0 = \varepsilon/(e^{\alpha_1\varepsilon} - e^{\alpha_0\varepsilon}) \text{ when } \varepsilon \neq 0, \text{ and } \gamma_0 = 1/(\alpha_1 - \alpha_0) \text{ when } \varepsilon = 0. \tag{3.1}$$

Equivalently, the heterogeneity scaling factor $\lambda = e^{\alpha}$ follows a *bounded Pareto distribution* on $[\lambda_0, \lambda_1]$, where $\lambda_0 = e^{\alpha_0}$, $\lambda_1 = e^{\alpha_1}$, with a density $f(\lambda)$ given by

$$f(\lambda) = \gamma_0 \lambda^{\varepsilon - 1} \; for \; \lambda \in [\lambda_0, \lambda_1], \; f(\lambda) = 0 \; for \; \lambda \notin [\lambda_0, \lambda_1] \; with \, also \, here$$
$$\gamma_0 = \varepsilon / (\lambda_1^{\varepsilon} - \lambda_0^{\varepsilon}) \; when \; \varepsilon \neq 0 \; and \; \gamma_0 = 1/(\log \lambda_1 - \log \lambda_0) \; when \; \varepsilon = 0.$$

3.1 *Uniform Distributions* ($\varepsilon = 0$)

In this benchmark configuration, we get immediately from (2.6)

$$b \frac{\partial S}{\partial b}(p, b) = \frac{s^*(p, e^{-\alpha_0} b) - s^*(p, e^{-\alpha_1} b)}{\alpha_1 - \alpha_0}. \tag{3.2}$$

It is then clear that one gets a locally increasing aggregate budget share, i.e. $b \frac{\partial S}{\partial b}(p, b) > 0$, when individual budget shares $s^*(p, \beta)$ are increasing "in the large", i.e. under assumptions *(H.0)* and *(H)*. Symmetrically, one gets $b \frac{\partial S}{\partial b}(p, b) < 0$ under assumptions *(H.1)* and *(H)*. On the other hand, if the support of the distribution, i.e. $|\alpha_1 - \alpha_0|$, becomes increasingly large, given β_0^*, β_1^*, aggregate budget shares becomes less sensitive to income variations. One can in fact show that the aggregate budget share income elasticity converges then uniformly to $\varepsilon = 0$.

Proposition 1 *Under assumption (H), given $p \gg 0$ and the agents' common income $e^{\alpha_0} \beta_1^* < b < e^{\alpha_1} \beta_0^*$,*

(a) under assumption (H.0), one has $b \frac{\partial S}{\partial b}(p, b) > 0$. Furthermore

$$0 < \frac{b \frac{\partial S}{\partial b}(p, b)}{S(p, b)} \leqq \frac{1}{s_1^*(\log(b/\beta_1^*) - \alpha_0)} \tag{3.3}$$

converges to 0 when α_0 tends to $-\infty$, uniformly on compact income subintervals $[b_m, b_M]$ of $(e^{\alpha_0} \beta_1^*, e^{\alpha_1} \beta_0^*)$.

(b) Under assumption (H.1), one has $b \frac{\partial S}{\partial b}(p, b) < 0$. Furthermore

$$\frac{1}{s_0^*(\log(b/\beta_0^*) - \alpha_1)} \leqq \frac{b \frac{\partial S}{\partial b}(p, b)}{S(p, b)} < 0 \tag{3.4}$$

converges to 0 when α_1 tends to $+\infty$, uniformly on compact subintervals $[b_m, b_M]$ of $(e^{\alpha_0} \beta_1^*, e^{\alpha_1} \beta_0^*)$.

A proof is given in the appendix. The implications for aggregate expenditures $W(p, b)$ are immediate through (2.7). Details are left to the reader.

3.2 Strictly Monotonic Exponential Distributions

We focus now on exponential distributions of the heterogeneity parameter, $g'(\alpha) = \varepsilon g(\alpha)$ for α in $[\alpha_0, \alpha_1]$ with $\varepsilon \neq 0$. We get from (2.6) and (3.1)

$$b\frac{\partial S}{\partial b}(p, b) = \varepsilon S(p, b) + \gamma_0[e^{\alpha_0\varepsilon}s^*(p, e^{-\alpha_0}b) - e^{\alpha_1\varepsilon}s^*(p, e^{-\alpha_1}b)]$$
$$with\ \gamma_0 = \varepsilon/(e^{\alpha_1\varepsilon} - e^{\alpha_0\varepsilon}). \tag{3.5}$$

When $\varepsilon > 0$, it is straightforward to verify here, as above in our heuristic discussion of (2.8), that the aggregate budget share is locally increasing, in fact $b\frac{\partial S}{\partial b}(p, b) - \varepsilon S(p, b) > 0$, if individual budget shares $s^*(p, \beta)$ increase strongly enough when comparing low incomes $\beta \leqq \beta_0^*$ to large incomes $\beta \geqq \beta_1^*$, as in assumption *(H.0)*. One can in fact show the much stronger result that the aggregate income elasticity $b\frac{\partial S}{\partial b}(p, b) \Big/ S(p, b)$ actually converges to $\varepsilon > 0$, uniformly on compact income intervals, when the support $[\alpha_0, \alpha_1]$ of the heterogeneity density becomes infinitely large, if individual budget shares $s^*(p, \beta)$ converge fast enough to 0 (faster than β^ε) when $\beta \leqq \beta_0^*$ goes to 0.

Symmetrically, when $\varepsilon < 0$, one verifies also easily that $b\frac{\partial S}{\partial b}(p, b) - \varepsilon S(p, b) < 0$ if individual budget shares $s^*(p, \beta)$ decrease strongly enough when comparing low incomes $\beta \leqq \beta_0^*$ to large incomes $\beta \geqq \beta_1^*$, as in assumption *(H.1)*. Here also, one can show that the aggregate income elasticity $b\frac{\partial S}{\partial b}(\beta, b) \Big/ S(p, b)$ converges to $\varepsilon < 0$, uniformly on compact income intervals, when the support $[\alpha_0, \alpha_1]$ of the heterogeneity density becomes infinitely large, if individual budget shares $s^*(p, \beta)$ go fast enough to 0 when $\beta \geqq \beta_1^*$ increases without bound.

Proposition 2 *Under assumption (H), given* $p \gg 0$ *and the agents' common income* $e^{\alpha_0}\beta_1^* < b < e^{\alpha_1}\beta_0^*$,

(a) If $\varepsilon > 0$, *under assumption (H.0), one has* $b\frac{\partial S}{\partial b}(p, b) > \varepsilon S(p, b) > 0$ *if*

$$[\log s_1^* - \log s_0^*]/(\alpha_1 - \alpha_0) > \varepsilon. \tag{3.6}$$

The aggregate income elasticity $b\frac{\partial S}{\partial b}(p, b) \Big/ S(p, b)$ *converges uniformly to* ε *when* $\alpha_0 \to -\infty, \alpha_1 \to +\infty$, *if*

$$0 \leqq s^*(p, \beta) \leqq a\beta^{\gamma\varepsilon} \ \mathit{for\ all}\ \beta \leqq \beta_0^*, \mathit{with}\ a \geqq 0, \gamma > 1. \tag{3.7}$$

(b) If $\varepsilon < 0$*, under assumption (H.1), one has* $b\dfrac{\partial S}{\partial b}(p, b) < \varepsilon S(p, b) < 0$ *if*

$$[\log s_1^* - \log s_0^*]/(\alpha_1 - \alpha_0) < \varepsilon. \tag{3.8}$$

The aggregate income elasticity $b\dfrac{\partial S}{\partial b}(p, b) \Big/ S(p, b)$ *converges uniformly to* ε *when* $\alpha_0 \to -\infty, \alpha_1 \to +\infty$, *if*

$$0 \leqq s^*(p, \beta) \leqq a\beta^{\gamma\varepsilon} \mathit{for\ all}\ \beta \geqq \beta_1^*, \mathit{with}\ a \geqq 0, \gamma > 1. \tag{3.9}$$

A proof is given in the appendix. The above has immediate implications for aggregate expenditures, through (2.7) and (3.5).

Corollary 1 *Under the assumptions of Proposition 2, income derivatives of aggregate expenditures are given by*

$$\frac{\partial W}{\partial b}(p, b) = (1 + \varepsilon)S(p, b) + [g(\alpha_0)s^*(p, e^{-\alpha_0}b) - g(\alpha_1)s^*(p, e^{-\alpha_1}b)]. \tag{3.10}$$

(a) When $\varepsilon > 0$, (H.0) and *(3.6)* imply $b\dfrac{\partial W}{\partial b}(p, b) > (1 + \varepsilon)W(p, b) > 0$. Under *(3.7)*, $b\dfrac{\partial W}{\partial b}(p, b) \Big/ W(p, b)$ converges uniformly to $1 + \varepsilon$ when $\alpha_0 \to -\infty$, $\alpha_1 \to +\infty$.

(b) When $\varepsilon < 0$, (H.1) and *(3.8)* imply $b\dfrac{\partial W}{\partial b}(p, b) < (1 + \varepsilon)W(p, b)$, while $b\dfrac{\partial W}{\partial b}(p, b) \Big/ W(p, b)$ converges uniformly to $1 + \varepsilon$ when $\alpha_0 \to -\infty, \alpha_1 \to +\infty$ under *(3.9)*. Aggregate expenditures are then decreasing in income if $1 + \varepsilon < 0$.

It may be noted that when $\varepsilon > 0$, the claimed results are valid in the limiting, but nevertheless relevant, configuration where $s^*(p, \beta) = 0$ for all $0 < \beta \leqq \beta_0^*$, which corresponds to the case where $\log s_0^* = -\infty$ in (3.6) and $a = 0$ in (3.7). In that case, convergence of the aggregate income elasticity to $\varepsilon > 0$ obtains when $\alpha_0 \to -\infty$, even when $\alpha_1 > \log(b/\beta_0^*)$ remains bounded above. When $s_0^* > 0$ and thus $a > 0$ in (3.7), one needs both $\alpha_0 \to -\infty$ and $\alpha_1 \to +\infty$. The distribution over the heterogeneity parameter becomes then more concentrated on large α, hence the scaled incomes $e^{-\alpha}b$ on low levels. Both the aggregate budget share $S(p, b)$ and its income derivative become small, while the aggregate income elasticity gets closer to $\varepsilon > 0$. An analogous symmetric argument applies to the case $\varepsilon < 0$.

4 Applications

We apply next the above results and methods to a few specific frameworks : aggregate market demand, portfolio choice, and concave aggregate expenditures.

4.1 Aggregate Market Demand

We consider the case where "base" behavior is the result of maximizing individual preferences, or utility $U(x)$, over commodity bundles $x = (x_1, \ldots, x_n) \geqq 0$ for low and large incomes. Given the commodity price vector $p = (p_1, \ldots, p_n) \gg 0$, "base" expenditures for each commodity h, $w_h^*(p, \beta)$, $s_h^*(p, \beta)$, are the result of maximizing $U(x)$ subject to $p.x = \beta$ for $\beta < \beta_m$ and for $\beta_M < \beta$, where $0 < \beta_m < \beta_M$ are given thresholds. We assume standard conditions (strict monotonicity and concavity, smoothness) ensuring in particular interior solutions $x_h > 0$ within this admissible income range, that can be characterized by standard FOC relations equating marginal rates of substitution of any pair of commodities h, k with the corresponding price ratios p_h/p_k. Expenditure functions $w_h(\alpha, p, b)$ and $s_h(\alpha, p, b)$ corresponding to the heterogeneity scaling factor $\lambda = e^\alpha$, as in (2.1), (2.2), are then obtained by maximizing $U(x/\lambda) = U(e^{-\alpha}x)$ subject to $p.x = b$, whenever $e^{-\alpha}b < \beta_m$ or $\beta_M < e^{-\alpha}b$. *Other (possibly non utility maximizing) behaviors are of course compatible with our analysis for medium range incomes* $\beta_m \leqq \beta \leqq \beta_M$.

We wish to see under which conditions on the utility function $U(x)$ one can get the configurations described in Propositions 1, 2 above. For the simplicity of the exposition, we focus on the "increasing" case *(H.0)* and $\varepsilon \geqq 0$. Similar arguments extend easily to the other case *(H.1)* and $\varepsilon \leqq 0$. Intuitively, one should get the individual "base" budget share $s^*(p, \beta)$ of a particular group of commodities $h = (1, \ldots, i)$ to be increasing in the "large" as in *(H.0)* if the marginal rates of substitution $\frac{\partial U}{\partial x_h}(x) \Big/ \frac{\partial U}{\partial x_k}(x)$ between commodities $h = 1, \ldots, i$ in that group and those outside $k = i + 1, \ldots, n$, become low for small income commodity bundles, and high for large income commodity bundles, in a sense we wish to make precise. And similarly for limiting conditions such as (3.7) for incomes going to 0 in Proposition 2 for the case $\varepsilon > 0$.

A Two Commodities CES Example

Given $p = (p_1, p_2) \gg 0$, let base expenditures be the result of maximizing the CES utility function

$$U(x_1, x_2) = a_1 \frac{x_1^{1-\rho_1}}{1-\rho_1} + a_2 \frac{x_2^{1-\rho_2}}{1-\rho_2}, \quad \rho_1 > 0, \rho_2 > 0 \tag{4.1}$$

with $a_1 > 0, a_2 > 0$, under the budget constraint $p.x = \beta$ for low incomes $0 < \beta < \beta_m$ and large ones $\beta_m < \beta_M < \beta$. The optimum base budget share of commodity $1, s = p_1 x_1/\beta = s^*(p, \beta)$ is then the unique solution of the FOC

$$\frac{s^*(p, \beta)^{\rho_1}}{[1 - s^*(p, \beta)]^{\rho_2}} = \beta^{\rho_2 - \rho_1} d \tag{4.2}$$

with $d = a_1 p_1^{\rho_1 - 1}/a_2 p_2^{\rho_2 - 1}$. When $\rho_2 > \rho_1, s^*(p, \beta)$ is increasing within these income ranges $\beta < \beta_m, \beta_M < \beta$. Then *(H.0)* applies: the right hand side of (4.2) *a)* goes to 0 ($s^*(p, \beta)$ decreases to 0) when $\beta < \beta_m$ goes down to 0, and *b)* goes to $+\infty$ ($s^*(p, \beta)$ goes up to 1) when $\beta_M < \beta$ tends to $+\infty$. Therefore, for any arbitrary thresholds $0 < s_0^* < s_1^* < 1$, there exist β_0^* low enough and β_1^* large enough such that *(H.0)* applies.

Then Proposition 1 applies for uniform distributions ($\varepsilon = 0$). For increasing exponential densities $\varepsilon > 0$, Proposition 2 applies as well provided that ρ_2 is significantly greater than ρ_1. Indeed one gets from (4.2) for any $\beta \leqq \beta_0^*$

$$s^*(p, \beta) \leqq \beta^{(\rho_2 - \rho_1)/\rho_1} d^{1/\rho_1}, \tag{4.3}$$

which is bounded above by $a\beta^{\gamma\varepsilon}$ for some $\gamma > 1$ as in (3.7) (provided that, without loss of generality, $\beta_0^* < 1$) if $(\rho_2 - \rho_1)/\rho_1 \geqq \gamma\varepsilon > \varepsilon$.

As an illustration of the impact of large supports of the behavioral heterogeneity parameter density $g(\alpha)$, one may note that the limit value $\varepsilon > 0$ in Proposition 2 for the aggregate income elasticity $b\dfrac{\partial S}{\partial b}(p, b) \Big/ S(p, b)$, can be somewhat different from the underlying individual base income elasticity of $s^*(p, \beta)$ for low incomes: the latter goes, from (4.2), to $(\rho_2 - \rho_1)/\rho_1 > \varepsilon$ when β goes to 0. This occurs despite the fact that the distributions considered in Proposition 2 give increasing weight on large heterogeneity parameters α, hence on low income $e^{-\alpha}b$, when $\varepsilon > 0$.

Multicommodities Markets

The core economic mechanism, in the foregoing two commodities CES example, rests on a simple property of the marginal rate of substitution $MRS_{12}(x_1, x_2) = \dfrac{\partial U}{\partial x_1}(x_1, x_2) \Big/ \dfrac{\partial U}{\partial x_2}(x_1, x_2)$ when expenditures are fixed proportions of variable income $\beta, x_1 = \mu_1\beta, x_2 = \mu_2\beta$, i.e. $\mu_1 = s/p_1, \mu_2 = (1 - s)/p_2$ where commodity 1's budget share s is fixed. If $\rho_2 > \rho_1$

$$MRS_{12}(\mu_1\beta, \mu_2\beta) = \beta^{\rho_2 - \rho_1}(a_1\mu_2^{\rho_2}/a_2\mu_1^{\rho_1}) \tag{4.4}$$

goes to 0 when $\beta < \beta_m$ goes to 0, to $+\infty$ when $\beta_M < \beta$ goes up to $+\infty$. Then for low incomes, one needs to lower commodity $1's$ consumption (decrease s) to reestablish the optimum equality of the marginal rate of substitution MRS_{12} with the given price ratio p_1/p_2, and a similar symmetric argument applies for large incomes. Hence the

(H.0) pattern that arises in such a case. We consider now a similar mechanism in a multicommodity context.

We use a simple two steps procedure. Every commodity bundle $x = (x_1, \ldots, x_n) \geqq 0$ is split in two groups $x = (x^1, x^2)$ with $x^1 = (x_1, \ldots, x_i)$, $x^2 = (x_{i+1}, \ldots, x_n)$. And similarly for the price vector $p = (p_1, \ldots, p_n) \gg 0$, i.e. $p = (p^1, p^2)$ with $p^1 = (p_1, \ldots, p_i)$, $p^2 = (p_{i+1}, \ldots, p_n)$. The base expenditures resulting from maximizing $U(x)$ subject to $p.x = \beta$ for $\beta < \beta_m$ and $\beta_M < \beta$, we focus on the budget share allocated to the first group of commodities, which is noted $s^*(p, \beta) = \sum_{h=1}^{i} s_h^*(p, \beta)$. Let β_1, β_2 be a preliminary spending allocation for the two groups, with $\beta_1 + \beta_2 < \beta_m$ or $\beta_M < \beta_1 + \beta_2$, and define the corresponding "indirect" utility by the constrained maximization

$$V(p, \beta_1, \beta_2) = Max\ U(x^1, x^2)/p^1 \cdot x^1 = \beta_1, \quad p^2 \cdot x^2 = \beta_2. \tag{4.5}$$

Under standard regularity conditions on $U(x)$ (strict monotonicity and concavity, smoothness, interior solutions), the "indirect" utility $V(p, \beta_1, \beta_2)$ will display the same regularity properties. Further, the original optimum allocation of total income β among the two groups will be the result of maximizing $V(p, \beta_1, \beta_2)$ under the total budget constraint $\beta_1 + \beta_2 = \beta$. The corresponding optimum condition is then

$$\frac{\partial V}{\partial \beta_1}(p, \beta_1, \beta_2) = \frac{\partial V}{\partial \beta_2}(p, \beta_1, \beta_2). \tag{4.6}$$

The interpretation of this two steps procedure follows standard lines. The first stage (4.5) allocates expenditures so as to equalize, *within each group of commodities*, marginal utilities of additional spending

$$\frac{\partial V}{\partial \beta_1}(p, \beta_1, \beta_2) = \frac{1}{p_h}\frac{\partial U}{\partial x_h}(x), \quad h = 1, \ldots, i,$$
$$\frac{\partial V}{\partial \beta_2}(p, \beta_1, \beta_2) = \frac{1}{p_k}\frac{\partial U}{\partial x_k}(x), \quad k = i+1, \ldots, n,$$

under the budget constraints $p^1 \cdot x^1 = \beta_1$, $p^2 \cdot x^2 = \beta_2$. These marginal utilities are then equalized, *accross the two groups*, in the second stage (4.6), so as to reach the original optimum allocation obtained by maximizing directly $U(x)$ subject to $p.x = \beta$. The outcome of (4.6), under the budget constraint $\beta_1 + \beta_2 = \beta$, yields then the optimum budget share, $\beta_1/\beta = s^*(p, \beta)$, $\beta_2/\beta = 1 - s^*(p, \beta)$.

One expects that, similarly to our two commodities CES example, the *(H.0)* pattern and the conditions of Propositions 1, 2 with $\varepsilon \geqq 0$, will obtain if the "indirect" utility's marginal rate of substitution

$$MRS_{12}(p, \beta_1, \beta_2) = \frac{\partial V}{\partial \beta_1}(p, \beta_1, \beta_2) \Big/ \frac{\partial V}{\partial \beta_2}(p, \beta_1, \beta_2) \tag{4.7}$$

becomes low for small incomes $\beta_1 + \beta_2 < \beta_m$ and large for high incomes $\beta_M < \beta_1 + \beta_2$, along lines analogous to (4.4).

Proposition 3 *Let the indirect utility function $V(p, \beta_1, \beta_2)$ be defined as in (4.5).*

(1) Assume that for any fixed budget share $0 < s < 1$ of the first group of commodities, the marginal rate of substitution $MRS_{12}(p, \beta_1, \beta_2)$ in (4.7), with $\beta_1 = s\beta$, $\beta_2 = (1-s)\beta$, tends: (a) to 0 when $\beta < \beta_m$ goes down to 0, (b) to $+\infty$ when $\beta_M < \beta$ increases without bound. Then for any thresholds $0 < s_0^* < s_1^* < 1$, there exist $0 < \beta_0^* < \beta_m < \beta_M < \beta_1^*$, with β_0^* small enough and β_1^* large enough, such that (H.0) is satisfied. Then Proposition 1 applies when $\varepsilon = 0$, as well as the first part of Proposition 2(a) when $\varepsilon > 0$, under condition (3.6).

(2) Assume in addition, when $\varepsilon > 0$

$$MRS_{12}(p, a\beta^{\gamma\varepsilon+1}, \beta) \to 0 \text{ when } \beta \leqq \beta_0^* \text{ tends to } 0 \tag{4.8}$$

for some $a > 0, \gamma > 1$. Then for β_0^* small enough, condition (3.7) is satisfied and the whole Proposition 2(a) applies.

The argument of the proof is straightforward. Details are given in the appendix. A similar symmetric argument can be easily designed to generate the *(H.1)* pattern, to be applied to decreasing densities of the heterogeneity parameter α, i.e. $\varepsilon \leqq 0$.

4.2 Aggregate Portfolio Choices

We consider the aggregate market behavior of a continuum of agents who have an identical income b to invest in several risky securities and a single riskfree asset. The issue we investigate is: under which conditions the aggregate amount invested in the risky assets is an increasing proportion of their income. Or equivalently, when the aggregate amount invested in the riskfree asset is a decreasing proportion of their income. It has been well known since the work of Arrow [1, Chap. 3], that in the case of a single risky security and of a single expected utility maximizing investor, such a behavior would obtain if the agent's relative degree of risk aversion was a decreasing function of her wealth. But it has been also known since then that such an investment behavior was no longer true in the case of several risky securities (Cass and Stiglitz [12, 13], Hart [26]). Without mentioning the impact of modern "behavorial" departures from expected utility maximization ...

We show that swithching to an aggregate viewpoint with heterogeneous agents allows to make the above portfolio property to be valid even with many risky assets when there are no personal outside income insurance motives (as in the standard frameworks quoted above), and when the available risky securities still imply incomplete markets. We require individual "base" behavior, as measured by the budget shares function $s_h^*(p, \beta)$ of every asset, to be the result of consistent expected utility maximization only for quite low incomes $\beta < \beta_m$, and quite large ones $\beta_M < \beta$, while allowing for any "behavioral" departure (loss aversion, ambiguity or whatever ...) in the medium income range $\beta_m \leqq \beta \leqq \beta_M$. Variations of relative degrees of risk aversion are needed only "in the large", i.e. we assume high relative degrees of risk aversion for small incomes β, and a very low one (risk neutrality) for quite large incomes β, while anything can happen in medium income ranges. The proportion of aggregate investment in the riskless security is then shown to be decreasing with the investors' (identical) wealth b for large heterogeneous populations as in Propositions 1, 2 above, in the *(H.1)* pattern, with $\varepsilon \leqq 0$.

Formally, there are n risky securities available today. A unit of any such asset $h = 1, \ldots, n$ yields $d_{hj} > 0$ units of income tomorrow, that may depend on the occurrence of some event $j = 1, \ldots, J$. Its price today is $q_h > 0$, the corresponding gross rates of return being $R_{hj} = d_{hj}/q_h > 0$. There is a single riskfree asset, a unit of which generates a sure unit of income tomorrow. Its current unit price is $q_0 > 0$, the corresponding sure gross rate of return being $R_0 = 1/q_0 > 0$. The current asset prices vector is noted $p = (q_0, q_1, \ldots, q_n) \gg 0$. We assume that every portfolio of the risky securities involves some possible loss by comparison to the riskfree asset.

(Incomplete asset markets) *There exists* $\Delta R_m < 0$ *such that for every* $\lambda_h \geqq 0$ *with* $\sum_{h=1}^{n} \lambda_h = 1$, *one has for some* $j = 1, \ldots, J$,

$$\sum_{h=1}^{n} \lambda_h R_{hj} - R_0 < \Delta R_m < 0. \tag{4.9}$$

The "base" behavior of individual investors is described by the non-negative proportions of the variable income $\beta > 0$ invested in the riskless asset, $s = s^*(p, \beta) \geqq 0$, and in the risky ones, $s_h = s_h^*(p, \beta) \geqq 0$, with $s + \sum_{h=1}^{n} s_h = 1$. As in Sect. 2, we consider a continuum of agents with a common income b, whose actual portfolio choices $s(\alpha, p, b)$, $s_h(\alpha, p, b)$, are affected by a behavioral heterogeneity scaling factor $\lambda = e^{\alpha}$, as in (2.2), and look at the proportion of their aggregate investments in the various assets $S(p, b)$, $S_h(p, b)$, defined as in (2.4). We investigate then, within the framework of Sect. 3, conditions ensuring that the aggregate budget share invested in the riskless asset, $S(p, b)$, decreases with income b. To do so, we need to focus on the case where the individual "base" budget share of the riskfree security, $s^*(p, \beta)$, decreases "in the large", as in the *(H.1)* pattern.

We assume that for low incomes, $0 < \beta < \beta_m$, and large ones, $\beta_M < \beta$, where $0 < \beta_m < \beta_M$ are given thresholds, individual "base" choices of a portfolio $s = s^*(p, \beta)$, $s_h = s_h^*(p, \beta)$, are driven by the maximization of a standard VNM expected utility $E[u(y_j)] = \sum_{j=1}^{J} \pi_j u(y_j)$, where $\pi_j > 0$ are the subjective probabilities of occurrence of each state j, with $\sum_{j=1}^{J} \pi_j = 1$, subject to $s + \sum_{h=1}^{n} s_h = 1$. The incomes generated by such a portfolio, are given by, for each state $j = 1, \ldots, J$ (there is no exogenous risky income, so no outside income insurance motive).

$$y_j = \left(sR_0 + \sum_{h=1}^{n} s_h R_{hj} \right) \beta = R_0\beta + \sum_{h=1}^{n} (R_{hj} - R_0) s_h \beta. \quad (4.10)$$

On the other hand, the corresponding budget shares $s(\alpha, p, b)$, $s_h(\alpha, p, b)$ associated to the heterogeneity scaling factor $\lambda = e^{\alpha}$, as in (2.2), can be viewed as resulting from the maximization of the expected utility $E[u(e^{-\alpha} y_j)]$ under the same constraints in similar income ranges. We make the standard "rationality" assumption for all low income ranges $y_j < y_m$ and large ones $y_M < y_j$ generated by arbitrary portfolios (s, s_h) as in (4.10) with $\beta < \beta_m$ and $\beta_M < \beta$ respectively:
(U) *The VNM utility function $u(y)$ is, on the relevant income ranges $y < y_m$ and $y_M < y$, continuous, twice continuously differentiable for $y > 0$, increasing ($u'(y) > 0$, with $\lim_{y \to 0} u'(y) = +\infty$), strictly concave ($u''(y) < 0$).*

As a counterpart of (4.9), the following assumption states that risky securities are nevertheless financially attractive.

There is at least one risky asset h such that

$$E[R_s - R_0] = \sum_{j=1}^{J} \pi_j (R_{hj} - R_0) > 0. \quad (4.11)$$

Proposition 4 *Let the assumptions (U), (4.9) and (4.11) hold.*

(1) Assume that investors are risk neutral for high incomes:

$$\text{There is } y_1^* > y_M \text{ such that } \rho(y) = -yu''(y)/u'(y) = 0 \text{ for } y > y_1^*. \quad (4.12)$$

Then there exists $\beta_1^* > \beta_M$ such that $s^*(p, \beta) = 0$ for all $\beta \geqq \beta_1^*$.

(2) Assume that investors are very much risk averse for low incomes:

$$\rho(y) \to +\infty \text{ when } y \to 0. \quad (4.13)$$

Then for any $0 < s_0^* < 1$, there exists $0 < \beta_0^* < \beta_m$ such that $s^*(p, \beta) > s_0^*$ for all $\beta \leqq \beta_0^*$.

(3) Under assumptions (4.12), (4.13), the (H.1) pattern applies with $1 > s_0^* > s_1^* = 0$ as above. For any exponential distribution of the heterogeneity scaling parameter α and any investors' common income b satisfying (H), the aggregate budget share invested in the riskless security, $S(p, b)$, is decreasing in b when $\varepsilon \leqq 0$. Its elasticity with respect to b tends to ε when $\alpha_1 \to +\infty$.

A proof is given in the appendix. As a concluding remark, it may be worth re-emphasizing that the standard "rationality" expected utility assumption was used here only for low and large incomes, as in *(U)*. And this only to assume formally risk neutrality for large incomes, as in (4.12), or a relative degree of risk aversion that tends to $+\infty$ (not necessarily monotonically!) for low incomes as in (4.13). As a matter of fact, a cursory look at the argument in the proof suggests strongly that the driving mechanism to derive the above results (1) and (2) (and hence the whole Proposition from Propositions 1(b), 2(b)) is the investors' *degree of loss aversion*, that is absent for high incomes and becomes infinite for low incomes. On the other hand, the whole analysis is compatible with any "behavioral rationality departure" in the medium range incomes.

4.3 Concave/Convex Aggregate Expenditures

We extend the results of Sect. 3 to higher order derivatives, namely concavity/convexity of aggregate budget shares $S(p, b)$ and expenditures $W(p, b)$ with respect to income b. We maintain throughout our assumption (3.1) of exponential distributions for the heterogeneity parameter $\alpha = \log \lambda$ (i.e. Pareto distributions of the scaling factor λ). It is convenient to reformulate the income derivative of aggregate expenditure (3.10) as

$$b\frac{\partial W}{\partial b}(p, b) = W(p, b)(1 + \varepsilon) - \gamma_0[e^{\alpha\varepsilon} w(\alpha, p, b)]_{\alpha_0}^{\alpha_1}, \tag{4.14}$$

with the maintained notation $[f(\alpha)]_{\alpha_0}^{\alpha_1} = f(\alpha_1) - f(\alpha_0)$.

Lemma 2 *Differentiation with respect to income b, of (3.5) and (4.14), yields*

$$b\frac{\partial^2 S}{\partial b^2}(p, b) = (\varepsilon - 1)\frac{\partial S}{\partial b}(p, b) - \gamma_0 \left[e^{\alpha(\varepsilon - 1)}\frac{\partial s^*}{\partial \beta}(p, e^{-\alpha} b)\right]_{\alpha_0}^{\alpha_1}, \tag{4.15}$$

$$b\frac{\partial^2 W}{\partial b^2}(p, b) = \varepsilon\frac{\partial W}{\partial b}(p, b) - \gamma_0 \left[e^{\alpha\varepsilon}\frac{\partial w^*}{\partial \beta}(p, e^{-\alpha} b)\right]_{\alpha_0}^{\alpha_1}. \tag{4.16}$$

In what follows, as in Sect. 3, we let assumption *(H)* hold, while the price system $p \gg 0$ and the agents' common income $e^{\alpha_0}\beta_1^* < b < e^{\alpha_1}\beta_0^*$ are given. We keep assumption *(H.0)*, or *(H.1)*, that enabled us to evaluate the sign and magnitude of $\frac{\partial S}{\partial b}$ from a comparison of individual budget shares $s^*(p, \beta)$ for low and large incomes, i.e. for $\beta \leq \beta_0^*$ versus $\beta_1^* \leq \beta$ for some appropriate thresholds $\beta_0^* < \beta_1^*$. Our strategy is also here to compare the income derivatives of individual "base" budget shares $\frac{\partial s^*}{\partial \beta}(p, \beta)$, or expenditures $\frac{\partial w^*}{\partial \beta}(p, \beta)$, for low and large incomes β. We focus on budget shares through assumptions *(D.0)*, or *(D.1)* below. The individual "base" income derivative $\frac{\partial s^*}{\partial \beta}(p, \beta)$ will play accordingly here, through these conditions, a role similar to that of $s^*(p, \beta)$ in Sect. 3 through *(H.0)*, *(H.1)*. The main technical difference being that, by construction, the budget share $s^*(p, \beta)$ is bounded since it lies in [0,1], whereas the income derivative needs not.

(D.0) *(Increasing individual budget share's income derivative "in the large") Given $p \gg 0$ and $\beta_0^* < \beta_1^*$ as in (H.0) or (H.1), there exist $\sigma_0^* < \sigma_1^*$ such that $\frac{\partial s^*}{\partial \beta}(p, \beta) \leqq \sigma_0^*$ for $0 < \beta < \beta_0^*$, while $\sigma_1^* \leqq \frac{\partial s^*}{\partial \beta}(p, \beta)$ for $\beta_1^* \leqq \beta$.*

(D.1) *(Decreasing individual budget share's income derivative "in the large") Given $p \gg 0$ and $\beta_0^* < \beta_1^*$ as in (H.0) or (H.1), there exist $\sigma_0^* > \sigma_1^*$ such that $\frac{\partial s^*}{\partial \beta}(p, \beta) \geqq \sigma_0^*$ for $0 < \beta \leqq \beta_0^*$, while $\sigma_1^* \geqq \frac{\partial s^*}{\partial \beta}(p, \beta)$ for $\beta_1^* \leqq \beta$.*

We focus essentially on one configuration in the main text here, where the aggregate budget share $S(p, b)$ is locally decreasing and convex in income (Proposition 5 below, under assumptions *(H.1)*, *(D.0)* and a decreasing exponential heterogeneity distribution $g(\alpha)$, with $\varepsilon \leqq 0$). Such a configuration seems indeed of specific potential interest, as it can generate aggregate expenditures $W(p, b)$ that are increasing and concave (Corollary 2 below, with $-1 < \varepsilon < 0$), i.e. an aggregate (multicommodities) "concave consumption function", a topic that has attracted a significant attention (see, e.g. Caroll and Kimball [11], Gourinchas and Parker (2002)). Two other interesting configurations are considered in the appendix, where the aggregate budget share $S(p, b)$ is (1) locally increasing and convex (Proposition 6, under assumptions *(H.0)*, *(D.0)* and a significantly increasing exponential heterogeneity distribution, with $\varepsilon \geqq$ (1) and (2) locally increasing and concave (Proposition 7, under assumptions *(H.0)*, *(D.1)* and a moderately increasing exponential heterogeneity distribution with $0 \leqq \varepsilon \leqq 1$).

Proposition 5 (Locally decreasing and convex aggregate budget share) *Let (H.1) hold, as well as (D.0) with $\sigma_0^* < \sigma_1^* \leqq 0$, and consider a decreasing exponential heterogeneity distribution with $\varepsilon \leqq 0$.*

(1) One has $b\frac{\partial^2 S}{\partial b^2}(p, b) > (\varepsilon - 1)\frac{\partial S}{\partial b}(p, b)$ if σ_1^* is close enough to 0, i.e.

$$[\log|\sigma_1^*| - \log|\sigma_0^*|]/(\alpha_1 - \alpha_0) < \varepsilon - 1. \tag{4.17}$$

The aggregate budget share is then locally strictly decreasing, $\frac{\partial S}{\partial b}(p, b) < 0$, and strictly convex, $b\frac{\partial^2 S}{\partial b^2}(p, b) > 0$, if either $\varepsilon = 0$, or under condition (3.8) when $\varepsilon < 0$.

(2) Assume that the individual "base" budget share is non-increasing with income, $\frac{\partial s^*}{\partial \beta}(p, \beta) \leqq 0$, for all $\beta > 0$. Then the aggregate income elasticity, $b\frac{\partial^2 S}{\partial b^2}(p, b) \Big/ \frac{\partial S}{\partial b}(p, b)$, converges uniformly to $\varepsilon - 1 \leqq -1$ when $\alpha_0 \to -\infty, \alpha_1 \to +\infty$, if $\left|\frac{\partial s^*}{\partial \beta}(p, \beta)\right|$

(a) goes fast enough to 0 when β tends to $+\infty$, and (b) is bounded above or does not diverge to $+\infty$ too fast when β goes down to 0:

$$(a)\ for\ \beta \geqq \beta_1^*,\ \left|\frac{\partial s^*}{\partial \beta}(p,\beta)\right| \leqq a_1 \beta^{\nu_1(\varepsilon-1)}\ with\ a_1 \geqq 0, \nu_1 > 1, \tag{4.18}$$

$$(b)\ for\ 0 \leqq \beta_0^*,\ \left|\frac{\partial s^*}{\partial \beta}(p,\beta)\right| \leqq a_0 \beta^{\nu_0(\varepsilon-1)}\ with\ a_0 > 0, 0 \leqq \nu_0 < 1. \tag{4.19}$$

A detailed proof is given in the appendix. On the other hand, it is immediate to verify that when the aggregate budget share income elasticity $b\frac{\partial^2 S}{\partial b^2}(p,b)\Big/\frac{\partial S}{\partial b}(p,b)$ converges to δ, the corresponding income elasticity for aggregate expenditure $b\frac{\partial^2 W}{\partial b^2}(p,b)\Big/\frac{\partial W}{\partial b}(p,b)$ converge to $1+\delta$. One should expect accordingly that a moderately decreasing exponential heterogeneity distribution with $-1 < \varepsilon < 0$, should lead to an increasing and concave (again, multidimentional) aggregate expenditure.

Corollary 2 (Locally increasing and concave aggregate expenditure) *Let the assumptions of Proposition 5 above and of Corollary 1 at the end of Sect. 3 hold, with $-1 < \varepsilon < 0$, and let $\alpha_0 \to -\infty, \alpha_1 \to +\infty$.*

(a) Under condition (3.9), the aggregate income elasticity
$$b\frac{\partial W}{\partial b}(p,b)\Big/W(p,b) \to 1+\varepsilon > 0.$$

(b) The aggregate income elasticity $b\frac{\partial^2 W}{\partial b^2}(p,b)\Big/\frac{\partial W}{\partial b}(p,b) \to \varepsilon < 0.$

5 Conclusion

The analysis presented in this paper shows that exploiting further the parametric structure of heterogeneity preference scaling factors, here by assuming Pareto distributions, may be a significantly fruitful research avenue to get sharp monotonicity properties of macroeconomic aggregates. This allowed us to go well beyond earlier results about possible "insensitivity" properties of aggregate expenditures obtained with less specific notions of "behavioral heterogeneity", recalled in the introduction. In particular, the application made here to the study of wealth effects on aggregate portfolio choices, shows that this approach enables us to get neat aggregate monotonicity properties that are usually hard, or even impossible, to get at the individual level from standard microeconomic (expected utility maximizing, ambiguity, ...) theory in financial markets.

The theoretical analysis of this paper focussed on a single group of agents displaying heterogeneous behavior generated from a single individual "base" expenditure function and a single homothetic heterogeneity scaling factor Pareto distribution. Clearly, it can be easily extended to encompass other possible sources of heterogeneity, e.g. by considering several groups of such individuals, these groups $i = 1, \dots, n$ having relative sizes $\mu_i > 0$ with $\sum_{i=1}^{n} \mu_i = 1$, and being endowed with individual incomes $b_i > 0$, individual "base" expenditure functions and heterogeneity factor distributions that are possibly different. For a given income distribution with $b = \Sigma_i \mu_i b_i$, the elasticity of the aggregate budget share with respect to aggregate income b, can be easily deduced from the results of this paper if each group fits our framework (of course, its relation with each group's aggregate elasticity with respect its income b_i will have to take into account covariance, as noted e.g. by Paluch, Kneip and Hildenbrand [41]). Further, the impact of a (local) change of the income distribution, given aggregate income, can also be easily derived within such an extended framework. Such an extended model is then able to cope with a wide variety of observable behavioral heterogeneities, that have been so much documented in the empirical literature, as reviewed in the introduction, in various frameworks (goods and services, portfolio choices, …).

Preliminary research work suggests also that the methodological approach of this paper, which focussed as a starting point on aggregate income elasticities, can presumably also be applied fruitfully to price elasticities, by using commodity specific scaling factors, as in Grandmont [21, 22]. This may open the possibility to extend the analysis to general equilibrium frameworks, with perfect or imperfect competition.

Appendix

1. Proof of Proposition 1

(a) Under *(H.0)*, (3.2) implies

$$b\frac{\partial S}{\partial b}(p, b) \geqq [s_1^* - s_0^*]/(\alpha_1 - \alpha_0) > 0.$$

Moreover $S(p, b) \geqq s_1^*(\log(\frac{b}{\beta_1^*}) - \alpha_0)/(\alpha_1 - \alpha_0) > 0$. Combining this inequality with (3.2) yields immediately (3.3), hence the result.

(b) Under *(H.1)*, (3.2) implies

$$b\frac{\partial S}{\partial b}(p, b) \leqq [s_1^* - s_0^*]/(\alpha_1 - \alpha_0) < 0.$$

Moreover $S(p, b) \geqq s_0^*(\alpha_1 - \log(\frac{b}{\beta_0^*}))/(\alpha_1 - \alpha_0) > 0$. Combining this inequality with (3.2) yields immediately (3.4), hence the result. Q.E.D.

2. Proof of Proposition 2

(a) Case $\varepsilon > 0$. Under *(H.0)*, (3.5) implies

$$b\frac{\partial S}{\partial b}(p, b) - \varepsilon S(p, b) \geqq \gamma_0[e^{\alpha_0 \varepsilon} s_1^* - e^{\alpha_1 \varepsilon} s_0^*].$$

Since $\gamma_0 > 0$, the right hand side will be positive if and only if (3.6) is satisfied (including the limiting case where $s_0^* = 0$, or $\log s_0^* = -\infty$).

When the support $[\alpha_0, \alpha_1]$ becomes large, with $\alpha_0 < \log(b/\beta_1^*)$, $\log(b/\beta_0^*) < \alpha_1$, a lower bound for $S(p, b)$ is given by

$$\begin{aligned} S(p, b) &\geqq \int_{\alpha_0}^{\log(b/\beta_1^*)} s_1^* \gamma_0 e^{\varepsilon \alpha} \, d\alpha \\ &\geqq s_1^* \frac{\gamma_0}{\varepsilon} \left[\left(\frac{b}{\beta_1^*} \right)^{\varepsilon} - e^{\alpha_0 \varepsilon} \right] > 0. \end{aligned} \quad \text{(A.1)}$$

One gets then from (3.5)

$$\left| \frac{b\frac{\partial S}{\partial b}(p, b)}{S(p, b)} - \varepsilon \right| \leqq \varepsilon \frac{|e^{\alpha_0 \varepsilon} s^*(p, e^{-\alpha_0} b) - e^{\alpha_1 \varepsilon} s^*(p, e^{-\alpha_1} b)|}{s_1^* \left[\left(\frac{b}{\beta_1^*} \right)^{\varepsilon} - e^{\alpha_0 \varepsilon} \right]} \quad \text{(A.2)}$$

Since $s^*(p, e^{-\alpha_0} b) \geqq s_1^* > 0$, one needs to have $\alpha_0 \to -\infty$ to make the right hand side of (A.2) go to 0. This condition is actually sufficient when $s_0^* = 0$, hence $s^*(p, e^{-\alpha_1} b) = 0$, even when $\alpha_1 > \log(b/\beta_0^*)$ remains bounded above. Under the more general condition (3.7), where one may have $a > 0$, one gets

$$\left| \frac{\frac{b\partial S}{\partial b}(p, b)}{S(p, b)} - \varepsilon \right| \leqq \varepsilon \frac{e^{\alpha_0 \varepsilon} + a e^{\alpha_1 \varepsilon (1-\gamma)} b^{\gamma \varepsilon}}{s_1^* \left[\left(\frac{b}{\beta_1^*} \right)^{\varepsilon} - e^{\alpha_0 \varepsilon} \right]} \quad \text{(A.3)}$$

which tends uniformly to 0 when $\alpha_0 \to -\infty$ and since $\gamma > 1$, when $\alpha_1 \to +\infty$.

(b) Case $\varepsilon < 0$. Under *(H.1)*, (3.5) implies

$$b\frac{\partial S}{\partial b}(p, b) - \varepsilon S(p, b) \leqq \gamma_0[e^{\alpha_0 \varepsilon} s_1^* - e^{\alpha_1 \varepsilon} s_0^*].$$

Since $\gamma_0 > 0$, the right hand side will be negative if and only if (3.8) is satisfied (including the limiting case where $s_1^* = 0$, or $\log s_1^* = -\infty$).

When the support $[\alpha_0, \alpha_1]$ becomes large, with $\alpha_0 < \log(b/\beta_1^*), \log(b/\beta_0^*) < \alpha_1$, a lower bound for $S(p, b)$ is given by

$$S(p,b) \geqq \int_{\log(b/\beta_0^*)}^{\alpha_1} s_0^* \gamma_0 e^{\varepsilon\alpha}\, d\alpha$$
$$\geqq s_0^* \frac{\gamma_0}{\varepsilon}\left[e^{\alpha_1\varepsilon} - \left(\frac{b}{\beta_0^*}\right)^{\varepsilon}\right] > 0. \tag{A.4}$$

One gets then from (3.5)

$$\left|\frac{b\dfrac{\partial S}{\partial b}(p,b)}{S(p,b)} - \varepsilon\right| \leqq |\varepsilon| \frac{|e^{\alpha_1\varepsilon} s^*(p, e^{-\alpha_1}b) - e^{\alpha_0\varepsilon} s^*(p, e^{-\alpha_0}b)|}{s_0^*\left[\left(\dfrac{b}{\beta_0^*}\right)^{\varepsilon} - e^{\alpha_1\varepsilon}\right]}. \tag{A.5}$$

Since $s^*(p, e^{-\alpha_1}b) \geqq s_0^* > 0$, one needs to have $\alpha_1 \to +\infty$ to make the right hand side of (A.5) go to 0. This condition is sufficient when $s_1^* = 0$, hence $s^*(p, e^{-\alpha_0}b) = 0$, even when α_0 remains bounded away from $-\infty$. Under the more general condition (3.9), where one may have $a > 0$, one gets

$$\left|\frac{b\dfrac{\partial S}{\partial b}(p,b)}{S(p,b)} - \varepsilon\right| \leqq |\varepsilon| \frac{e^{\alpha_1\varepsilon} + a e^{\alpha_0\varepsilon(1-\gamma)} b^{\gamma\varepsilon}}{s_0^*\left[\left(\dfrac{b}{\beta_0^*}\right)^{\varepsilon} - e^{\alpha_1\varepsilon}\right]} \tag{A.6}$$

which tends uniformly to 0 when $\alpha_1 \to +\infty$ and, since $\varepsilon(1-\gamma) > 0$, when $\alpha_0 \to -\infty$. Q.E.D.

3. Proof of Proposition 3

We consider the standard case where $MRS_{12}(p, \beta_1, \beta_2)$ in (4.7) is a decreasing function of β_1 that tends to $+\infty$ when β_1 goes to 0, and an increasing function of β_2 that goes down to 0 when β_2 tends to 0, in the small income configuration $\beta_1 + \beta_2 < \beta_m$. Symmetrically $MRS_{12}(p, \beta_1, \beta_2)$ tends to 0 when $\beta_M < \beta_1$ goes up to $+\infty$, and tends to $+\infty$ when $\beta_M < \beta_2$ increases without bound. For $\beta < \beta_m$ and $\beta_M < \beta$, the optimum individual base budget share $s^*(p, \beta)$ of the first group of commodities is then the unique solution, given β, of

$$MRS_{12}(p, s\beta, (1-s)\beta) = 1, \tag{A.7}$$

where the left hand side of (A.7) is a decreasing function of s that tends to $+\infty$ when s goes down to 0, and tends to 0 when s goes up to 1.

The assumption made in Proposition 3.1 states that $MRS_{12}(p, s\beta, (1-s)\beta)$ is an increasing function of β, given s, that tends to 0 when β goes down to 0. Thus for any $0 < s_0^* < 1$ fixed, there exists β_0^* small enough such that $MRS_{12}(p, s_0^*\beta, (1-s_0^*)\beta)$ is uniformly small, hence less than 1, for all $\beta \leqq \beta_0^*$. Then one needs to lower s to reestablish the optimum of (A.7), i.e. $s^*(p, \beta) < s_0^*$ for all $\beta \leqq \beta_0^*$. A symmetric argument applies for an arbitrary $1 > s_1^* > s_0^*$ and $\beta_1^* > \beta_M$, so that $s_1^* < s^*(p, \beta)$, as in *(H.0)*.

Thus Proposition 1 applies for the case $\varepsilon = 0$. When $\varepsilon > 0$, Proposition 2*(a)* applies as well if condition (3.6) is satisfied. Assume now (4.8) for some $a > 0$ and $\gamma > 1$. One has for $\beta \leqq \beta_0^*$ going to 0

$$MRS_{12}(p, a\beta^{\gamma\varepsilon+1}, (1-a^{\gamma\varepsilon})\beta) < MRS_{12}(p, a\beta^{\gamma\varepsilon+1}, \beta) \to 0. \tag{A.8}$$

So if β_0^* is chosen small enough, the above is small, hence less than 1, which implies that $s^*(p, \beta) < a\beta^{\gamma\varepsilon}$, i.e. (3.7), for all $\beta \leqq \beta_0^*$. Thus Proposition 2*(a)* applies fully for the case $\varepsilon > 0$. Q.E.D.

4. Proof of Proposition 4

(1) Under (4.12), let $\beta_1^* > \beta_M$ be large enough so that all incomes generated by arbitrary portfolios as in (4.10) satisfy $y_j > y_1^*$. Then for any $\beta \geqq \beta_1^*$, if the "base" budget share invested in the riskfree security at the optimum portfolio $(s, (s_h))$, is positive, $s = s^*(p, \beta) > 0$, one should have $E[(R_{hj} - R_0)u'(y_j)] \leqq 0$ for every risky asset h, with equality whenever $s_h > 0$, where (y_j) is the resulting optimum income stream (4.10). That would not be possible for the risky security h mentioned in (4.11), since risk neutrality means that $u'(y_j)$ is a constant independent of y_j. Thus $s^*(p, \beta) = 0$ for all $\beta \geqq \beta_1^*$.

(2) Let $0 < s_0^* < 1$, an arbitrary "large" budget share invested in the riskfree asset, be fixed. Consider an arbitrarily small income level $0 < \beta < \beta_m$. Let $(s_h(\beta))$ be the budget shares of risky securities that maximize expected utility $E[u(y_j)]$ with $s = s_0^*$ and $\sum_{h=1}^{n} s_h(\beta) = 1 - s_0^*$. Let the resulting income stream $y_j^*(\beta)$ be given by (4.10):

$$y_j^*(\beta) = R_0\beta + \sum_{h=1}^{n} s_h(\beta)\beta(R_{hj} - R_0) > 0. \tag{A.9}$$

From the corresponding FOC, marginal expected utilities $E[R_{hj}u'(y_j^*(\beta))] > 0$ for additional investment in risky assets h must all be equalized for assets with $s_h(\beta) > 0$, while marginal expected utilities must be lower for those other assets k with $s_k(\beta) = 0$.

Consider now the optimum portfolio budget shares when the riskless security proportion s is free to vary. The necessary and sufficient condition for $s^*(p, \beta) > s_0^*$

is that marginal expected utility of investing in the riskfree asset exceeds that of investing in the risky securities, when these are evaluated at the optimum portfolio $(s_h(\beta))$ obtained when s is fixed at s_0^*:

$$E\left[\sum_{h=s}^{n} s_h(\beta)\beta(R_{hj} - R_0)u'(y_j^*(\beta))\right] < 0. \tag{A.10}$$

Taking an exact first order Taylor development of $u'(y_j^*(\beta))$ at $R_0\beta$ and using (A.9) generates the equivalent inequality

$$E[(y_j^*(\beta) - R_0\beta)u'(R_0\beta)] < E[(y_j^*(\beta) - R_0\beta)^2\rho(\theta_j(\beta))u'(\theta_j(\beta))/\theta_j(\beta)]$$

where the income $\theta_j(\beta) > 0$ lies in $[R_0\beta, y_j^*(\beta)]$. Dividing both sides by $\beta u'(R_0\beta) > 0$ and introducing the notation $\Delta R_j(\beta) = \sum_{h=1}^{n} s_h(\beta)(R_{hj} - R_0)$, leads to the equivalent inequality

$$E[\Delta R_j(\beta)] < E[(\Delta R_j(\beta))^2 \frac{\beta}{\theta_j(\beta)} \frac{u'(\theta_j(\beta))}{u'(R_0\beta)} \rho(\theta_j(\beta))]. \tag{A.11}$$

We now make $\beta > 0$ tend to 0 and prove that the right hand side of (A.11) tends to $+\infty$, which implies the existence of a small income level $\beta_0^* > 0$ such that $s^*(p, \beta > s_0^*$ for all $\beta \leqq \beta_0^*$, as claimed. Since $\sum_{h=1}^{n} s_h(\beta) = 1 - s_0^* > 0$, the income $\theta_j(\beta) > 0$ lies in the interval $\left[R_0\beta, \left(\sum_{h=1}^{n} s_h(\beta)R_{hj} + s_0^* R_0\right)\beta\right]$ and tends thus to 0, but the ratio $\beta/\theta_j(\beta)$ remains bounded away from 0. On the other hand, (4.9) implies the existence of a state j (that may depend on β) such that $\Delta R_j(\beta)(1 - s_0^*) < \Delta R_m < 0$. For that state j, one has $y_j^*(\beta) < R_0\beta$, hence $\theta_j(\beta) \leqq R_0\beta$ in which case $u'(\theta_j(\beta))/u'(R_0\beta) \geqq 1$. If $\pi > 0$ is the minimum of all state probabilities, the right hand side of (A.11) is larger than

$$\pi \left(\frac{\Delta R_m}{1 - s_0^*}\right)^2 \frac{\beta}{\theta_j(\beta)} \rho(\theta_j(\beta)).$$

Since $\theta_j(\beta)$ goes to 0, $\rho(\theta_j(\beta))$ tends to $+\infty$ by (4.13), so the above goes to $+\infty$ when β goes to 0, which proves (2).

(3) The claims made in (3) of Proposition 4 follow immediately, from Proposition 1*(b)* when $\varepsilon = 0$ and Proposition 2*(b)* when $\varepsilon < 0$. Q.E.D.

5. Proof of Proposition 5

(1) Under *(D.0)*, (4.5) implies

$$b\frac{\partial^2 S}{\partial b^2}(p,b) - (\varepsilon - 1)\frac{\partial S}{\partial b}(p,b) \geqq \gamma_0\left[e^{\alpha_0(\varepsilon-1)}\sigma_1^* - e^{\alpha_1(\varepsilon-1)}\sigma_0^*\right]. \tag{A.12}$$

If $\sigma_0^* < \sigma_1^* \leqq 0$, since $\gamma_0 > 0$, the right hand side is positive if and only if (4.17) holds (including the limiting case where $\sigma_1^* = 0$, or $\log|\sigma_1^*| = -\infty$). To complete the proof, one needs to prove that $\frac{\partial S}{\partial b}(p,b) < 0$. Under *(H.1)*, this is true if $\varepsilon = 0$ (Proposition 1), and under condition (3.8) when $\varepsilon < 0$ (Proposition 1), and under condition (3.8) when $\varepsilon < 0$ (Proposition 2*(b)*.

(2) Let $[\alpha_0, \alpha_1]$ become large, with $\alpha_0 < \log(b/\beta_1^*), \log(b/\beta_0^*) < \alpha_1$. Under the assumption $\frac{\partial s^*}{\partial \beta}(p,\beta) \leqq 0$ for all $\beta > 0$, a lower bound for

$$\left|\frac{\partial S}{\partial b}(p,b)\right| = \int_{\alpha_0}^{\alpha_1}\left|\frac{\partial s^*}{\partial \beta}(p, e^{-\alpha}b)\right| e^{-\alpha} g(\alpha)\, d\alpha \tag{A.13}$$

is given by

$$\begin{aligned}\left|\frac{\partial S}{\partial b}(p,b)\right| &\geqq \int_{\log(b/\beta_0^*)}^{\alpha_1} \gamma_0|\sigma_0^*|e^{\alpha(\varepsilon-1)}\, d\alpha \\ &\geqq |\sigma_0^*|\frac{\gamma_0}{\varepsilon - 1}\left[e^{\alpha_1(\varepsilon-1)} - (b/\beta_0^*)^{\varepsilon-1}\right].\end{aligned} \tag{A.14}$$

One gets then from (4.15)

$$\begin{aligned}&\left|\frac{b\frac{\partial^2 S}{\partial b^2}(p,b)}{\frac{\partial S}{\partial b}(p,b)} - (\varepsilon - 1)\right| \\ &\leqq \left|\frac{\varepsilon - 1}{\sigma_0^*}\right| \frac{\left|e^{\alpha_0(\varepsilon-1)}\frac{\partial s^*}{\partial \beta}(p, e^{-\alpha_0}b) - e^{\alpha_1(\varepsilon-1)}\frac{\partial s^*}{\partial \beta}(p, e^{-\alpha_1}b)\right|}{(b/\beta_0^*)^{\varepsilon-1} - e^{\alpha_1(\varepsilon-1)}}.\end{aligned} \tag{A.15}$$

Since $\frac{\partial s^*}{\partial \beta}(p, e^{-\alpha_1}b) \leqq \sigma_0^* < 0$, one needs to have $\alpha_1 \to +\infty$, together with (4.19), to make $e^{\alpha_1(\varepsilon-1)}\left|\frac{\partial s^*}{\partial \beta}(p, e^{-\alpha_1}b)\right|$ go to 0. This condition is actually sufficient to make the right hand side of (A.15) go to 0 when $\sigma_1^* = 0$, hence $\frac{\partial s^*}{\partial \beta}(p, e^{-\alpha_0}b) = 0$, even when α_0 is bounded below. Under the additional more general condition (4.18), where one may have $a_1 > 0$ when $\sigma_1^* < 0$, one verifies that the absolute value of the numerator of the right hand side of (A.15) is bounded above by

$$a_1 b^{\nu_1(\varepsilon-1)} e^{\alpha_0(\varepsilon-1)(1-\nu_1)} + a_0 b^{\nu_0(\varepsilon-1)} e^{\alpha_1(\varepsilon-1)(1-\nu_0)}, \tag{A.16}$$

which tends to 0 when $\alpha_0 \to -\infty, \alpha_1 \to +\infty$ under the conditions specified in (4.18), (4.19). Q.E.D.

6. Proposition 6 (Locally increasing and concave aggregate budget share) *Let (H.0) hold, as well as (D.0) with $\sigma_1^* > \sigma_0^* \geqq 0$, and consider an increasing exponential heterogeneity distribution with $\varepsilon \geqq 1$.*

(1) *One has $b\frac{\partial^2 S}{\partial b^2}(p,b) > (\varepsilon - 1)\frac{\partial S}{\partial b}(p,b)$ if*

$$[\log \sigma_1^* - \log \sigma_0^*]/(\alpha_1 - \alpha_0) > \varepsilon - 1 \geqq 0. \tag{A.17}$$

The aggregate budget share is then locally strictly increasing, $\frac{\partial S}{\partial b}(p,b) > 0$, and strictly convex, $b\frac{\partial^2 S}{\partial b^2}(p,b) > 0$, under condition (3.6).

(2) *Assume that the individual "base" budget share is non-decreasing with income, $\frac{\partial s^*}{\partial \beta}(p,\beta) \geqq 0$ for all $\beta > 0$. Then the aggregate income elasticity $b\frac{\partial^2 S}{\partial b^2}(p,b)\Big/\frac{\partial S}{\partial b}(p,b)$ converges uniformly to $\varepsilon - 1 \geqq 0$ when $\alpha_0 \to -\infty, \alpha_1 \to +\infty$, if $\frac{\partial s^*}{\partial \beta}(p,b)$ (a) is bounded above or does not tend to $+\infty$ too fast when $\beta_1^* \leqq \beta \to +\infty$, and (b) goes to 0 fast enough when $\beta_0^* \geqq \beta \to 0$:*

$$(a)\ 0 < \sigma_1^* \leqq \frac{\partial s^*}{\partial \beta}(p,\beta) \leqq a_1 \beta^{\nu_1(\varepsilon-1)} \text{ for } \beta_1^* \leqq \beta, \text{ with } a_1 > 0, 0 \leqq \nu_1 < 1, \tag{A.18}$$

$$(b)\ 0 \leqq \frac{\partial s^*}{\partial \beta}(p,\beta) \leqq a_0 \beta^{\nu_0(\varepsilon-1)} \text{ for } 0 < \beta \leqq \beta_0^*, \text{ with } a_0 > 0, \nu_0 > 1. \tag{A.19}$$

Proof. (1) As seen in the proof of Proposition 5, *(D.0)* implies (A.12). Since $\gamma_0 > 0$ and $0 \leqq \sigma_0^* < \sigma_1^*$ here, the right hand side is positive if and only if (A.17) holds. Since $\frac{\partial S}{\partial b}(p,b) > 0$ under condition (3.6) (Proposition 2*(a)*, the proof is complete.

(2) Let $[\alpha_0, \alpha_1]$ become large, with $\alpha_0 < \log(b/\beta_1^*), \log(b/\beta_0^*) < \alpha_1$. Under the assumption $\frac{\partial s^*}{\partial \beta}(p,\beta) \geqq 0$ for all $\beta > 0$, from (A.13), a lower bound for $\frac{\partial S}{\partial b}(p,b)$ is given by

$$\frac{\partial S}{\partial b}(p,b) \geqq \int_{\alpha_0}^{\log(b/\beta_1^*)} \gamma_0 \sigma_1^* e^{\alpha(\varepsilon-1)}\, d\alpha,$$

which is equal to $\sigma_1^*[\gamma_0/(\varepsilon-1)][(b/\beta_1^*)^{(\varepsilon-1)} - e^{\alpha_0(\varepsilon-1)}] > 0$ when $\varepsilon > 1$, and to $\sigma_1^*\gamma_0$
$[\log(b/\beta_1^*) - \alpha_0] > 0$ when $\varepsilon = 1$.

One gets then from (4.15) when $\varepsilon = 1$

$$\left|\frac{b\dfrac{\partial^2 S}{\partial b^2}(p,b)}{\dfrac{\partial S}{\partial b}(p,b)}\right| \leqq \frac{1}{\sigma_1^*}\frac{\left|\dfrac{\partial s^*}{\partial \beta}(p,e^{-\alpha_0}b) - \dfrac{\partial s^*}{\partial \beta}(p,e^{-\alpha_1}b)\right|}{\log(b/\beta_1^*) - \alpha_0}.$$

Assumption (A.18) implies that $\dfrac{\partial s^*}{\partial \beta}(p,e^{-\alpha_0}b)$ is bounded above by $a_1 > 0$ when $\varepsilon = 1$, while $0 \leqq \dfrac{\partial s^*}{\partial \beta}(p,e^{-\alpha_1}b) \leqq \sigma_0^*$ from *(D.0)*. So the above income elasticity goes to $\varepsilon - 1 = 0$ when $\alpha_0 \to -\infty$ (even if α_1 is bounded above).

When $\varepsilon > 1$, (4.15) implies

$$\left|\frac{b\dfrac{\partial^2 S}{\partial b^2}(p,b)}{\dfrac{\partial S}{\partial b}(p,b)} - (\varepsilon-1)\right| \leqq \frac{\varepsilon-1}{\sigma_1^*}\frac{\left|e^{\alpha_0(\varepsilon-1)}\dfrac{\partial s^*}{\partial \beta}(p,e^{-\alpha_0}b) - e^{\alpha_1(\varepsilon-1)}\dfrac{\partial s^*}{\partial \beta}(p,e^{-\alpha_1}b)\right|}{(b/\beta_1^*)^{\varepsilon-1} - e^{\alpha_0(\varepsilon-1)}}. \tag{A.20}$$

Since $0 < \sigma_1^* \leqq \dfrac{\partial s^*}{\partial \beta}(p,e^{-\alpha_0}b)$, one needs to make α_0 go to $-\infty$, together with (A.18), to make $e^{\alpha_0(\varepsilon-1)}\dfrac{\partial s^*}{\partial \beta}(p,e^{-\alpha_0}b)$ go to 0. The condition is in fact sufficient to make the right hand side of (A.20) go to 0 when $\sigma_0^* = 0$, hence $\dfrac{\partial s^*}{\partial \beta}(p,e^{-\alpha_1}b) = 0$, even when α_1 is bounded above. Under the additional more general condition (A.19), where one may have $\sigma_0^* > 0$ and $a_0 > 0$, the absolute value of the numerator of the right hand side of (A.20) is bounded above by an expression identical to (A.16). This expression goes to 0 under conditions (A.18), (A.19) when $\alpha_0 \to -\infty, \alpha_1 \to +\infty$. Q.E.D.

7. Proposition 7 (Locally increasing and concave aggregate budget share) *Let (H.0) hold, as well as (D.1) with $\sigma_0^* > \sigma_1^* \geqq 0$, and consider an increasing exponential heterogeneity distribution with $0 \leqq \varepsilon \leqq 1$.*

(1) *One has $b\dfrac{\partial^2 S}{\partial b^2}(p,b) < (\varepsilon-1)\dfrac{\partial S}{\partial b}(p,b)$ if*

$$[\log\sigma_1^* - \log\sigma_0^*]/(\alpha_1 - \alpha_0) < \varepsilon - 1 \leqq 0. \tag{A.21}$$

The aggregate budget share is then locally strictly increasing, $\frac{\partial S}{\partial b}(p, b) > 0$, *and strictly concave,* $b\frac{\partial^2 S}{\partial b^2}(p, b) < 0$, *if either* $\varepsilon = 0$, *or under condition* (3.6) *when* $\varepsilon > 0$.

(2) *Assume that the individual "base" budget share is non-decreasing with income,* $\frac{\partial s^*}{\partial \beta}(p, \beta) \geqq 0$, *for all* $\beta > 0$. *The aggregate income elasticity* $b\frac{\partial^2 S}{\partial b}(p, b) \Big/ \frac{\partial S}{\partial b}(p, b)$ *converges uniformly to* $\varepsilon - 1 \leqq 0$ *when* $\alpha_0 \to -\infty, \alpha_1 \to +\infty$, *if* $\frac{\partial s^*}{\partial \beta}(p, \beta)$ *(a) goes to 0 fast enough when* $\beta_1^* \leqq \beta$ *tends to* $+\infty$, *and (b) is bounded above or does not go too fast to* $+\infty$ *when* $\beta_0^* \geqq \beta$ *tends to* 0:

$$(a)\ 0 \leqq \frac{\partial s^*}{\partial \beta}(p, \beta) \leqq a_1 \beta^{\nu_1(\varepsilon-1)}\ \textit{for}\ \beta \geqq \beta_1^*,\ \textit{with}\ a_1 \geqq 0, \nu_1 > 1, \tag{A.22}$$

$$(b)\ 0 < \sigma_0 \leqq \frac{\partial s^*}{\partial \beta}(p, \beta) \leqq a_0 \beta^{\nu_0(\varepsilon-1)}\ \textit{for}\ \beta \geqq \beta_0^*,\ \textit{with}\ a_0 > 0, 0 \leqq \nu_0 < 1. \tag{A.23}$$

Proof. (1) Under *(D.1)*, (4.15) implies

$$b\frac{\partial^2 S}{\partial b^2}(p, b) - (\varepsilon - 1)\frac{\partial S}{\partial b}(p, b) \leqq \gamma_0 \left[e^{\alpha_0(\varepsilon-1)}\sigma_1^* - e^{\alpha_1(\varepsilon-1)}\sigma_0^*\right].$$

If $\sigma_0^* > \sigma_1^* \geqq 0$, since $\gamma_0 > 0$, the right hand side is negative if and only if (A.21) holds. To complete the proof, one needs to show $\frac{\partial S}{\partial b}(p, b) > 0$. Under *(H.0)*, this is true if $\varepsilon = 0$ (Proposition 1), and under condition (3.6) when $\varepsilon > 0$ (Proposition 2*a*).

(2) Let $[\alpha_0, \alpha_1]$ become large with $\alpha_0 < \log(b/\beta_1^*)$, $\log(b/\beta_0^*) < \alpha_1$. Under the assumption $\frac{\partial s^*}{\partial \beta}(p, \beta) \geqq 0$ for all $\beta > 0$, from (A.13), a lower bound for $\frac{\partial S}{\partial b}(p, b)$ is given by

$$\frac{\partial S}{\partial b}(p, b) \geqq \sigma_0^* \frac{\gamma_0}{1 - \varepsilon}[(b/\beta_0^*)^{\varepsilon-1} - e^{\alpha_1(\varepsilon-1)}]$$

when $0 \leqq \varepsilon < 1$, and by

$$\frac{\partial S}{\partial b}(p, b) \geqq \sigma_0^* \gamma_0[\alpha_1 - \log(b/\beta_0^*)]$$

when $\varepsilon = 1$.

One gets then from (4.15) when $\varepsilon = 1$

$$\left|\frac{b\dfrac{\partial^2 S}{\partial b^2}(p,b)}{\dfrac{\partial S}{\partial b}(p,b)}\right| \leqq \frac{1}{\sigma_0^*}\frac{\left|\dfrac{\partial s^*}{\partial \beta}(p,e^{-\alpha_0}b)-\dfrac{\partial s^*}{\partial \beta}(p,e^{-\alpha_1}b)\right|}{\alpha_1-\log(b/\beta_0^*)}.$$

Assumption (A.23) implies that $0<\sigma_0^*\leqq\dfrac{\partial s^*}{\partial\beta}(p,e^{-\alpha_1}b)\leqq a_0$ when $\varepsilon=1$ while $0\leq\dfrac{\partial s^*}{\partial\beta}(p,e^{-\alpha_0}b)\leq\sigma_1^*$ from *(D.1)*. So the above income elasticity goes to $0=\varepsilon-1$ when α_1 tends to $+\infty$ (even when α_0 is bounded below).

When $\varepsilon<1$, (4.15) implies

$$\left|\frac{b\dfrac{\partial^2 S}{\partial b^2}(p,b)}{\dfrac{\partial S}{\partial b}(p,b)}-(\varepsilon-1)\right|$$
$$\leqq \frac{|\varepsilon-1|}{\sigma_0^*}\frac{\left|e^{\alpha_0(\varepsilon-1)}\dfrac{\partial s^*}{\partial \beta}(p,e^{-\alpha_0}b)-e^{\alpha_1(\varepsilon-1)}\dfrac{\partial s^*}{\partial \beta}(p,e^{-\alpha_1}b)\right|}{(b/\beta_0^*)^{\varepsilon-1}-e^{\alpha_1(\varepsilon-1)}}. \quad \text{(A.24)}$$

Since $\dfrac{\partial s^*}{\partial\beta}(p,e^{-\alpha_1}b)\geqq\sigma_0^*>0$, one needs to make $\alpha_1\to+\infty$, together with (A.23), to make $e^{\alpha_1(\varepsilon-1)}\dfrac{\partial s^*}{\partial\beta}(p,e^{-\alpha_1}b)$ go to 0. This condition is in fact sufficient to make the right hand side of (A.24) go to 0 when $\sigma_1^*=0$, hence $\dfrac{\partial s^*}{\partial\beta}(p,e^{-\alpha_0}b)=0$, even when α_0 is bounded below. Under the additional more general condition (A.22), where one may have $\sigma_1^*>0$, $a_1>0$, the absolute value of the numerator of the right hand side of (A.24) is bounded above by the same expression stated in (A.16). Under the conditions (A.22), (A.23), this expression goes to 0 when $\alpha_0\to-\infty$, $\alpha_1\to+\infty$. Q.E.D.

References

1. Arrow, K.J.: Essays in the Theory of Risk Bearing, Chap. 3, Markham, Chicago (1970)
2. Barten, A.P.: Family composition, prices and expenditure patterns. In: Hart, P., Mills, G., Whitaker, J.D. (eds.) Econometric Analysis for National Economic Planning: 16th Symposium of the Colston Society, pp. 277–292. Butterworth, London (1964)
3. Barten, A.P.: The systems of consumer demand functions approach?: A review. In: Part, A., Intriligator, M.D. (eds.) Frontiers of Quantitative Economics, vol. 3, pp. 23–58. North-Holland, Amsterdam (1977)
4. Becker, G.S.: Irrational behavior and economic theory. J. Polit. Econ. **70**, 1–13 (1962)
5. Blundell, R.W., Lewbel, A.: The information content of equivalence scales. J. Econom. **50**, 49–68 (1991)

6. Blundell, R.W., Stoker, T.M.: Heterogeneity and aggregation. J. Econ. Lit. **43**(2), 347–391 (2005)
7. Calvet, L.E., Campbell, J.Y., Sodini, P.: Down or out?: Assessing the welfare cost of households investment mistakes. J. Polit. Econ. **115**(5), 707–747 (2007)
8. Calvet, L.E., Campbell, J.Y., Sodini, P.: Measuring the financial sophistication of households. Am. Econ. Rev. Papers Proc. **99**(2), 393–398 (2009)
9. Calvet, L.E., Comon, E.: Behavioral heterogeneity and the income effect. Rev. Econ. Stat. **85**, 653–669 (2003)
10. Calvet, L.E., Sodini, P.: Twin picks?: Disentangling the determinants of risk-taking in households portfolio. J. Finance **69**(2), 867–906 (2014)
11. Carroll, C.D., Kimball, M.S.: On the concavity of the consumption function. Econometrica **64**, 981–992 (1996)
12. Cass, D., Stiglitz, J.E.: The structure of investor preferences and asset returns, and separability in portfolio allocation?: A contribution to the pure theory of mutual funds. J. Econ. Theory **2**(2), 122–160 (1970)
13. Cass, D., Stigliz, J.E.: Risk aversion and wealth effects on portfolios with many assets. Rev. Econ. Stud. **39**(3), 331–354 (1972)
14. Chan, Y.L., Kogan, L.: Catching up with the Joneses?: Heterogeneous preferences and the dynamics of asset prices. J. Polit. Econ. **110**(6), 1255–1285 (2002)
15. Curcuru, S., Heaton, J., Lucas, D., Moore, D.: Heterogeneity and portfolio choice: theory and evidence. In: Ait Sahalia, Y., Hansen, L.P. (eds.) Handbook of Financial Econometrics, vol. 11, Chap. 6, pp. 337–382. North Holland, Amsterdam (2010)
16. Deaton, A., Muellbauer, J.: Economics and Consumer Behavior. Cambridge University Press, Cambridge (1980)
17. Dierker, H., Dierker, E., Trockel, W.: Price-dispersed preferences and C^1 mean demand. J. Math. Econ. **13**, 11–42 (1984)
18. Gabaix, X.: Power law in economics and finance. Ann. Rev. Econ. **1**, 255–293 (2009)
19. Giraud, G., Maret, I.: Behavioral heterogeneity in large economies. BETA DP 2002-04, University of Strasbourg (2002)
20. Grandmont, J.-M.: Distributions of preferences and the law of demand. Econometrica **55**, 155–161 (1987)
21. Grandmont, J.-M.: Transformations of the commodity space, behavioral heterogeneity and the aggregation problem. J. Econ. Theory **57**, 1–35 (1992)
22. Grandmont, J.-M.: Behavioral heterogeneity and Cournot oligopoly equilibrium. Ric. Econ. **47**, 167–187 (1993)
23. Hildenbrand, W.: On the law of demand. Econometrica **51**, 997–1019 (1983)
24. Hildenbrand, W.: Market Demand: Theory and Empirical Evidence. Princeton University Press, Princeton (1994)
25. Hildenbrand, W., Kneip, A.: On behavioral heterogeneity. Econ. Theor. **25**, 155–169 (2005)
26. Hart, O.D.: Some negative results on the existence of comparative statics results in portfolio theory. Rev. Econ. Stud. **42**(4), 615–621 (1974)
27. Jerison, M.: Dispersed excess demands, the weak axiom and uniqueness of equilibrium. J. Math. Econ. **31**, 15–48 (1999)
28. Jorgenson, D.W., Slesnik, D.T.: Aggregate consumer behavior and household equivalence scales. J. Bus. Econ. Stat. **5**, 219–232 (1987)
29. Kirman, A.: Heterogeneity in economics. J. Econ. Interact. Coord. **1**(1), 89–117 (2006)
30. Kneip, A.: Behavioral heterogeneity and structural properties of aggregate demand. J. Math. Econ. **31**, 49–79 (1999)
31. Lewbel, A.: Consumer demand systems and household equivalence scales. In: Pesaran, M.H., Schmidt, P. (eds.) Handbook of Applied Econometrics: Vol. II: Microeconomics, pp. 167–201. Blackwell, Oxford (1997)
32. Lewbel A.: Modeling heterogeneity. In: Blundell, R., Newey, W., Persson, T. (eds.) Advances in Economics and Econometrics, Theory and Applications: 9th World Congress of the Econometric Society, vol. 3, Chap. 5, pp. 111–119. Econometric Society Monographs, 43. Cambridge University Press, Cambridge (2007)

33. Lewbel, A., Pendakur, K.: Equivalence Scales. In: McMillan, Durlauf, S., Blume, L.E. (eds.) Entry for the New Palgrave Dictionary of Economics, 2nd ed. (2008)
34. Lewbel, A., Pendakur, K.: Unobserved preference heterogeneity in demand using generalized random coefficients. J. Polit. Econ. **125**(4), 1100–1148 (2017)
35. Mas-Colell, A., Neuefind, W.: Some generic properties of aggregate excess demand and an application. Econometrica **45**, 591–599 (1977)
36. Matzkin, R.L.: Heterogeneous choice. In: Blundell, R., Newey, W., Persson, T. (eds.) Advances in Economics and Econometrics, Theory and Applications: 9th Word Congress of the Econometric Society, vol. 3, Chap. 4, pp. 75–110. Econometric Society Monographs, 43. Cambridge University Press, Cambridge (2007)
37. McFadden, D.: Economic choices. Am. Econ. Rev. **91**(3), 351–378 (2001)
38. Muellbauer, J.: Testing the Barten model of household composition effects and the cost of children. Econ. J. **87**, 460–487 (1977)
39. Muellbauer, J.: The estimation of the Prais-Houthakker model of equivalence scales. Econometrica **48**, 153–176 (1980)
40. Prais, S.J., Houthakker, H.S.: The Analysis of Family Budgets, 2nd ed. Cambridge University Press, New York (1971)
41. Paluch, M., Kneip, A., Hildenbrand, W.: Individual versus aggregate income elasticities for heterogeneous populations. J. Appl. Economet. **27**, 847–869 (2012)
42. Quah, J.: The law of demand when income is price dependent. Econometrica **65**, 1421–1142 (1997)
43. Quah, J.: Demand is heterogeneous in Grandmont's model. Nuffield College, Working Paper, W12 (2001)

Linear Exchange Economies

Bernard Cornet

Abstract We consider a pure exchange economy with finitely many commodities and consumers. Each consumer has a **linear** utility function defined on the closed positive orthant of the commodity space and is endowed with a nonnegative quantity of each commodity. This linear exchange economy was extensively studied by Gale [9–11], who proved the existence of equilibria and the uniqueness of equilibria utility levels under the so-called assumption of irreducibility. Computational aspects have also been investigated by Eaves [6] who proposed a finite algorithm to compute an equilibrium (provided an equilibrium exists) for any linear exchange economy. The goal of this paper is to show that the set of equilibria of a linear exchange economy can be defined as the set of solution of a concave minimization problem, thus allowing to show directly convex properties of the set of equilibria, together with the uniqueness and existence properties of equilibria. Moreover, the existence of equilibria is proved without using a fixed point theorem, a situation which makes linear exchange economies one of the rare examples of economies for which this is actually possible.

1 Introduction

We consider a pure exchange economy $\mathcal{E}$ with finite numbers L of commodities and, I, of consumers. Each consumer has a **linear** utility function defined on the closed positive orthant of the commodity space $\mathbb{R}^L$ and is endowed with a nonnegative quantity of each commodity. Linear exchange economies have been extensively studied in the seminal papers by Gale [9–11], who proved the existence of equilibria and the uniqueness of equilibria utility levels under the so-called assumption of irreducibility, then by Bonnisseau et al. [1, 2], Florig [8], and Lahiri [12] in the finite case, and by Florenzano and Moreno-Garcia [7] in the case of a continuum of agents. Computational aspects have also been investigated by Eaves [6] who proposed a finite algorithm to compute a Walras equilibrium (provided an equilibrium exists)

B. Cornet (✉)
Department of Economics, University of Kansas, Lawrence, KS, USA
e-mail: cornet@ku.edu

T. Maruyama (ed.), *Marginal Revolution in Economics*, Monographs in Mathematical Economics 6, https://doi.org/10.1007/978-981-99-4342-5_8

for any linear exchange economy. New developments of such algorithms, following the scheme of polyhedral complementarity have been extensively studied recently by Shmyrev [13–16].

The aim of this paper is to show that the set of equilibria $(x_1^*, \ldots, x_I^*, p^*) \in (\mathbb{R}^L)^I \times \mathbb{R}^L$ of a linear exchange economy is identical to the set of solutions of a maximization problem whose set of constraints is convex, and whose objective function $\varphi(x, p)$ is concave with respect to the variable $x = (x_1, \ldots, x_I)$ and also with respect to the variable p up to the change of variable $p_\ell = \exp(q_\ell)$ $(\ell = 1, \ldots, L)$. A previous result was proved by Cornet [4] and a related result was obtained by Gale [10] but for a particular distribution of the endowments of the consumers. Consequences of this equivalence result are then given to get properties of the equilibrium set of $\mathcal{E}$ and to the existence and uniqueness of equilibria. In particular, the existence of equilibria for a linear exchange economy can be proved without using a fixed point theorem, a situation which makes linear exchange economies one of the rare examples of economies for which this is actually possible.

Firstly, our equivalence result allows us to deduce that the set of equilibrium allocations of $\mathcal{E}$ is convex [Corollary 1] and that the set of equilibrium prices is homeomorphic to a convex set [Corollary 2]. Also of interest for the structure of the equilibrium set of $\mathcal{E}$, from our equivalence result we deduce that if $(x_1^*, \ldots, x_I^*, p^*)$ and $(x_1^{**}, \ldots, x_I^{**}, p^{**})$ are two equilibria of $\mathcal{E}$, and if $\alpha \in (0, 1)$, then $(\alpha x_1^* + (1 - \alpha)x_1^{**}, \ldots, \alpha x_I^* + (1 - \alpha)x_I^{**}, q)$ is also an equilibrium of $\mathcal{E}$, where $q \in \mathbb{R}^L_{++}$ is defined by $q_\ell = (p_\ell^*)^\alpha (p_\ell^{**})^{1-\alpha}$ for every ℓ. Secondly, our equivalence result allows to give a direct proof of Gale's existence result [Corollary 2] without the use of a fixed-point argument. Finally, we provide a result on uniqueness of equilibrium prices under the assumption that at least one consumer has a positive endowment of each commodity.

The precise statements of our results are given in the Sect. 2 and the proofs of the results are given in Sect. 3.

2 The Model and the Results

We[1] consider a linear exchange economy $\mathcal{E}$ with positive finite numbers, L, of commodities and, I, of consumers. The consumption set of the ith consumer $(i = 1, \ldots, I)$ is the closed positive orthant $\mathbb{R}^L_+$ of $\mathbb{R}^L$, her initial endowment is a vector $\omega_i \in \mathbb{R}^L$ and her utility function $u_i : \mathbb{R}^L_+ \to \mathbb{R}$ is defined by $u_i(x_i) = a_i \cdot x_i$ for some given vector $a_i \in \mathbb{R}^L$.

[1] If $x = (x_\ell)$ and $y = (y_\ell)$ are two vectors of $\mathbb{R}^L$, we denote by $x \cdot y = \sum_{\ell=1}^L x_\ell y_\ell$, the scalar product of $\mathbb{R}^L$, by $\|x\| = (x \cdot x)^{1/2}$ the Euclidean norm. The notation $x \geq y$ (resp. $x \gg y$, resp. $x > y$) means that, for every coordinate ℓ, $x_\ell \geq y_\ell$ (resp. for every ℓ, $x_\ell > y_\ell$, resp. $x \geq y$ and $x \neq y$); we let $\mathbb{R}^L_+ = \{x \in \mathbb{R}^L : x \geq 0\}$ be the closed positive orthant of $\mathbb{R}^L$ and $\mathbb{R}^\ell_{++} = \{x \in \mathbb{R}^L : x \gg 0\}$ be the interior of $\mathbb{R}^L_+$.

We define the set $\mathcal{A}$ of attainable allocations by:

$$\mathcal{A} = \{(x_1, \dots, x_I) \in (\mathbb{R}^L_+)^I \ : \ \sum_{i=1}^I x_i \leq \sum_{i=1}^I \omega_i\}$$

and we recall that $(x_1^*, \dots, x_I^*, p^*) \in (\mathbb{R}^L_+)^I \times \mathbb{R}^L_+$ is a (Walras) equilibrium of $\mathcal{E}$ if the two following conditions hold:

(a) *For every* i, x_i^* *maximizes* $u_i(x_i) = a_i \cdot x_i$ *over her budget set*: $\{x_i \geq 0 \ : \ p^* \cdot x_i \leq p^* \cdot \omega_i\}$;
(b) $\sum_{i=1}^I x_i^* \leq \sum_{i=1}^I \omega_i$.

In the following, we assume that:
Assumption A1: For every $i = 1, \dots, I$, $a_i > 0$, and $\omega_i > 0$.
Assumption A2: $\sum_{i=1}^I a_i \gg 0$ and $\sum_{i=1}^I \omega_i \gg 0$.

The first assumption **A1** is standard. The second one states that every good ℓ is desired by at least some consumer and owned by at least some (possible other) consumer. This assumption is clearly not limiting for, if it was not satisfied, one would simply remove the goods ℓ which do not satisfy Assumption **A2**. Moreover, we recall that, under Assumptions **A1** and **A2**, if p^* is an equilibrium price of $\mathcal{E}$, then all its components are positive, that is, $p^* \gg 0$.[2]

Hereafter, we denote by $v_i : \mathbb{R}^L_{++} \to \mathbb{R}$ the indirect utility function of the ith consumer, i.e.,

$$v_i(p) := \max\{a_i \cdot x_i \ : \ x_i \geq 0, \ p \cdot x_i \leq p \cdot \omega_i\} \text{ for } p \in \mathbb{R}^L_{++}.$$

We consider the following maximization problem in which a_i and ω_i $(i = 1, \dots, I)$ are fixed parameters and the variable of the problem is the vector $(x_1, \dots, x_I, p) \in (\mathbb{R}^L)^I \times \mathbb{R}^L$.

$$(\mathcal{P}) : \text{maximize} \min_{i=1,\dots,I} [a_i \cdot x_i - v_i(p)]$$

$$\text{subject to} : \sum_{i=1}^I x_i \leq \sum_{i=1}^I \omega_i$$

$$x_i \geq 0 \ \ (i = 1, \dots, I)$$

$$p \gg 0.$$

We make a change of variable by defining, for $q = (q_1, \dots, q_L) \in \mathbb{R}^L$:

$$V_i(q) = v_i(\exp(q_1), \dots, \exp(q_L)) \ \ (i = 1, \dots, I)$$

and we consider the following related problem in which, as previously, a_i and ω_i $(i = 1, \dots, I)$ are fixed parameters and the variable of the problem is the vector $(x_1, \dots, x_I, q) \in (\mathbb{R}^L)^I \times \mathbb{R}^L$.

---

[2] Indeed, suppose on the contrary that for some good ℓ, say $\ell = 1$, one has $p_1^* \leq 0$. Then, for every i, the vector $x_i = x_i^* + (1, 0, \dots, 0)$ satisfies $x_i \geq 0$ and $p^* \cdot x_i \leq p^* \cdot x_i^* \leq p^* \cdot \omega_i$, hence x_i belongs to the budget set of consumer i. Thus, $a_i \cdot x_i^* \geq a_i \cdot x_i$ from the Equilibrium Condition of Consumer i. But Assumption **A2** implies that for some consumer i, one has $a_{i1} > 0$. Consequently, $a_i \cdot x_i > a_i \cdot x_i^*$, which contradicts $a_i \cdot x_i^* \geq a_i \cdot x_i$ from above.

$$\begin{aligned}(\mathcal{Q}) : \text{maximize } &\min_{i=1,\ldots,I} [a_i \cdot x_i - V_i(q)]\\ \text{subject to : } &\sum_{i=1}^{I} x_i \leq \sum_{i=1}^{I} \omega_i\\ &x_i \geq 0 \ \ (i = 1, \ldots, I)\\ &q \in \mathbb{R}^L.\end{aligned}$$

The objective function of the first problem $(\mathcal{P})$ is not concave, in general, but the next result shows that the objective function of the modified problem $(\mathcal{Q})$ is concave (and its constraints are linear). Thus, $(\mathcal{Q})$ is a concave programming problem.

We now state our first result.

Theorem 1 *Under Assumptions A1 and A2 the following holds.*
(a) The objective function of the problem $\mathcal{Q}$ *is concave.*
(b) Let $(x_1^*, \ldots, x_I^*, p^*) \in (\mathbb{R}^L)^I \times \mathbb{R}^L_{++}$, *and let* $q^* \in \mathbb{R}^L$ *be defined by* $q_\ell^* = \ln p_\ell^*$ *for every* ℓ. *Then the following assertions are equivalent:*

(i) $(x_1^*, \ldots, x_I^*, p^*)$ *is an equilibrium of* $\mathcal{E}$;
(ii) $(x_1^*, \ldots, x_I^*, p^*)$ *is a solution of problem* $\mathcal{P}$;
(iii) $(x_1^*, \ldots, x_I^*, q^*)$ *is a solution of problem* $\mathcal{Q}$.

We now give a first consequence of Theorem 1. Hereafter, we denote by $W(\mathcal{E})$ the set of (Walras) equilibria of the economy $\mathcal{E}$, by $W_x(\mathcal{E})$ the set of equilibrium allocations and by $W_p(\mathcal{E})$ the set of equilibrium prices of $\mathcal{E}$, that is:

$$W_x(\mathcal{E}) = \{(x_1, \ldots, x_I) \in (\mathbb{R}^L)^I \ : \ \exists p \in \mathbb{R}^L_{++} : (x_1, \ldots, x_I, p) \in W(\mathcal{E})\},$$
$$W_p(\mathcal{E}) = \{p \in \mathbb{R}^L_{++} \ : \ \exists (x_1, \ldots, x_I) \in (\mathbb{R}^L)^I : (x_1, \ldots, x_I, p) \in W(\mathcal{E})\}.$$

Corollary 1 *Under Assumptions A1 and A2, the set* $W(\mathcal{E})$ *of equilibria of* $\mathcal{E}$ *is homeomorphic to a convex set, more precisely to the convex set:*

$$W'(\mathcal{E}) = \{(x_1^*, \ldots, x_I^*, \ln p_1^*, \ldots, \ln p_L^*) \ : \ (x_1^*, \ldots, x_I^*, p^*) \in W(\mathcal{E})\}.$$

Hence, the set $W_x(\mathcal{E})$ *of equilibrium allocations of* $\mathcal{E}$ *is convex and the set* $W_p(\mathcal{E})$ *of equilibrium prices of* $\mathcal{E}$ *is homeomorhic to the convex set:*

$$W'_p(\mathcal{E}) = \{(\ln p_1^*, \ldots, \ln p_L^*) \ : \ p^* \in W_p(\mathcal{E})\}.$$

The proof of Corollary 1 is a direct consequence of Theorem 1 and of the definitions of problems $\mathcal{P}$ and $\mathcal{Q}$. It is worth noticing that the first part of the previous corollary can be equivalently reformulated as follows.

Corollary 1'. *Under Assumptions A1 and A2, let* $(x_1^*, \ldots, x_I^*, p^*)$ *and* $(x_1^{**}, \ldots, x_I^{**}, p^{**})$ *be two equilibria of* $\mathcal{E}$, *let* $\alpha \in (0, 1)$ *and define* $r^* \in \mathbb{R}^L_{++}$ *by* $r_\ell = (p_\ell^*)^\alpha (p_\ell^{**})^{1-\alpha}$ *for every* ℓ.

Then $(\alpha x_1^* + (1-\alpha)x_1^{**}, \dots, \alpha x_I^* + (1-\alpha)x_I^{**}, r^*)$ *is also an equilibrium of* $\mathcal{E}$.

Under the assumptions made previously, the set $W(\mathcal{E})$ may be empty, e.g., see Gale [11] for an example. The next theorem guarantees the existence of an equilibrium under the following additional assumption introduced by Gale [9].

Irreducibility Assumption A3: *There do not exist a proper subset* $\boldsymbol{I}$ *of the set of consumers* $\{1, \dots, I\}$ *and a proper subset* $\mathbf{L}$ *of the set of commodities* $\{1, \dots, L\}$ *such that:*

$\omega_{i\ell} = 0$ *whenever* $(i, \ell) \in \boldsymbol{I} \times \mathbf{L}$ *and* $a_{i\ell} = 0$ *whenever* $i \notin \boldsymbol{I}$ *and* $\ell \notin \mathbf{L}$.

Let us note that the above assumption **A3** is satisfied in particular, under Assumptions **A1** and **A2** together with one of the following additional assumptions (i) $a_i \gg 0$ for every i, or (ii) $\omega_i \gg 0$ for every i.

Corollary 2 (Gale [11]) *Under Assumptions* **A1**, **A2**, *and* **A3**, *there exists a solution of problem* $\mathcal{P}$, *hence there exists an equilibrium of the economy* $\mathcal{E}$.

The proof we give hereafter is different from Gale's original one and it does not involve any fixed-point theorem. In other words, the class of linear exchange economies is one of the very rare classes of economies for which the existence of equilibria can be proved without invoking a fixed-point theorem.

Proof of Corollary 2. The objective function $\varphi : (\mathbb{R}^L)^I \times \mathbb{R}^L_{++} \to \mathbb{R}$ of problem $\mathcal{P}$ is clearly continuous but its set of constraints is not compact. Without any loss of generality, we can add to problem $\mathcal{P}$, the constraint $\sum_{\ell=1}^L p_\ell = 1$, since the objective function of $\mathcal{P}$ is positively homogeneous of degree 0 with respect to the price vector p. Then the new set of constraints $C = \{(x, p) \in \mathcal{A} \times \mathbb{R}^L_{++} : \sum_{\ell=1}^L p_\ell = 1\}$ is nonempty and bounded (but is not closed).

To show the existence of a solution of the objective function φ over the set C, it clearly suffices to show that, for every $\bar{x} \in \mathcal{A}$, and every $\bar{p} \in \partial\mathbb{R}^L_{++}$, $\bar{p} \neq 0$, when $(x, p) \to (\bar{x}, \bar{p})$ then $\varphi(x, p) \to -\infty$. For this to be true, it suffices to show that:

- $\exists i$, $\forall \bar{p} \in \partial\mathbb{R}^L_{++}$, $\bar{p} \neq 0$, $v_i(p) \to +\infty$ when $p \to \bar{p}$ and $p \in \mathbb{R}^L_{++}$.

Indeed, let $\boldsymbol{I} := \{i = 1, \dots, I \,:\, \bar{p} \cdot \omega_i = 0\}$, and let $\mathbf{L} := \{\ell = 1, \dots, L \,:\, \bar{p}_\ell > 0\}$. Then $\boldsymbol{I}$ and $\mathbf{L}$ are proper subsets of consumers and goods, respectively, since on one hand, $\sum_{i=1}^I \omega_i \gg 0$ by **A2** and, on the other hand, $\bar{p} \in \partial\mathbb{R}^L_{++}$, and $\bar{p} \neq 0$. For every $(i, \ell) \in \boldsymbol{I} \times \mathbf{L}$, one clearly has $\omega_{i\ell} = 0$. Thus, the Irreducibility Assumption **A3** implies that there exist $i \in \{1, \dots, I\} \setminus \boldsymbol{I}$ and $\ell \in \{1, \dots, L\} \setminus \mathbf{L}$ such that $a_{i\ell} > 0$. Let $p \gg 0$, from the definition of $v_i(p)$, taking $x_i = \mathbf{1}_\ell(p \cdot \omega_i / p_\ell)$ (where $\mathbf{1}_\ell$ is the vector with all its coordinates equal to 0 but the ℓ-th one equal to 1), one deduces that:

$$v_i(p) \geq a_i \cdot x_i \geq p \cdot \omega_i \frac{a_{i\ell}}{p_\ell}.$$

But $a_{i\ell} > 0$, $\bar{p} \cdot \omega_i > 0$ (since $i \notin \boldsymbol{I}$), $\bar{p}_\ell = 0$ (since $h \notin \mathbf{L}$) imply that $p \cdot \omega_i (a_{i\ell}/p_\ell) \to +\infty$ when $p \to \bar{p}$ and $p \in \mathbb{R}^L_{++}$. Hence $v_i(p) \to +\infty$, when $p \to \bar{p}$ and $p \in \mathbb{R}^L_{++}$. □

The end of this section investigates the uniqueness of equilibria in the linear model. The next theorem guarantees that the utility level of every consumer is uniquely determined at equilibria.

Theorem 2 (Gale [11]) *Under Assumptions A1 and A2, the equilibrium utility level of every consumer is uniquely determined, that is, if* $(x_1^*, \dots, x_I^*, p^*)$ *and* $(x_1^{**}, \dots, x_I^{**}, p^{**})$ *are two equilibria of* $\mathcal{E}$*, then* $a_i \cdot x_i^* = a_i \cdot x_i^{**}$ *for every* i.

In Sect. 3.4, we provide a simple proof of Theorem 2, different from Gale's original one.

Our last result is concerned with the uniqueness of the equilibrium price up to the multiplication by a positive scalar.

Theorem 3 *Assume that A1, A2, and* $\omega_i \gg 0$ *for some consumer* i*. Then the equilibrium price is uniquely defined, up to the multiplication by a positive scalar, that is, if* p^* *and* p^{**} *are two equilibrium prices of* $\mathcal{E}$*, then* $p^* = tp^{**}$ *for some positive real number* $t > 0$.

The proof of Theorem 3 is given in Sect. 3.4. This uniqueness property of the equilibrium price vector is of fundamental importance in the study of Walrasian exchange processes by Champsaur and Cornet [3].

3 Proofs

3.1 Proof of Part (a) of Theorem 1

• The proof of Part (a) of Theorem 1 is a direct consequence of the following lemma. We denote by $\psi : (\mathbb{R}^L)^I \times \mathbb{R}^L \to \mathbb{R}$ the objective function of problem $\mathcal{Q}$, that is:

$$\psi(x, q) = \min_{i=1,\dots,I}[a_i \cdot x_i - V_i(q)].$$

Lemma 1 *(a)* $v_i(p) = p \cdot \omega_i \max_{\ell=1,\dots,L}(a_{i\ell}/p_\ell)$ *for every* $p \gg 0$.
(b) For every i*, the function* $V_i : \mathbb{R}^L \to R$ *is convex.*
(c) The function $\psi : (\mathbb{R}^L)^I \times \mathbb{R}^L \to R$ *is concave.*

Proof *Part (a).* For every i and every $p \gg 0$, one has:

$$\begin{aligned} v_i(p) &\stackrel{\text{def}}{=} \max\{a_i \cdot x_i \;:\; p \cdot x \le p \cdot \omega_i,\ x \ge 0\} \\ &= \min\{\lambda p \cdot \omega_i \;:\; \lambda p_\ell \ge a_{i\ell}\ (\ell = 1, \dots, L),\ \lambda \ge 0\}. \end{aligned}$$

This follows from the duality theorem of linear programming [see, for example, Gale (1960)], noticing that $v_i(p)$ is defined as the value of a Linear Programming Problem whose set of constraints is nonempty (it contains 0) and compact (since $p \gg 0$), thus the value $v_i(p)$ is finite. Hence, $v_i(p)$ is equal to the value of the dual problem, which is exactly the above statement. Consequently, one has:

$v_i(p) = p \cdot \omega_i \max_{\ell=1,\dots,L}(a_{i\ell}/p_\ell)$ for every $p \gg 0$.

Part (b). From Part (a) one deduces that, for every $q \in \mathbb{R}^L$,

$V_i(q) = v_i(\exp(q_1), \dots, \exp(q_\ell)) = \max_{\ell=1,\dots,L}[\sum_{k=1}^{L} \omega_{ik} a_{i\ell} \exp(q_k - q_\ell)]$.

The functions $q \to \exp(q_k - q_\ell)$ are clearly convex for every k and every ℓ. Since $a_{i\ell} \geq 0$, $\omega_{i\ell} \geq 0$ and since the supremum of convex functions is also convex, then V_i is also convex. □

Part (c). This a direct consequence of Part (b) and of the fact that the infimum of concave functions is also concave. □

3.2 *Proof of Part (b) of Theorem 1*

• $[(i) \Rightarrow (ii)]$. Let $(x_1^*, \dots, x_I^*, p^*)$ be an equilibrium of $\mathcal{E}$. Then $a_i \cdot x_i^* = v_i(p^*)$ for every i, hence $0 = \min_{i=1,\dots,I}[a_i \cdot x_i^* - v_i(p^*)]$. In view of this equality, it suffices to show that:

$$0 \geq \min_{i=1,\dots,I}[a_i \cdot x_i - v_i(p)] \quad \text{for every } (x, p) \in \mathcal{A} \times \mathbb{R}^L_{++}.$$

Indeed, suppose on the contrary that, for some $(x, p) \in A \times \mathbb{R}^L_{++}$ one has $0 < \min_i[a_i \cdot x_i - v_i(p)]$. Then, for every i, $a_i \cdot x_i > v_i(p)$, which implies that $p \cdot x_i > p \cdot \omega_i$. Summing up these inequalities, one gets $p \cdot \sum_{i=1}^{I} x_i > p \cdot \sum_{i=1}^{I} \omega_i$, which contradicts the facts that $\sum_{i=1}^{I} x_i \leq \sum_{i=1}^{I} \omega_i$ and $p \gg 0$. □

• $[(ii) \Rightarrow (iii)]$. The proof is straightforward. □

• $[(iii) \Rightarrow (i)]$. Let $(x_1^*, \dots, x_I^*, q^*)$ be a solution of $\mathcal{Q}$, we first prove the following claim.

Claim: $u^* \stackrel{\text{def}}{=} \min_{i=1,\dots,I}[a_i \cdot x_i^* - V_i(q^*)] \geq 0$.

We notice that $(u^*, x_1^*, \dots, x_I^*, q^*)$ is a solution of the following problem, in which the variable is the vector $(u, x_1, \dots, x_I, q) \in \mathbb{R} \times (\mathbb{R}^L)^I \times \mathbb{R}^L$.

$$\begin{aligned}
(\mathcal{Q}') : &\text{ maximize } u \\
&\text{subject to: } u \leq a_i \cdot x_i - V_i(q) \quad (i = 1, \dots, I) \\
&\sum_{i=1}^{I} x_i \leq \sum_{i=1}^{I} \omega_i \\
&u \in \mathbb{R}, x_i \geq 0 \quad (i = 1, \dots, I), q \in \mathbb{R}^L.
\end{aligned}$$

The above problem is clearly a Concave Programming Problem, that is, its objective is concave (in fact it is linear) and its constraints are defined by convex functions since V is convex by Lemma 1. Moreover, the Slater Constraint Qualification is satisfied since there clearly exists $(u, x_1, \dots, x_I, q) \in \mathbb{R} \times (\mathbb{R}^L)^I \times \mathbb{R}^L$ satisfying all the constraints of $(\mathcal{Q}')$ and satisfying $u < a_i \cdot x_i - V_i(q) \quad (i = 1, \dots, I)$.

Consequently, there exist Kuhn-Tucker's multipliers $\lambda_i^* \geq 0 \ \ (i = 1, \dots, I)$, $\mu_\ell^* \geq 0 \ \ (\ell = 1, \dots, L)$ such that the following three assertions hold:

(3.1) $u^* - a_i \cdot x_i^* + V_i(q^*) \leq 0$ and $(u^* - a_i \cdot x_i^* + V_i(q^*))\lambda_i^* = 0 \ \ (i = 1, \dots, I)$,

(3.2) $\sum_{i=1}^I x_i^* - \sum_{i=1}^I \omega_i \leq 0$ and $\left(\sum_{i=1}^I x_i^* - \sum_{i=1}^I \omega_i\right) \cdot \mu^* = 0$,

(3.3) $(u^*, x_1^*, \dots, x_I^*, q^*)$ is a solution of the following problem:

maximize $L(u, x, q) := u - \sum_{i=1}^I \lambda_i^*[u - a_i \cdot x_i + V_i(q)] - \mu^* \cdot \left(\sum_{i=1}^I x_i - \sum_{i=1}^I \omega_i\right)$

subject to: $(u, x, q) \in \mathbb{R} \times (\mathbb{R}_+^L)^I \times \mathbb{R}^L$.

But, from (3.3), letting $x = x^*$ and $q = q^*$, one gets $\sum_{i=1}^I \lambda_i^* = 1$ so that:

$$L(u, x, q) = \sum_{i=1}^I \lambda_i^*[a_i \cdot x_i - V_i(q)] - \mu^* \cdot \left(\sum_{i=1}^I x_i - \sum_{i=1}^I \omega_i\right).$$

The first order necessary (and sufficient) conditions of the maximisation problem (3.3) with respect to the variables $x = (x_1, \dots, x_I)$ [keeping $u = u^*$ and $q = q^*$ fixed] over the set $(\mathbb{R}_+^L)^I$ imply that

(3.4) $\lambda_i^* a_i - \mu^* \leq 0$, $\left(\lambda_i^* a_i - \mu^*\right) \cdot x_i^* = 0$, $x_i^* \geq 0 \ \ (i = 1, \dots, I)$.

Recalling that $\sum_{i=1}^I \lambda_i^* = 1$ and summing up (over i) the conditions (3.1) one gets $u^* = \sum_{i=1}^I \lambda_i^*\left(a_i \cdot x_i^* - V_i(q^*)\right)$. From this equality and (3.2) one deduces that

(3.5) $u^* = L(u^*, x^*, q^*)$.

Let $\varepsilon \in \mathbb{R}_{++}^L$ be arbitrary, we define the vector $q \in \mathbb{R}^L$ by

$$q = \left(\ln(\mu_1^* + \varepsilon_1), \dots, \ln(\mu_L^* + \varepsilon_L)\right).$$

But conditions (3.2), (3.3), (3.4) and (3.5) imply that:

$$\begin{aligned} u^* = L(u^*, x^*, q^*) \geq L(u^*, x^*, q) & \\ &= \sum_{i=1}^I \lambda_i^*[a_i \cdot x_i^* - V_i(q)] + 0 \\ &= \sum_{i=1}^I \lambda_i^*[a_i \cdot x_i^* - v_i(\mu^* + \varepsilon)] \\ &= \sum_{i=1}^I [\mu^* \cdot x_i^* - \lambda_i^* v_i(\mu^* + \varepsilon)] \\ &= \sum_{i=1}^I [\mu^* \cdot \omega_i - \lambda_i^* v_i(\mu^* + \varepsilon)] \end{aligned}$$

From Lemma 1, for every $p \gg 0$, $v_i(p) = p \cdot \omega_i \max_\ell a_{i\ell}/p_\ell$, so that:

$$u^* \geq \sum_{i=1}^{I}(\mu^* + \varepsilon)\cdot\omega_i - \varepsilon\cdot\sum_{i=1}^{I}\omega_i - \sum_{i=1}^{I}\lambda_i^*(\mu^* + \varepsilon)\cdot\omega_i \max_{\ell}\frac{a_{i\ell}}{\mu_\ell + \varepsilon_\ell}.$$

But (3.4) implies that, for every i, $\lambda_i^* a_{i\ell} \leq \mu_\ell^* \leq \mu_\ell^* + \varepsilon_\ell$. Consequently:

$$\begin{aligned} u^* &\geq \sum_{i=1}^{I}(\mu^* + \varepsilon)\cdot\omega_i[1 - \lambda_i^* \max_{\ell}\frac{a_{i\ell}}{\mu_\ell + \varepsilon_\ell}] - \varepsilon\cdot\sum_{i=1}^{I}\omega_i \\ &\geq -\varepsilon\cdot\sum_{i=1}^{I}\omega_i. \end{aligned} \tag{3.1}$$

Taking the limit when $\varepsilon \to 0$, one gets $u^* \geq 0$, which ends the proof of the claim. □

We now end the proof of the theorem by showing that $(x_1^*, \ldots, x_I^*, p^*)$ is an equilibrium of $\mathcal{E}$, where $p^* \in \mathbb{R}_{++}^L$ is defined $p_\ell^* = \exp(q_\ell^*)$ for every ℓ. From the above claim, we deduce that

$$a_i \cdot x_i^* \geq V_i(q^*) = v_i(p^*) \quad \text{for every } i.$$

We also know that the equilibrium Market Clearing Condition (b) : $\sum_{i=1}^{I} x_i^* \leq \sum_{i=1}^{I}\omega_i$ holds since $(x_1^*, \ldots, x_I^*)$ satisfies the constraints of problem $(\mathcal{Q})$. Thus the proof will be complete if we show that for every i, x_i^* belongs to the budget set of the i-th consumer, that is, if

$$p^* \cdot x_i^* \leq p^* \cdot \omega_i \quad \text{for every } i.$$

The proof of this last assertion is standard but is given for the sake of completeness. Indeed, for all i, the above inequality $a_i \cdot x_i^* \geq v_i(p^*)$ implies that $p^* \cdot x_i^* \geq p^* \cdot \omega_i$ [see Debreu (1959)].[3] But these inequalities, together with the equilibrium market clearing condition (b), imply that $p^* \cdot x_i^* = p^* \cdot \omega_i$ for every i. This ends the proof of Part (b) of the theorem. □

3.3 *Proof of Theorem 2*

We prepare the proof of Theorem 2 with a lemma.

Lemma 2 *Under A1, A2, let $p \gg 0$, $q \gg 0$, and let $r \gg 0$ be defined by $r_\ell = \sqrt{p_\ell q_\ell}$ for all ℓ.*

(a) $\frac{1}{2}v_i(p) + \frac{1}{2}v_i(q) \geq v_i(r)$ *for every i.*

(b) $\frac{1}{2}v_i(p) + \frac{1}{2}v_i(q) > v_i(r)$ *whenever $v_i(p) \neq v_i(q)$.*

---

[3] A direct proof of this assertion can be given in the linear case. For every integer n, $a_i \cdot [x_i^* + (1/n)a_i] > a_i \cdot x_i^* \geq v_i(p^*)$, which implies that $p^* \cdot [x_i^* + (1/n)a_i] > p^* \cdot \omega_i$. Hence, taking the limit, when $n \to +\infty$, one gets $p^* \cdot x_i^* \geq p^* \cdot \omega_i$.

***Proof of Lemma* 2.** *Part (a)* It follows from the convexity of V_i proved in Lemma 1.

Part (b) Let i such that $v_i(p) \neq v_i(q)$. We choose $\ell \in \{1, \ldots, L\}$ so that

$$v_i(r) = a_{i\ell} \frac{r \cdot \omega_i}{r_\ell},$$

and we notice that $a_{i\ell} > 0$. Indeed if $a_{i\ell} = 0$, then $v_i(r) = 0$ which implies that $a_{ik} r \cdot \omega_i \leq 0$ for every $k = 1, \ldots, L$; but $r \cdot \omega_i > 0$ since $r \gg 0$ and $\omega_i > 0$. Hence $a_i = 0$, a contradiction.

Recalling that the arithmetic mean is greater than or equal to the geometric mean (and the equality holds only when the two numbers are equal) one deduces that:

$$\frac{1}{2}\frac{p_k \omega_{ik}}{p_\ell} + \frac{1}{2}\frac{q_k \omega_{ik}}{q_\ell} \geq \sqrt{\frac{p_k \omega_{ik}}{p_\ell}}\sqrt{\frac{q_k \omega_{ik}}{q_\ell}} = \frac{r_k \omega_{ik}}{r_\ell} \text{ for every } k = 1, \ldots, L.$$

The end of the proof will distinguish the two following cases (i): one of the above inequalities is strict for some k, and (ii): the above inequalities are in fact equalities for every k. In the first case, summing up (over k) one gets

$$\frac{1}{2}\frac{p \cdot \omega_i}{p_\ell} + \frac{1}{2}\frac{q \cdot \omega_i}{q_\ell} > \frac{r \cdot \omega_i}{r_\ell}.$$

Multiplying this inequality by $a_{i\ell} > 0$ one deduces that

$$\frac{1}{2} v_i(p) + \frac{1}{2} v_i(q) \geq \frac{1}{2} a_{i\ell} \frac{p \cdot \omega_i}{p_\ell} + \frac{1}{2} a_{i\ell} \frac{q \cdot \omega_i}{q_\ell} > a_{i\ell} \frac{r \cdot \omega_i}{r_\ell} = v_i(r).$$

We now consider the second case (ii) for which one must have:

$$\frac{p_k \omega_{ik}}{p_\ell} = \frac{q_k \omega_{ik}}{q_\ell} = \frac{r_k \omega_{ik}}{r_\ell} \quad \text{for every } k = 1, \ldots, L.$$

Summing up (over k) and multiplying by $a_{i\ell} > 0$ one gets:

$$v_i(p) \geq a_{i\ell} \frac{p \cdot \omega_i}{p_\ell} = a_{i\ell} \frac{r \cdot \omega_i}{r_\ell} = v_i(r)$$

and similarly $v_i(q) \geq v_i(r)$. Hence $\frac{1}{2}\big(v_i(p) + v_i(q)\big) \geq v_i(r)$. Recalling that $v_i(p) \neq v_i(q)$, one deduces that: $\frac{1}{2} v_i(p) + \frac{1}{2} v_i(q) > v_i(r)$. □

***Proof of Theorem* 2.** Let $(x_1, \ldots, x_I, p)$ and $(y_1, \ldots, y_I, q)$ be two Walras equilibria of $\mathcal{E}$, we want to show that $v_i(p) = v_i(q)$ for every i. Suppose on the contrary that for some i, say $i = 1$, one has $v_1(p) \neq v_1(q)$. From Lemma 2 we deduce that

$$\text{for every } i = 1: \quad a_1 \cdot (\tfrac{1}{2}x_1 + \tfrac{1}{2}y_1) = \tfrac{1}{2}v_1(p) + \tfrac{1}{2}v_1(q) > v_1(r)$$

$$\text{and for every } i > 1: \quad a_i \cdot (\tfrac{1}{2}x_i + \tfrac{1}{2}y_i) = \tfrac{1}{2}v_i(p) + \tfrac{1}{2}v_i(q) \geq v_i(r).$$

Consequently,

for $i = 1$, $r \cdot (\frac{1}{2}x_1 + \frac{1}{2}y_1) > r \cdot \omega_1$,

and for $i > 1$, $r \cdot (\frac{1}{2}x_i + \frac{1}{2}y_i) \geq r \cdot \omega_i$ [see Footnote 3].

Summing up these inequalities (over i) one deduces that

$$r \cdot \sum_{i=1}^{I} \omega_i = r \cdot \left(\sum_{i=1}^{I} \frac{1}{2} x_i + \sum_{i=1}^{I} \frac{1}{2} y_i \right) > r \cdot \sum_{i=1}^{I} \omega_i,$$

a contradiction. This ends the proof of Theorem 2. □

3.4 Proof of Theorem 3

We prove the result by contradiction. Suppose there exist two Walras equilibria $(x_1, \ldots, x_I, p)$ and $(y_1, \ldots, y_I, q)$ of $\mathcal{E}$ such that, for every $t > 0$, $p \neq tq$. We define the vector $r \in \mathbb{R}^L_{++}$ by $r_\ell = \sqrt{p_\ell q_\ell}$ for every ℓ. From the assumptions made in the theorem, there exists i, say $i = 1$, such that $\omega_1 \gg 0$. We claim that

$$\frac{1}{2}v_1(p) + \frac{1}{2}v_1(q) > v_1(r).$$

Suppose for the moment that the claim is proved. Then the proof of Theorem 3 goes as follows. The above claim implies that $a_1 \cdot (\frac{1}{2}x_1 + \frac{1}{2}y_1) = \frac{1}{2}v_1(p) + \frac{1}{2}v_1(q) > v_1(r)$ and, from Lemma 2a, for every $i > 1$, $a_i \cdot (\frac{1}{2}x_i + \frac{1}{2}y_i) = \frac{1}{2}v_i(p) + \frac{1}{2}v_i(q) \geq v_i(r)$. But, as in the proof of Theorem 2, these inequalities lead to a contradiction.

We now prove the above claim. In view of the definition of v_i, it suffices to show that for $i = 1$, one has:

$$\frac{1}{2}(p \cdot \omega_i / p_\ell + q \cdot \omega_i / q_\ell) > r \cdot \omega_i / r_\ell \quad \text{for every } \ell.$$

Indeed, since the arithmetic mean is greater than or equal to the geometric mean (and the equality holds only when the two numbers are equal) one deduces that, for every ℓ

$$\frac{1}{2}\frac{p_k\omega_{ik}}{p_\ell} + \frac{1}{2}\frac{q_k\omega_{ik}}{q_\ell} \geq \sqrt{\frac{p_k\omega_{ik}}{p_\ell}}\sqrt{\frac{q_k\omega_{ik}}{q_\ell}} = \frac{r_k\omega_{ik}}{r_\ell} \quad \text{for every } k = 1, \ldots, L.$$

Furthermore, one of the above inequality is strict for some k. Indeed, suppose on the contrary that the equality holds for every k, then $p_k\omega_{ik}/p_\ell = q_k\omega_{ik}/q_\ell$ for every k. Recalling that $\omega_{ik} > 0$ for every k, one deduces that $p = (p_\ell/q_\ell)q$, a contradiction.

Summing up (over k) the above inequalities, one deduces that

$$\frac{1}{2}(p \cdot \omega_i / p_\ell + q \cdot \omega_i / q_\ell) > r \cdot \omega_i / r_\ell \quad \text{for every } \ell.$$

This ends the proof of the claim and the proof of the theorem. □

References

1. Bonnisseau, J.-M., Florig, M., Jofre, A.: Continuity and uniqueness of equilibria for linear exchange economies. J. Optim. Theory Appl. **109**(2), 237–263 (2001)
2. Bonnisseau, J.-M., Florig, M., Jofre, A.: Differentiability of equilibria for linear exchange economies. J. Optim. Theory Appl. **109**(2), 265–288 (2001)
3. Champsaur, P., Cornet, B.: Walrasian exchange processes. In: Gabszewicz J.J. et al. (eds.) Economic Decision-Making: Games, Econometrics and Optimisation, Contributions in Honor of Jacques H. Drèze, pp. 3–13. North Holland, Amsterdam (with Paul Champsaur) (1990)
4. Cornet, B.: Linear exchange economies. In: Cahiers Eco-Maths. CERMSEM, Université Paris I Panthéon-Sorbonne (1989)
5. Debreu, G.: Theory of Value. Wiley, New York (1959)
6. Eaves, C.: A finite algorithm for the linear exchange model. J. Math. Econ. **3**, 117–203 (1976)
7. Florenzano, M., Moreno-Garcia, E.: Linear exchange economies with a continuum of agents. Span. Econ. Rev. **3**, 253–272 (2001)

8. Florig, M.: Equilibrium correspondence of linear exchange economies. J. Optim. Theory Appl. **120**(1), 97–109 (2004)
9. Gale, D.: Price equilibrium for linear models of exchange. Technical Report P-1156, The Rand Corporation (1957)
10. Gale, D.: The Theory of Linear Economic Models. McGraw-Hill. New York (1960)
11. Gale, D.: The linear exchange model. J. Math. Econ. **3**, 205–209 (1976)
12. Lahiri, S.: Uniqueness of competitive equilibrium price in linear exchange economies: a note (2012). SSRN: https://ssrn.com/abstract=2150039
13. Shmyrev, V.I.: On an approach to the determination of equilibrium in elementary exchange models. Soviet Math. - Doklady **27**(1), 230–233 (1983)
14. Shmyrev, V.I.: Polyhedral complementarity algorithms for finding equilibrium in linear models of competitive economics. Diskretn. Anal. Issled. Oper. **21**(2), 84–101 (2014). (in Russian)
15. Shmyrev, V.I.: Polyhedral complementarity approach to equilibrium problem in linear exchange models. In: Valdman J. (ed.) Optimization algorithms: examples, pp. 27–46. IntechOpen, London, United Kingdom (2018). https://doi.org/10.5772/intechopen.77206
16. Shmyrev, V.I.: Duality in linear economic models of exchange. Trudy Inst. Mat. i Mekh. UrO RAN **26**(3):258–274 (2020). https://doi.org/10.21538/0134-4889-2020-26-3-258-274 (in Russian)

Name Index

T. Maruyama (ed.), *Marginal Revolution in Economics*, Monographs in Mathematical Economics 6, https://doi.org/10.1007/978-981-99-4342-5

Subject Index

T. Maruyama (ed.), *Marginal Revolution in Economics*, Monographs in Mathematical Economics 6, https://doi.org/10.1007/978-981-99-4342-5

SPRINGER NATURE

GPSR Compliance

The European Union's (EU) General Product Safety Regulation (GPSR) is a set of rules that requires consumer products to be safe and our obligations to ensure this.

If you have any concerns about our products, you can contact us on ProductSafety@springernature.com

In case Publisher is established outside the EU, the EU authorized representative is:

Springer Nature Customer Service Center GmbH
Europaplatz 3
69115 Heidelberg, Germany

The manufacturer's authorised representative in the EU is Springer Nature Customer Service Centre GmbH, Europaplatz 3, 69115 Heidelberg, Germany. If you have any concerns regarding our products, please contact ProductSafety@springernature.com

Printed and bound by CPI Group (UK) Ltd, Croydon, CR0 4YY
30/09/2026
02225311-0020